High Throughput Protein Expression and Purification

METHODS IN MOLECULAR BIOLOGY™

John M. Walker, Series Editor

475. Cell Fusion: *Overviews and Methods, edited by Elizabeth H. Chen, 2008*
474. Nanostructure Design: *Methods and Protocols, edited by Ehud Gazit and Ruth Nussinov, 2008*
473. Clinical Epidemiology: *Practice and Methods, edited by Patrick Parfrey and Brendon Barrett, 2008*
472. Cancer Epidemiology, Volume 2: *Modifiable Factors, edited by Mukesh Verma, 2008*
471. Cancer Epidemiology, Volume 1: *Host Susceptibility Factors, edited by Mukesh Verma, 2008*
470. Host-Pathogen Interactions: *Methods and Protocols, edited by Steffen Rupp and Kai Sohn, 2008*
469. Wnt Signaling, Volume 2: *Pathway Models, edited by Elizabeth Vincan, 2008*
468. Wnt Signaling, Volume 1: *Pathway Methods and Mammalian Models, edited by Elizabeth Vincan, 2008*
467. Angiogenesis Protocols: *Second Edition, edited by Stewart Martin and Cliff Murray, 2008*
466. Kidney Research: *Experimental Protocols, edited by Tim D. Hewitson and Gavin J. Becker, 2008*
465. Mycobacteria, Second Edition, edited by *Tanya Parish and Amanda Claire Brown, 2008*
464. The Nucleus, Volume 2: *Physical Properties and Imaging Methods, edited by Ronald Hancock, 2008*
463. The Nucleus, Volume 1: *Nuclei and Subnuclear Components, edited by Ronald Hancock, 2008*
462. Lipid Signaling Protocols, edited by *Banafshe Larijani, Rudiger Woscholski, and Colin A. Rosser, 2008*
461. Molecular Embryology: *Methods and Protocols, Second Edition, edited by Paul Sharpe and Ivor Mason, 2008*
460. Essential Concepts in Toxicogenomics, edited by *Donna L. Mendrick and William B. Mattes, 2008*
459. Prion Protein Protocols, edited by *Andrew F. Hill, 2008*
458. Artificial Neural Networks: Methods and Applications, edited by *David S. Livingstone, 2008*
457. Membrane Trafficking, edited by *Ales Vancura, 2008*
456. Adipose Tissue Protocols, Second Edition, edited by *Kaiping Yang, 2008*
455. Osteoporosis, edited by *Jennifer J. Westendorf, 2008*
454. SARS- and Other Coronaviruses: *Laboratory Protocols*, edited by *Dave Cavanagh, 2008*
453. Bioinformatics, Volume II: *Structure, Function and Applications*, edited by *Jonathan M. Keith, 2008*
452. Bioinformatics, Volume I: *Data, Sequence Analysis and Evolution*, edited by *Jonathan M. Keith, 2008*
451. Plant Virology Protocols: *From Viral Sequence to Protein Function*, edited by *Gary Foster, Elisabeth Johansen, Yiguo Hong, and Peter Nagy, 2008*
450. Germline Stem Cells, edited by *Steven X. Hou and Shree Ram Singh, 2008*
449. Mesenchymal Stem Cells: *Methods and Protocols*, edited by *Darwin J. Prockop, Douglas G. Phinney, and Bruce A. Brunnell, 2008*
448. Pharmacogenomics in Drug Discovery and Development, edited by *Qing Yan, 2008*
447. Alcohol: *Methods and Protocols*, edited by *Laura E. Nagy, 2008*
446. Post-translational Modification of Proteins: *Tools for Functional Proteomics*, Second Edition, edited by *Christoph Kannicht, 2008*
445. Autophagosome and Phagosome, edited by *Vojo Deretic, 2008*
444. Prenatal Diagnosis, edited by *Sinhue Hahn and Laird G. Jackson, 2008*
443. Molecular Modeling of Proteins, edited by *Andreas Kukol, 2008.*
442. RNAi: Design and Application, edited by *Sailen Barik, 2008*
441. Tissue Proteomics: *Pathways, Biomarkers, and Drug Discovery*, edited by *Brian Liu, 2008*
440. Exocytosis and Endocytosis, edited by *Andrei I. Ivanov, 2008*
439. Genomics Protocols, Second Edition, edited by *Mike Starkey and Ramnanth Elaswarapu, 2008*
438. Neural Stem Cells: *Methods and Protocols*, Second Edition, edited by *Leslie P. Weiner, 2008*
437. Drug Delivery Systems, edited by *Kewal K. Jain, 2008*
436. Avian Influenza Virus, edited by *Erica Spackman, 2008*
435. Chromosomal Mutagenesis, edited by *Greg Davis and Kevin J. Kayser, 2008*
434. Gene Therapy Protocols: Volume II: *Design and Characterization of Gene Transfer Vectors*, edited by *Joseph M. LeDoux, 2008*
433. Gene Therapy Protocols: Volume I: *Production and In Vivo Applications of Gene Transfer Vectors*, edited by *Joseph M. LeDoux, 2008*
432. Organelle Proteomics, edited by *Delphine Pflieger and Jean Rossier, 2008*
431. Bacterial Pathogenesis: *Methods and Protocols*, edited by *Frank DeLeo and Michael Otto, 2008*
430. Hematopoietic *Stem Cell Protocols*, edited by *Kevin D. Bunting, 2008*
429. Molecular Beacons: *Signalling Nucleic Acid Probes, Methods and Protocols*, edited by *Andreas Marx and Oliver Seitz, 2008*
428. Clinical Proteomics: *Methods and Protocols*, edited by *Antonia Vlahou, 2008*
427. Plant Embryogenesis, edited by *Maria Fernanda Suarez and Peter Bozhkov, 2008*
426. Structural Proteomics: *High-Throughput Methods*, edited by *Bostjan Kobe, Mitchell Guss, and Huber Thomas, 2008*
425. 2D PAGE: *Sample Preparation and Fractionation*, Volume II, edited by *Anton Posch, 2008*
424. 2D PAGE: *Sample Preparation and Fractionation*, Volume I, edited by *Anton Posch, 2008*
423. Electroporation Protocols: *Preclinical and Clinical Gene Medicine*, edited by *Shulin Li, 2008*
422. Phylogenomics, edited by *William J. Murphy, 2008*
421. Affinity Chromatography: *Methods and Protocols*, Second Edition, edited by *Michael Zachariou, 2008*

METHODS IN MOLECULAR BIOLOGY™

High Throughput Protein Expression and Purification

Methods and Protocols

Edited by

Sharon A. Doyle

GlaxoSmithKline, Walnut Creek CA, USA

Humana Press

Editor
Sharon A. Doyle
GlaxoSmithKline
Walnut Creek CA
USA

Series Editor
John M. Walker
School of Life Sciences
University of Hertfordshire
Hatfield, Hertfordshire, AL10 9AB, UK

ISBN: 978-1-58829-879-9 e-ISBN: 978-1-59745-196-3
ISSN: 1064-3745 e-ISSN: 1940-6029
DOI: 10.1007/978-1-59745-196-3

Library of Congress Control Number: 2008931181

Cover illustration:

Printed on acid-free paper

9 8 7 6 5 4 3 2 1

springer.com

Preface

Advances in genome sequencing have changed the scope of scientific inquiry for the biological sciences. Entire genome sequences can now be determined rapidly, providing researchers with the raw data that encode all proteins produced by an organism. One can now devise experiments to discover the structures, functions, and interactions of these proteins on a cellular or organismal level that previously was not possible. Scientists in a wide range of fields including structural biology, functional genomics, and drug discovery are poised to benefit greatly from these advances.

This exciting opportunity has brought with it new challenges because proteins do not behave in a uniform and predictable manner as do their DNA counterparts. There is no universal method to isolate a protein from its expression system in its native form, due to potentially large variations in size, charge, shape, and external chemical moieties. In addition, the conditions required for proper folding and stability, for example, vary greatly among proteins.

Yet for experiments to yield large numbers of proteins, parallel processing is a necessity. Biochemists, instrumentation engineers, and bioinformaticists have recognized this challenge and over the last few years have made tremendous progress on developing technologies to enable high-throughput protein production. This includes not only instrumentation to handle large numbers of samples in parallel, but also strategies that create and exploit common features to enable simplified cloning, stable expression, and purification of many types of proteins.

This comprehensive volume presents current methodologies including various high-throughput cloning schemes, protein expression analysis, and production protocols. Methods are described that utilize *E. coli*, insect, and mammalian cells, as well as cell-free systems for the production of a wide variety of proteins, including glycoproteins and membrane proteins. This volume provides details of the most successful protocols currently in use by leading scientists, presented in a manner that we hope will be useful to those with training in protein biochemistry, as well as to those new to laboratory-based protein biochemistry who plan to apply these protocols to related fields.

Walnut Creek, CA | Sharon Doyle

Contents

Contributors

JAMES M. ABDULLAH • *Biosciences Division, Argonne National Laboratory, Lemont, IL, USA*

DAVID ALDERTON • *Oxford Protein Production Facility, Welcome Trust Centre for Human Genetics, Oxford, UK*

ERIN S. ARROYO • *Biosciences and Biotechnology Division, Lawrence Livermore National Laboratory, Livermore, CA, USA*

DOUGLAS A. AUSTEN • *Gene Expression, Vertex Pharmaceuticals, Cambridge, MA, USA*

BRIAN P. AUSTIN • *Macromolecular Crystallography Laboratory, Center for Cancer Research, National Cancer Institute at Frederick, Frederick, MD, USA*

MAHMOUD REZA BANKI • *Department of Chemical Engineering, Princeton University, Princeton, NJ, USA*

GRAHAM BENCH • *Atmospheric, Earth and Energy Division, Lawrence Livermore National Laboratory, Livermore, CA, USA*

NICK S. BERROW • *Oxford Protein Production Facility, Welcome Trust Centre for Human Genetics, Oxford, UK*

CRAIG D. BLANCHETTE • *Biosciences and Biotechnology Division, Lawrence Livermore National Laboratory, Livermore, CA, USA*

PAUL G. BLOMMEL • *Center for Eukaryotic Structural Genomics, Department of Biochemistry, University of Wisconsin-Madison, Madison, WI, USA*

MIRJAM BUCHS • *Biologics Center, Novartis Institutes for Biomedical Research, Basel, Switzerland*

JENNY A. CAPPUCCIO • *Biosciences and Biotechnology Division, Lawrence Livermore National Laboratory, Livermore, CA, USA*

CHACKO S. CHAKIATH • *Protein Expression Laboratory, SAIC-Frederick, Inc., National Cancer Institute at Frederick, MD, USA*

STEPHEN P. CHAMBERS • *Gene Expression, Vertex Pharmaceuticals, Cambridge, MA, USA*

DEB K. CHATTERJEE • *Protein Expression Laboratory, SAIC-Frederick, Inc., National Cancer Institute at Frederick, MD, USA*

SCOTT CHERRY • *Macromolecular Crystallography Laboratory, Center for Cancer Research, National Cancer Institute at Frederick, Frederick, MD, USA*

BRETT A.CHROMY • *Biosciences and Biotechnology Division, Lawrence Livermore National Laboratory, Livermore, CA, USA*

MATTHEW A. COLEMAN • *Biosciences and Biotechnology Division, Lawrence Livermore National Laboratory, Livermore, CA, USA*

FRANK R. COLLART • *Biosciences Division, Argonne National Laboratory, Lemont, IL, USA*

MICHELE H. CORZETT • *Biosciences and Biotechnology Division, Lawrence Livermore National Laboratory, Livermore, CA, USA*

MARK I. DONNELLY • *Biosciences Division, Argonne National Laboratory, Lemont, IL, USA*
WILLIAM H. ESCHENFELDT • *Biosciences Division, Argonne National Laboratory, Lemont, IL, USA*
SAID ESHAGHI • *Division of Biophysics, Department of Medical Biochemistry and Biophysics, Karolinska Instiutet, Stockholm, Sweden*
DOMINIC ESPOSITO • *Protein Expression Laboratory, National Cancer Institute, SAIC-Frederick, Frederick, MD, USA*
JULIA FLETCHER • *Cloning and Protein Expression, R&D, Invitrogen Corporation, Carlsbad, CA, USA*
BRIAN G. FOX • *Center for Eukaryotic Structural Genomics, Department of Biochemistry, University of Wisconsin-Madison, Madison, WI, USA*
JOHN R. FULGHUM • *Gene Expression, Vertex Pharmaceuticals, Cambridge, MA, USA*
LESLIE A. GARVEY • *Protein Expression Laboratory, SAIC-Frederick, Inc., National Cancer Institute at Frederick, MD, USA*
SABINE GEISSE • *Biologics Center, Novartis Institutes for Biomedical Research, Basel, Switzerland*
ALISON R. GILLIES • *Department of Chemical Engineering, Princeton University, Princeton, NJ, USA*
PAUL T. HENDERSON • *Biosciences and Biotechnology Division, Lawrence Livermore National Laboratory, Livermore, CA, USA*
ANGELA K. HINZ • *Biosciences and Biotechnology Division, Lawrence Livermore National Laboratory, Livermore, CA, USA*
PAUL D. HOEPRICH JR. • *Biosciences and Biotechnology Division, Lawrence Livermore National Laboratory, Livermore, CA, USA*
IAN HUNT • *Protein Structure Unit, Centre for Proteomic Chemistry, Novartis Institutes for Biomedical Research, Cambridge, MA, USA*
ANDRZEJ JOACHIMIAK • *Biosciences Division, Argonne National Laboratory, Lemont, IL, USA*
FEDERICO KATZEN • *Cloning and Protein Expression, R&D, Invitrogen Corporation, Carlsbad, CA, USA*
BRIAN K. KAY • *Department of Biological Sciences, University of Illinois at Chicago, Chicago, IL, USA*
VLADIMIR KERY • *Biomimetic Therapeutics, Franklin, TN, USA*
ERNIE KIM • *Protein Structure Unit, Centre for Proteomic Chemistry, Novartis Institutes for Biomedical Research, Cambridge, MA, USA*
HEATH KLOCK • *The Genomics Institute of the Novartis Research Foundation, San Diego, CA, USA*
JIM KOEHN • *Protein Structure Unit, Centre for Proteomic Chemistry, Novartis Institutes for Biomedical Research, Cambridge, MA, USA*
WIESLAW KUDLICKI • *Cloning and Protein Expression, R&D, Invitrogen Corporation, Carlsbad, CA, USA*
EDWARD A. KUHN • *Biosciences and Biotechnology Division, Lawrence Livermore National Laboratory, Livermore, CA, USA*

RICHARD J. LAW • *Biosciences and Biotechnology Division, Lawrence Livermore National Laboratory, Livermore, CA, USA*

SCOTT A. LESLEY • *The Genomics Institute of the Novartis Research Foundation, San Diego, CA, USA*

CHIANN-TSO LIN • *Molecular Biosciences, Pacific Northwest National Laboratory, Richland, WA, USA*

FAN LU • *Gene Expression, Vertex Pharmaceuticals, Cambridge, MA, USA*

MARION MAHNKE • *Biologics Center, Novartis Institutes for Biomedical Research, Basel, Switzerland*

PETER A. MARTIN • *Center for Eukaryotic Structural Genomics, Department of Biochemistry, University of Wisconsin-Madison, Madison, WI, USA*

CYNTHIA SANVILLE MILLARD • *Biosciences Division, Argonne National Laboratory, Lemont, IL, USA*

PRISCILLA A. MOORE • *Molecular Biosciences, Pacific Northwest National Laboratory. Richland, WA, USA*

SREEDEVI NALLAMSETTY • *Macromolecular Crystallography Laboratory, Center for Cancer Research, National Cancer Institute at Frederick, Frederick, MD, USA*

JOANNE E. NETTLESHIP • *Oxford Protein Production Facility, Welcome Trust Centre for Human Genetics, Oxford, UK*

RAYMOND J. OWENS • *Oxford Protein Production Facility, Welcome Trust Centre for Human Genetics, Oxford, UK*

JOSEPH B. PESAVENTO • *Biosciences and Biotechnology Division, Lawrence Livermore National Laboratory, Livermore, CA, USA*

TODD PETERSON • *Cloning and Protein Expression, R&D, Invitrogen Corporation, Carlsbad, CA, USA*

YANN POULIQUEN • *Biologics Center, Novartis Institutes for Biomedical Research, Basel, Switzerland*

NAHID RAHMAN-HUQ • *Oxford Protein Production Facility, Welcome Trust Centre for Human Genetics, Oxford, UK*

MICHAEL SACHS • *Protein Structure Unit, Centre for Proteomic Chemistry, Novartis Institutes for Biomedical Research, Cambridge, MA, USA*

KORY D. SEDER • *Center for Eukaryotic Structural Genomics, Department of Biochemistry, University of Wisconsin-Madison, Madison, WI, USA*

BRENT W. SEGELKE • *Biosciences and Biotechnology Division, Lawrence Livermore National Laboratory, Livermore, CA, USA*

KALAVATHY SITARAMAN • *Protein Expression Laboratory, SAIC-Frederick, Inc., National Cancer Institute at Frederick, MD, USA*

LUCY STOLS • *Biosciences Division, Argonne National Laboratory, Lemont, IL, USA*

TODD A. SULCHEK • *Biosciences and Biotechnology Division, Lawrence Livermore National Laboratory, Livermore, CA, USA*

SUSANNE E. SWALLEY • *Gene Expression, Vertex Pharmaceuticals, Cambridge, MA, USA*

SANG THAI • *Department of Biological Sciences, University of Illinois at Chicago, Chicago, IL, USA*

Joseph E. Tropea • *Macromolecular Crystallography Laboratory, Center for Cancer Research, National Cancer Institute at Frederick, Frederick, MD, USA*
Veronica V. Volgina • *Department of Biological Sciences, University of Illinois at Chicago, Chicago, IL, USA*
Vicki L. Walsowrth • *Biosciences and Biotechnology Division, Lawrence Livermore National Laboratory, Livermore, CA, USA*
David S. Waugh • *Macromolecular Crystallography Laboratory, Center for Cancer Research, National Cancer Institute at Frederick, Frederick, MD, USA*
David W. Wood • *Department of Chemical Engineering, Princeton University, Princeton, NJ, USA*
Russell L. Wrobel • *Center for Eukaryotic Structural Genomics, Department of Biochemistry, University of Wisconsin-Madison, Madison, WI, USA*

Chapter 1

High-Throughput Protein Production (HTPP): A Review of Enabling Technologies to Expedite Protein Production

Jim Koehn and Ian Hunt

Summary

Recombinant protein production plays a crucial role in the drug discovery process, contributing to several key stages of the pathway. These include exploratory research, target validation, high-throughput screening (HTS), selectivity screens, and structural biology studies. Therefore the quick and rapid production of high-quality recombinant proteins is a critical component of the successful development of therapeutic small molecule inhibitors. This chapter will therefore attempt to provide an overview of some of the current "best-in-class" cloning, expression, and purification strategies currently available that enhance protein production capabilities and enable greater throughput. As such the chapter should also enable a reader with limited understanding of the high-throughput protein production (HTPP) process with the necessary information to set up and equip a laboratory for multiparallel protein production.

Key words: Deep-well block protein expression; Miniaturized protein purification; High throughput

1.1. Introduction

Over the last few years a number of technologies have been developed to expedite the production of recombinant proteins for therapeutic studies. These include the use of rapid cloning systems, miniaturization of cell growth conditions, and a variety of innovative automation systems for expression and purification of recombinant proteins. A quick Web search easily identifies several companies offering complete high-throughput protein production (HTPP) systems. However true HTPP is the preserve of only a few laboratories and/or consortia worldwide whose focus is the rapid generation of many hundreds of proteins simultaneously. What technologies are available to those labs that require only a modest, but significant improvement in throughput?

Sharon A. Doyle (ed.), *Methods In Molecular Biology: High Throughput Protein Expression and Purification, vol. 498*

Book doi: 10.1007/978-1-59745-196-3

This question is probably relevant to most labs where target prioritization and interest necessitate they work on specific protein targets that can often be difficult to express. This review will therefore attempt to provide an overview of some of the current "best–in-class" cloning, expression, and purification strategies currently available. Specifically, the review will focus on some of the developments in *E. coli*- and baculovirus-mediated protein expression systems that have enhanced protein production within the context of the drug discovery environment.

1.2. High-Throughput Cloning Methods

One of the standard procedures when setting out to express a recombinant protein is to screen a series of constructs to identify the most viable protein for the generation of sufficient soluble material for downstream purposes. This may include expressing full-length proteins or perhaps specific domains or chimeric proteins. A series of fusion partners may also be investigated for their effects on driving enhanced expression or their capacity to capture and purify the target protein quickly with minimal impurities. Using traditional cloning methodologies, the generation of the many possible combinations and their analysis in different expression systems would be so labor intensive and time consuming as to make a parallel strategy of expression screening impractical. However, over the past few years many of the limitations relating to the generation of multiple expression plasmids (and constructs) have been addressed by a number of elegant recombinatorial cloning systems that enable the rapid cloning of potentially hundreds of genes and constructs simultaneously (*see* **Table 1.1**). By far the most popular are Gateway® (Invitrogen) and LIC (Novagen, EMD Biosciences), although the former does result in the expression of extraneous coding regions that may be deleterious to downstream applications (e.g., crystallography). Moreover, the long-term cost in adopting these technologies can be substantial. Therefore, several reports have detailed low-cost alternatives that offer comparable capabilities. Most notable are those described by Klock et al. *(1)* and Benoit et al. *(2)*. In both these cases, the resulting expression plasmids contain no extraneous coding regions and thus offer significant advantages to those approaches which are commercially available. In addition, both systems are easily adapted to liquid-handling workstations.

Irrespective of the method employed for the construction of the expression plasmid(s), propagation of sufficient recombinant DNA for expression studies necessitates plasmid selection and purification. Nowadays (even for large numbers of constructs)

Table 1.1
Comparison of rapid direction cloning strategies

	Cloning system					
	Traditional	TOPO	Gateway	Creator	In-fusion	LIC
Commercial supplier		Invitrogen http://www.invitrogen.com	Invitrogen	Clontech http://www.clontech.com	Clontech	Novagen http://www.emdbiosciences.com
Mechanism of gene insertion	Classic sticky end ligation using T4 DNA ligase	TOPO cloning technology	Lambda bacteriophage	Cre-loxP		T4 DNA polymerase
Efficiency	Varies	Varies; insertion of large DNA fragments (1,500 bp) can be problematic	>90%	>90%	>90%	>90%
Potential use in multiparallel expression strategies	Limited	Yes	Yes	Yes	Yes	Yes
Incorporation of additional amino acids	Yes, typically the addition of 2 residues corresponding to restriction sites	No	Addition of 8 amino acid residues	Introduction of Cre-LoxP regions	No	Yes, addition of 5 amino acid residue. Can be removed by enterokinase cleavage
Comments		Potentially any expression vector can be adapted. Conversion to TOPO systems is however currently costly			Following ligation requires transfer to Creator. Cloning vehicle only	

this is a relatively straightforward process; following single-plate transformation (*see* **Note 1**), positive clones can be selected, propagated, and purified using a variety of standard commercial high-throughput kits that are easily adapted to liquid-handling workstations (QIAwell Ultra Plasmid BioRobot Kit, Qiagen; Wizard SV 96 Plasmid DNA Purification Kit, Promega; Perfect Prep Plasmid 96 Vac Direct Bind Kit, Eppendorf and Nucleospin Robot-96 Plasmid Kit, Macherey-Nagel). Most of these commercial kits can either be used in a standalone format with minimal equipment requirements (96-single tube well magnet, Qiagen Cat no. 36915 or vacuum manifold), or adapted to a variety of liquid-handling systems (*see* **Section 1.6** for further details). For each specific robot, it is recommended to refer to the kit manufacturer's Web site for detailed information on plate/machine compatibility since this can vary markedly.

1.3. Expression Systems

1.3.1. *E. coli*

Bacterial expression is the most commonly employed expression vehicle for the production of recombinant proteins. It is relatively simple to manipulate, inexpensive to culture, and the amount of time necessary to generate a recombinant protein is relatively short. Recombinant expression of proteins is normally achieved through the induction of a strong promoter system. Several are commonly used including T7, lambda Pl, and *ara*B. Perhaps the most popular is the T7-based pET expression plasmids (commercially available from Novagen). However, while this system leads to the generation of large amounts of mRNA and concomitant protein expression, the high levels of mRNA can cause ribosome destruction and cell death. Furthermore, leaky expression of T7 RNA polymerase may result in plasmid or expression instability. Use of the lac operator and T7 lysozyme (pLysS) can, however, provide an extra level of repression. A more complete review of *E. coli* expression vectors can be found elsewhere *(3–5)*.

1.3.1.1. HT *E. coli* Autoinduction

A very useful tool in the expression testing of multiple proteins and conditions, autoinduction was first described by Studier *(6)*. Autoinduction refers to bacterial cultures grown in media containing specific components that after an initial period of tightly regulated, uninduced growth, automatically induces target protein expression without IPTG. Typically, autoinduced expression produces a greater proportion of soluble target protein than does IPTG-induced expression. This method can provide higher protein yields and greater convenience compared to standard IPTG induction, in particular, when working with large numbers of parameters and or constructs. Several autoinduction reagents are sold commercially including Overnight Express™ autoinduction from Novagen (http://www.emdbiosciences.com/novagen).

1.3.1.2. Expression Strains

Many *E. coli* strains optimized for protein expression purposes are commercially available from suppliers such as Invitrogen, Novagen, and Stratagene. The Origami™ and Origami B™ strains (Novagen) have been developed to express proteins that require disulfide bonds to achieve their active, correctly folded conformation. These strains with mutations in both the thioredoxin reductase (*trx*B) and glutathione reductase (*gor*) genes greatly enhance disulfide bond formation in the cytoplasmic space. Furthermore, strains such as BL21 CodonPlus™ (Stratagene) and Rosetta-2™ derived from BL21 (Novagen) have been reported to enhance expression of gene sequences that contain rarely used *E. coli* codons in the expression of heterologous protein in *E. coli*. Strains that have been engineered to express the *Oleispire anrarctica* chaperones Cpn10 and Cpn60 have also been shown to enhance expression. Commercially available as ArticExpress™ (Stratagene), this strain is particularly useful in conferring improved protein expression levels at lower temperatures, potentially increasing the yield of active, soluble recombinant proteins. Many of these strains are sold in 8-well strips and 96-well plate formats, allowing convenient transfer of protocols to HT formats using liquid-handling workstations.

1.3.2. Baculovirus

A detailed review of the different baculovirus systems and their application to HT protein expression is covered elsewhere *(7)*; however, for completeness we briefly review some of the developments that have enhanced (at least in part) the streamlining and speed of baculovirus-mediated insect cell expression. In this regard it is noteworthy that while the Bac-to-Bac® system (Invitrogen) is perhaps the most popular insect cell expression system, it is one of the least amenable to HT applications. The large quantities of recombinant bacmid that is required for transfection make the system very time consuming and cumbersome, especially when working with 24–96 different constructs. Conversely transient insect cell systems (Insect Direct™, Novagen) and some of the newer baculovirus-mediated systems (FlashBAC™, NextGen Sciences and BacMagic™, Novagen) that do not require bacmid propagation are actually more amenable to HTPP and automation. Therefore, it will be interesting to see protein science evolve toward higher throughput, if we see a switch in the popularity of the respective systems.

1.3.2.1. HT Bacmid Propagation (Bac-to-Bac Only)

The generation of recombinant bacmid can be performed by the use of several commercial kits and a detailed protocol is described elsewhere in this book. Both the R.E.A.L Prep 96 Kit (Qiagen) and PerfectPrep™ BAC 96 (Eppendorf/Brinkmann) kits are amenable for HT purification of bacmid DNA by both manual and liquid-handling strategies.

1.3.2.2. HT Transfection

The use of the Amaxa Nucleotransfection system (*see* http://Amaxa.com) provides a very fast and highly efficient method of performing large numbers of suspension-based transfections of mammalian and insect cells. The system can simultaneously transfect 8–96 constructs in approximately 5 min into 50–100 μL suspension culture volumes. Since the process is extremely efficient and can be conducted in suspension-based cultures, protein expression studies can be completed in a relatively short period of time (following a single round of viral application). This represents a significant advantage over traditional (manual) approaches in which transfection can take 2–3 h to complete and which are extremely laborious when working with 24 or more different constructs. Taken together, the system represents a significant breakthrough in streamlining the HT insect cell expression process. Moreover, coupled with the use of transient or non-bacmid based baculovirus expression strategies, this approach offers a very powerful system for insect-driven HTPP.

1.3.3. Other Expression Systems

Several other systems are also used extensively in the expression of recombinant proteins for structural, functional, and high-throughput assays. However, for the sake of brevity this review has focused on those expression systems most routinely encountered in HTPP laboratories.

1.3.3.1. Mammalian Cell Expression

The expression of heterologous proteins in a mammalian background offers many clear advantages to their generation in *E. coli* or insect cells, including correct post-translational modification and folding. However, while the use of mammalian cells such as CHO or HEK293 is well documented, the process of creating stable mammalian cell lines can often be laborious and time consuming *(8)*. Transient expression systems that utilize suspension cell lines therefore may provide a viable alternative, producing high quantities of recombinant protein in a very short period of time. Of particular interest is the use of HEK293-EBNA cells for rapid transient expression studies *(9–11)*. The system utilizes episomally replicating plasmids featuring the Epstein–Barr virus (EBV) *oriP* driven by EBNA-1 protein generated from a gene integrated into the HEK293 genome. In most cases transient protein expression is driven by the strong CMV promoter. Furthermore, the system is highly amenable to automation, and many groups have begun to adapt the system to miniaturized high-throughput strategies in a similar fashion to those described for *E. coli* and baculovirus incorporating FLAG and polyhistidine tags to facilitate rapid HT-purification and quantitation.

1.3.3.2. Cell-Free Expression

Traditional cell-free expression systems, while providing an attractive "quick" route to the production of proteins, have always been marred by low expression levels. However, several improvements have been made that now enable expression of 5–10 mg/mL. Changes include the optimization of lysate composition, introduction of semi-continuous reactions, and energy regeneration systems *(12–15)*. The use of a eukaryotic-based approach coupled with the rapid, cell-free utility offered by the system has a number of clear advantages and offers a viable alternate to traditional approaches. Indeed, some groups, notably RIKEN Structural Genomics Initiative, appear almost exclusively to generate recombinant protein in this way *(16, 17)*. Several commercial cell-free system are currently available and include the Rapid Translation System™ (RTS) platform (Roche), EasyXpress™ (Qiagen), and Expressway™ (Invitrogen), the former offering a complete scalable system for small-scale PCR-mediated screening to large-scale (10 mL) production in a variety of cell lysates (*E. coli* and wheat-germ bases). However, the expression of proteins that require SH-bond formation or membrane expression is not possible in either case. Furthermore, the commercial systems, while very quick and amenable to HT strategies, are very expensive and therefore are cost prohibitive for many laboratories.

1.4. Protein Expression and Purification Strategies

1.4.1. Miniaturized Protein Expression and Purification Screening

E. coli being a very robust organism is highly amenable for growth in 96- and 24-deep-well block formats (1–3 mL) using standard shaking incubators (*see* **Note 2**). Conversely, the growth of deep-well block insect cell cultures is somewhat more problematic. Indeed, while insect cells have been shown to grow successfully in 4–10 mL volumes using 24-deep-well blocks, reliable growth using 48- and 96-well blocks is difficult to achieve. Notwithstanding, by careful optimization of agitation speeds and humidity of either conventional shakers or those designed specifically for deep-well blocks, insect cells can be cultivated and used for transfection, viral amplification, and protein expression studies *(18–20)*.

Deep-well block expression requires a concomitant method of analyzing the resulting recombinant protein expression data. To this end, several groups have developed reporter assays to quickly assess the solubility of recombinant protein directly from the culture supernatant, thereby avoiding time-consuming and labor-intensive extraction and purification of the protein of interest *(21–24)*. However, in many cases additional information regarding the expressed protein is required which necessitates the lysis, purification, and analysis of the recombinant protein. Several commercial lysis buffers are available and are listed. Some that do not require the harvesting of cellular debris by centrifugation have also been developed. Sold commercially as POP Culture™ (Novagen) and FastBreak™ (Promega) their use greatly facilitates the downstream analysis of protein expression, since they allow in situ analysis of cell lysates without the need for cellular clarification via centrifugation (*see* **Note 3**). This can be extremely advantageous when setting up an HTPP strategy on a liquid-handling workstation, since centrifugation steps can add significant time onto the process.

1.4.1.1. Affinity Tags for Solubility and HT Purification

The use of such a multiparallel expression strategy to conduct optimization and screening studies (whether conducted in a prokaryotic or eukaryotic host) necessitates the adoption of a generic purification approach that is quick, simple, and cheap to perform. In addition to offering a convenient one-step generic purification strategy, several fusion partners have also been shown to enhance solubility of the target protein. An excellent summary of the findings of various studies was presented by Braun and LaBaer *(25)*. Clearly, the use of many tags in a multiparallel expression strategy is impractical since the requirements for the different methods of purification and analysis would make for a logistically huge undertaking. Therefore typically most groups have their favorites based on experience [and success] and will preferentially use these in their work, often in concert with each other. Indeed, when used in tandem, affinity, solubility-enhancing, or reporter tags can provide very useful generic purification tools. For example, two affinity tags can be used in concert to produce high-purity protein via a two-step chromatography procedure, while GFP, S-tags, Trx, or NusA can be used in tandem with His6 and MBP to rapidly analyze and quantify expression levels and also allow downstream purification.

However, it is always important to remember that fusion tags (irrespective of size) can potentially interfere with folding, function, or crystallization capabilities. Conversely, there are many instances of proteins whose expression has been enhanced by the use of tags but which have become unfolded and precipitate following removal of the tag by enzymatic cleavage at an engineered recognition site adjacent the tag.

Deciding which to use is therefore difficult, since they all appear to offer various advantages over each other. Furthermore, some are far more amenable to adaptation to high-throughput purification strategies than others (most notably the His6 tag). A list of some of the commonly used tags that are amenable to rapid single-step purification of proteins that are commercially available is shown in **Table 1.2**. However as mentioned earlier,

Table 1.2
Comparison of commonly used tags for purification and expression enhancement

Tag	Residues	Size (kDa)	Use	Matrix/elution	Commercial supplier	Comment
Polyhistidine	2–10	0.84	Purification	Divalent metal/imidazole or low pH	Qiagen, Invitrogen, Novagen	Tag or elution may affect protein properties
Maltose binding protein	396	40	Purification and enhanced solubility	Amylose resin, elution with maltose	New England Biolabs, Novagen	One-step purification of relatively pure protein. Matrix compatible with nonionic detergents and high slat, but not reducing agents
Glutathione S-transferase	211	26	Purification and enhanced solubility	Glutathione agarose, elution with glutathione	Amersham Biosciences, Novagen	GST dimerization and glutathione elution may affect fusion protein properties
Intein CBD	51	5.59	Purification	Chitin	IMPACT (Novagen)	Compatible with high salt and nonionic detergents; purification must be done in the absence of thiol-containing reducing agents until elution step
Strep-tag II	8	1.20	Purification	Strep-Tactin/biotin or desthiobiotin	Sigma	
S-tag	15	1.75	Detection and purification	S-fragment/low pH	Novagen	Used in RNase S assay for quantitative determination of expression levels. Harsh elution conditions (pH 2.0), typically used in conjunction with other tags (His6)

(continued)

Table 1.2 (continued)

Tag	Residues	Size (kDa)	Use	Matrix/elution	Commercial supplier	Comment
NusA	495	54.87	Increased expression and solubility	NA	Novagen, Invitrogen	Must be used in conjunction with another affinity tag for protein purification; large tag, may affect properties of fusion protein
Trx	109	11.67				
DsbA	208	23.10				
Ubiquitin (Ub)	128	14.73	Solubility	NA		
SUMO (Smt3)	101	11.60				
GFP	239	27	Detection	NA	Clontech, Invitrogen	Allows *in vivo* visualization of expression

Additional sources: References *(26–28)*

undoubtedly the most popular and widely used strategy for purification of proteins is via the hexa-histidine epitope, and as such this popularity is reflected in the large number of strategies available for high-throughput His6-mediated purification (**Table 1.3**). Indeed, the development of plate-based purification strategies that allow the rapid capture of 24–96 different proteins and their

Table 1.3
Commercial kits and reagents for analyzing histidine-tagged recombinant proteins

Product	Company	Comment
Filtration plates	Clontech; BD TALON™ HT 96-Well purification plate (635622)	Kits include plates which contain charged Nickel resin in a filter-plate format ready for use
	GE Healthcare; His MultiTrap™ HP (28-4009-89), His MultiTrap™ FF (28-4009-90)	
	Novagen (Kit); RoboPop™ Ni-NTA His-Bind (71788-3)	
	Promega (Kit); HisLink™ 96 Protein purification system (V3680)	
	Qiagen; Ni-NTA Superflow 96 BioRobot Kit (969261)	
	Sigma; His-Select Nickel 96-Well filter plates (H 0413)	
Nickel-coated plates	Qiagen; Ni-NTA HisSorb™ plates (35081)	Intended for high-throughput capture and analysis of His-tagged protein which bind to Nickel-coated plates Limited binding capacities
	Pierce; HisGrab™ Nickel-coated plates (15242)	
	Sigma; His-Select Nickel-coated plates (S 5563)	
Special Plates	Pierce; SwellGel Nickel-Chelated Disks, Plate (75824)	Disks are dehydrated nickel-chelated agarose gel which rehydrate during capture and his-tag protein purification
	Sigma; His-Select iLAP™ HC Nickel-Coated Plate (H 9412)	Integrates cell lysis and his-tagged protein capture surface of the plate. Limited binding capacity and protein recovery have been observed in our studies
Magnetic bead reagents	Clontech; TALON™ Magnetic Affinity Resin (635502)	
	Dynal (Invitrogen); Dynabeads® TALON™ (101.01)	Large particle size settles quickly
	Qiagen; Ni-NTA Magnetic Agarose Beads (36111)	Lower binding capacity when compared in our studies

(continued)

Table 1.3 (continued)

Product	Company	Comment
	Promega; MagneHis™ Ni-Particles (V8560)	Laboratory choice for binding capacity and low background
	Sigma; His-Select HC Nickel Magnetic Beads (H 1786)	
Magnetic bead Kits	Qiagen; Ni-NTA Superflow 96 BioRobot Kit (969261)	All you need except the robot
	Promega; MagneHis™ Protein Purification System (V8500)	

subsequent purification by affinity resins has greatly expedited multiparallel applications. The use of magnetic beads (Novagen, Invitrogen, and Qiagen) has also been another major contribution for high-throughput protein screening *(29, 30)*. Many of the systems can be automated onto liquid-handling systems or carried out on the bench using standalone vacuum manifolds or magnets (*see* **Note 4** below).

As mentioned earlier, the removal of large fusion proteins can be facilitated by the incorporation of protease cleavage recognition sites at the N-terminus of the fusion partner. Cleavage can be conducted either in solution following purification or immediately after enzyme capture "in situ" on the chromatography resin itself. Of particular note in regard to high-throughput applications, the AKTA Express (GE Healthcare) provides a very elegant procedure for on-column cleavage coupled to multidimensional chromatography that is highly amenable to HT protein purification.

Many different protease cleavage enzymes are commercially available, with some of the more popular listed in **Table 1.4**. However, there are a number of problems associated with their use; in many cases cleavage can be incomplete and cause a reduction in protein yields, leave undesirable amino acid extensions at the N- or C-terminal ends, or act nonspecifically within the recombinant protein itself that may have significant deleterious repercussions to downstream processes such as structural or functional study of the recombinant protein. Excision of an affinity tag can however be achieved with minimal nonspecific side effects by the careful selection of the protease. In our experience PreScission™ (GE Healthcare) and TEV (Invitrogen) appear to offer the best solutions with regard to minimal nonspecific secondary effects. As such they therefore offer an excellent generic protease cleavage system for high-throughput purposes.

Table 1.4
Comparison of cleavage sites used for the removal of fusion partners

Excision site	Cleavage enzyme	Commercial provider
Asp Asp Asp Asp Lys↓	Enterokinase	New England Biolabs
Ile Glu/Asp Gly Arg↓	Factor Xa protease	Different distributors
Leu Val Pro Arg↓Gly Ser	Thrombin	Different distributors
Glu Asn Leu Tyr Phe Gln↓Gly	TEV protease	Invitrogen
Leu Glu Val Leu Phe Gln↓Gly Pro	PreScission™ protease	GE Healthcare

Adapted from *(26)*

1.4.2. Medium-Scale Protein Expression and Purification Optimization

Following the HTPP evaluation of multiple constructs for suitability in downstream biochemical applications (e.g., X-ray crystallography, HTS, biophysical analysis), several constructs are typically selected for a more thorough protein biochemical analysis at 100–500 mL cell culture volumes. Traditionally this type of analysis has been done sequentially. However, there is huge utility in purifying and characterizing proteins in a parallel fashion, since information gathered collectively on these proteins can provide quite insightful information that can be used to cycle back for additional rounds of experiments. Moreover, the quantities generated at this level (100–500 mL) may be sufficient for protein characterization studies that are not practical at the smaller cell culture volumes (2–10 mL). For these purposes cell cultures expressing several constructs can easily be grown in standard laboratory shakers, with the caveat of limits in rates and controllability of oxygenation, mixing, and temperature control. However, shakers can easily be overwhelmed when several laboratory members are conducting medium-scale optimization of several different constructs simultaneously. Several adaptations have therefore been made to enhance capacity. A novel and cheap method of *E. coli* growth using 2-L polyethylene terephthalate (PET) bottles was described recently *(31)*. These bottles fit in a standard shaker and their design offers sufficient aeration to promote excellent growth. Disposal of the bottles afterward reduces the risk of crosscontamination of subsequent cultures. An alternative approach that utilizes presterilized, disposable 1-L tissue-culture grade roller bottles stood on end allows 12–24 separate culture volumes to be handled simultaneously using a single standard shaking incubator (Hunt, unpublished data). However, the most elegant solution to the problem is the LEX fermentation system from Harbinger Biotech (Toronto, ON, Canada). The LEX system replaces traditional medium-low throughput mechanical shakers with an ultra high-throughput bench-top bioreactor system. It was originally developed at the Structural Genomics Consortium (Toronto, Canada) and is extremely well suited for laboratories operating multiple HT – expression screening campaigns.

Concomitantly, several automated medium-scale systems are commercially available that can be used to expedite multiparallel protein purification. Most notable are the AKTA Express (GE Healthcare), Profina Purification System (BioRad). and Protein Maker (Emerald Biosystems), all of which provide very versatile solutions for multiparallel purification of proteins identified by small-scale HTPP screening.

1.5. Protein Analytics

Analysis of the results from HT expression and purification is still somewhat of a bottleneck. While e-gels and multimodule gel electrophoresis apparatus are available (E-PAGE 96 and XCell MultiCell, Invitrogen), analysis still requires staining and visual examination of the various samples. When dealing with many samples this can be rather laborious; moreover, data collection and storage is extremely cumbersome. Two microfluidic systems that offer high-throughput and cherry picking capabilities are Agilent Technologies lab-on-a-chip 5100 ALP and Calipers LabChip90 automated electrophoresis system. However, both are very expensive and ideally require dedicated users to operate the systems. They are therefore not particularly compatible for open access or multiuser laboratories. Nevertheless, they offer significant benefits for HTPP and have become popular additions to many laboratories.

1.6. HTPP Automation

The miniaturization of expression systems such as *E. coli* and baculovirus to a 24- to 48-deep-well plate format, the introduction of recombinatorial cloning systems, and the proliferation of a myriad of different microtiter-based protein purification strategies (filtration and magnetic bead mediated) have greatly expedited the capabilities of a single laboratory to rapidly analyze many constructs and expression conditions for optimization of large-scale protein production. In tandem with these developments has been the proliferation of several automated systems that offer solutions to some part or all of the process. A summary of some of the commercial systems currently available for HTPP is listed in **Table 1.5**. Interestingly, initial

Table 1.5
Commercial automation solutions for HT protein production

Equipment	Company and Web page	Comments
General liquid handling		
Biorobot 8000	Qiagen	Standard liquid-handling workstation that can be configured to customer requirements Both cloning and protein purification processes can be easily adapted to most systems Can be somewhat laborious to set up
Microlab 4200	Hamilton	
TECAN Genesis	TECAN	
Biomek 2000, 3000	Beckman	
Expression workstation	NexGen Sciences	Convenient plug-and-play system requiring minimal user intervention and training
High-end automation		
Dedicated HTPP commercial systems		
Piccolo	TAP	Very high-end HTPP system(s). Currently employed at only a handful of sites
Expression factory	NextGen Sciences	
Biorobot protein expression system	Qiagen	Provides a robust, HT strategy for colony selection and arraying capabilities
QPix2 Colony picker	Genetix	
Low-end automation		
Low-throughput bench-top systems		
QIAcube	Qiagen	Low-medium throughput systems for 12–24-DNA/protein purification
Maxwell 16	Promega	
Kingfisher	Fisher Thermo Electron	Rapid 24- to 96-microwell protein purification using magnetic bead technology
Biosprint	Qiagen	
LEX 48L HT Bench-top bioreactor	Harbinger Biotech.	Low-cost approach for HT *E. coli* fermentation
Baculoworkstation	NextGen Sciences	Semiautomated liquid-handling system for 24- to 48-well insect cell culture transfections. Fits easily into BSC

commercial solutions were based on the premise that customers were looking for high-end automation solutions where hundreds of clones and expression conditions were de rigueur (e.g., Piccolo, TAP, Cambridge, UK; and Expression Factory, NextGen Sciences, Cambridge, UK). However, this has been shown to be a premise/domain of only a handful of companies. Therefore, several bench-top systems that allow multi-parallel analysis of 24–36 constructs have been developed. These systems while not only appealing to most laboratories also provide a much more cost-effective entry into HT-expression testing. For example, an elegant and relatively cost-effective tool for rapid histidine-tagged protein purification makes use of KingFisher (Fisher Thermo Electron) or Biosprint (Qiagen) systems and magnetic Dynabeads® beads (Invitrogen). In less than 60 min you can purify your recombinant protein from up to 96 crude-cell lysates. This strategy is a relatively cost-effective system in which to conduct HTPP and with its small footprint can easily be accommodated in a standard laboratory. Other commercial systems include the QIAcube (Qiagen) and Maxwell 16® (Promega), both of which offer medium-throughput solutions for DNA and protein purification, respectively. Moreover, most HTPP protocols can be easily automated using standard laboratory liquid-handling workstations (Hamilton, Tecan, and Beckman Coulter). However, this adaptation can represent a significant amount of resource to adapt to refine the protocols to the specific system. Therefore, in this regard commercial systems that are "plug and play" offer attractive alternatives. For example NextGen Sciences Expression Workstation™ offers a very convenient entry in HTPP with preprogrammed protocols and technical support. The same company also offers an automated liquid-handling system for insect cell seeding into 24-well plates, transfections, infections, viral dilutions, and parallel expression screening. As such the Baculoworkstation™ is the first high-throughput recombinant baculovirus system commercially offered (*see* http://www.nextgensciences.com).

Taken together, many automation solutions are now available for performing aspects of HTPP and at prices that make it affordable to many laboratories. With this capability, the generation of "difficult-to-express" recombinant proteins for biochemical study becomes a much more tractable possibility for many more laboratories. However, although greatly enhancing throughput and the ability to interrogate more conditions, the technology should not be used as a replacement for sensible experimental design.

1.7. Notes

1. HT transformations can be achieved manually using a 48-well Vented QTray with Divider (Genetix). Briefly, following heat shock, up to 48 different transformants can be plated and spread using Coli Roller Plating Beads (Novagen). Other standard molecular biology plasticware required includes 24- and 48-well blocks and 96-well microplates (we find ABgene, Epsom, UK; or Qiagen both excellent sources

for HTPP applications), HT96 Competent cells (Novagen), and shaker incubator (e.g., from Infors AG; New Brunswick, Edison, NJ; Schel Lab Shaker, Schel USA).

2. Incubators suitable for HT *E. coli* and insect cell expression: Growth of both *E. coli* and insect cell cultures in deep-well blocks can be achieved in either conventional shakers (Infor AG, Bottmingen, CH, New Brunswick, Edison, NJ) or those designed specifically for that purpose following careful optimization of agitation speeds (Shel Shaker Incubators, Schel, USA; GlasCol, Terre Haute, IN; Thomson, San Diego, CA; HiGro, GeneMachines, San Carlos, CA).
3. Several in situ lysis buffers that offer convenient tools for HT-expression studies are commercially available Pierce: B-PER™ (http://www.pierce.com), Novagen: BugBuster™ HTP extraction reagent, POP™ culture reagent, RoboPOP™ purification kit, Insect POP™ culture reagent (http://www.emdbiosciences.com), and Promega: FastBreak™ cell lysis reagent (http://www.promega.com).
4. There are several companies that offer metal affinity resins in a variety of formats (immobilized chromatography, filtration and spin plates and columns, and magnetic beads). These include nitrilotriacetic acid agarose (Ni-NTA™ – Qiagen) and nickel sepharose™ high performance (GE Healthcare). Another material commonly used is TALON™ resin (Clontech). Composed of cobalt carboxymethylaspartate coupled to a solid support resin it exhibits far less nonspecific protein binding than Ni-NTA resin, resulting in higher purity of the protein preparation.

References

1. Klock, H. E., White, A., Koesema, E., & Lesley, S. A. (2005) Methods and results for semi-automated cloning using integrated robotics. *J. Struct. Funct. Genomics* **6,** 89–94.
2. Benoit, R. M., Wilhelm, R. N., Scherer-Becker, D., & Ostermeier, C. (2006) An improved method for fast, robust, and seamless integration of DNA fragments into multiple plasmids. *Protein Expr. Purif.* **45,** 66–71.
3. Baneyx, F. (1999) Recombinant protein expression in *Escherichia coli. Curr. Opin. Biotechnol.* **10,** 411–421.
4. Hannig, G. & Makrides, S. C. (1998) Strategies for optimizing heterologous protein expression in *Escherichia coli. Trends Biotechnol.* **16,** 54–60.
5. Makrides, S. C. (1996) Strategies for achieving high-level expression of genes in *Escherichia coli. Microbiol. Rev.* **60,** 512–538.
6. Studier, F. W. (2005) Protein production by auto-induction in high-density shaking cultures. *Protein Expr. Purif.* **41,** 207–234.
7. Hunt, I. (2005) From gene to protein: a review of new and enabling technologies for multi-parallel protein expression. *Protein Expr. Purif.* **40,** 1–22.
8. Wurm, F. & Bernard, A. (1999) Large-scale transient expression in mammalian cells for recombinant protein production. *Curr. Opin. Biotechnol.* **10,** 156–159.
9. Durocher, Y., Perret, S., & Kamen, A. (2002) High-level and high-throughput recombinant protein production by transient transfection of suspension-growing human 293-EBNA1 cells. *Nucleic Acids Res.* **30,** E9.
10. Geisse, S. & Kocher, H. P. (1999) Protein expression in mammalian and insect cell systems. *Methods Enzymol.* **306,** 19–41.

11. Pham, P. L., Perret, S., Doan, H. C., Cass, B., St Laurent, G., Kamen, A., & Durocher, Y. (2003) Large-scale transient transfection of serum-free suspension-growing HEK293 EBNA1 cells: peptone additives improve cell growth and transfection efficiency. *Biotechnol. Bioeng.* **84,** 332–342.
12. Hino, M., Shinohara, Y., Kajimoto, K., Terada, H., & Baba, Y. (2002) Requirement of continuous transcription for the synthesis of sufficient amounts of protein by a cell-free rapid translation system. *Protein Expr. Purif.* **24,** 255–259.
13. Kawasaki, T., Gouda, M. D., Sawasaki, T., Takai, K., & Endo, Y. (2003) Efficient synthesis of a disulfide-containing protein through a batch cell-free system from wheat germ. *Eur. J. Biochem.* **270,** 4780–4786.
14. Madin, K., Sawasaki, T., Ogasawara, T., & Endo, Y. (2000) A highly efficient and robust cell-free protein synthesis system prepared from wheat embryos: plants apparently contain a suicide system directed at ribosomes. *Proc. Natl. Acad. Sci. USA.* **97,** 559.
15. Sawasaki, T., Seki, M., Sinozaki, K., & Endo, Y. (2002) High-throughput expression of proteins from cDNAs catalogue from Arabidopsis in wheat germ cell-free protein synthesis system. *Tanpakushitsu Kakusan Koso* **47,** 1003–1008.
16. Kigawa, T. & Yokoyama, S. (2002) High-throughput cell-free protein expression system for structural genomics and proteomics studies. *Tanpakushitsu Kakusan Koso* **47,** 1014–1019.
17. Stevens, R. C., Yokoyama, S., & Wilson, I. A. (2001) Global efforts in structural genomics. *Science* **294,** 89–92.
18. Chambers, S. P., Austen, D. A., Fulghum, J. R., & Kim, W. M. (2004) High-throughput screening for soluble recombinant expressed kinases in *Escherichia coli* and insect cells. *Protein Expr. Purif.* **36,** 40–47.
19. Malde, V. & Hunt, I. (2004) Calculation of baculovirus titer using a microfluidic-based Bioanalyzer. *Biotechniques* **36,** 942–945.
20. Mccall, E. J., Danielsson, A., Hardern, I. M., Dartsch, C., Hicks, R., Wahlberg, J. M., & Abbott, W. M. (2005) Improvements to the throughput of recombinant protein expression in the baculovirus/insect cell system. *Protein Expr. Purif.* **42,** 29–36.
21. Lesley, S. A., Graziano, J., Cho, C. Y., Knuth, M. W., & Klock, H. E. (2002) Gene expression response to misfolded protein as a screen for soluble recombinant protein. *Protein Eng.* **15,** 153–160.
22. Waldo, G. S. (2003) Improving protein folding efficiency by directed evolution using the GFP folding reporter. *Methods Mol. Biol.* **230,** 343–359.
23. Waldo, G. S. (2003) Genetic screens and directed evolution for protein solubility. *Curr. Opin. Chem. Biol.* **7,** 33–38.
24. Wigley, W. C., Stidham, R. D., Smith, N. M., Hunt, J. F., & Thomas, P. J. (2001) Protein solubility and folding monitored in vivo by structural complementation of a genetic marker protein. *Nature Biotechnol.* **19,** 131–136.
25. Braun, P. & LaBaer, J. (2003) High throughput protein production for functional proteomics. *Trends Biotechnol.* **21,** 383–388.
26. Stevens, R. C. (2000) Design of high-throughput methods of protein production for structural biology. *Struct. Fold Des.* **8,** R177–R185.
27. Terpe, K. (2003) Overview of tag protein fusions: from molecular and biochemical fundamentals to commercial systems. *Appl. Microbiol. Biotechnol.* **60,** 523–533.
28. Kimple, M. E. & Sondek, J. (2004) *Affinity Purification, 9.9.1–9.9.19.* In: Current Protocols in Protein Science, Wiley, NY, USA.
29. Drees, J., Smith, J., Schafer, F., & Steinert, K. (2004) High-throughput expression and purification of 6xHis-tagged proteins in a 96-well format. *Methods Mol. Med.* **94,** 179–190.
30. Lanio, T., Jeltsch, A., & Pingoud, A. (2003) High-throughput purification of polyHis-tagged recombinant fusion proteins. *Methods Mol. Biol.* **205,** 199–203.
31. Millard, C. S., Stols, L., Quartey, P., Kim, Y., Dementieva, I., & Donnelly, M. I. (2003) A less laborious approach to the high-throughput production of recombinant proteins in *Escherichia coli* using 2-liter plastic bottles. *Protein Expr. Purif.* **29,** 311–320.

Chapter 2

Designing Experiments for High-Throughput Protein Expression

Stephen P. Chambers and Susanne E. Swalley

Summary

The advent of high-throughput protein production and the vast amount of data it is capable of generating has created both new opportunities and problems. Automation and miniaturization allow experimentation to be performed more efficiently, justifying the cost involved in establishing a high-throughput platform. These changes have also magnified the need for effective statistical methods to identify trends and relationships in the data. The application of quantitative management tools to this process provides the means of ensuring maximum efficiency and productivity.

Key words: Protein expression optimization; Quantitative analysis; Experimental design; Screening; Statistics

2.1. Introduction

The amount of protein, particularly when recombinant, is more often described in qualitative than quantitative terms. Frequently, proteins are visualized on a gel to characterize amount and purity. The once-mandatory protein purification tables, describing protein production efficiencies, are now rarely found in publications. Protein yields are frequently described subjectively as estimates or percentages. This over-reliance on qualitative measurement reflects the difficulties encountered in accurately determining amounts of protein. The problem is only aggravated when working in a high-throughput protein production environment. This bottleneck in generating quantitative data has now effectively been removed with the development of the LabChip®90 protein assay system capable of analyzing >288 samples (three

Sharon A. Doyle (ed.), *Methods in Molecular Biology: High Throughput Protein Expression and Purification, vol. 498*

Book doi: 10.1007/978-1-59745-196-3

96-well plates) per chip priming *(1)*. When integrated into a high-throughput protein production platform, a vast amount of data can be generated, thereby, requiring effective quantitative and statistical methods to identify trends and relationships. In this chapter, we attempt to illustrate some of the advantages in using experimental design and show how, when combined with high-throughput expression, it can be used to optimize protein production. We encourage readers who are interested in this area to consult further references for a more detailed introduction to statistical analysis in experimentation *(2)*.

2.1.1. Design of Experiments

Statistical design of experiments (DOE), or simply experimental design, is a proven technique used extensively today in many industrial-manufacturing processes. Considering that this method was originally conceived to identify genetic variation in crops, it has not, until recently, been widely taken up by life scientists. As more research disciplines are using automation and microfluidics to obtain faster results, an increasing number of scientists are now recognizing the assistance that experimental design can provide. Consequently, this technique is finding increasing acceptance in many areas beyond its origins in genetics.

Among the advantages that DOE can provide is the increased amount of information per experiment compared to an ad hoc approach. The second benefit occurs in providing an organized approach toward analysis and interpretation of results, thus facilitating communication. Another advantage is the ability to identify interactions among factors, leading to more reliable prediction of response in areas not directly covered by experimentation. The fourth benefit is in the assessment of information reliability in light of experimental and analytical variation. The uptake of this mathematical technique has been greatly aided by the availability of DOE software packages, like JMP (*see* **Note 1**), making it accessible to the nonstatistician.

2.1.2. Optimization of Protein Production

Optimization of protein production using a conventional one-factor-at-a-time approach is a very labor-intensive endeavor, due to the large number of potential factors and their interactions that can affect expression. Interactions make it difficult to optimize factors independently, increasing the number of experiments required to cover the variable space to identify the maximum response. Through DOE techniques the total number of experiments can be reduced, by evaluating the more relevant interactions among variables, and through the use of partial factorial experimental models. Even then, however, the throughput of traditional protein expression is insufficient to perform the required number of experiments in a reasonable period of time and at a viable cost. Only now through the recent development of high-throughput protein expression platforms is it possible to take full advantage of DOE optimization of protein production.

2.2. Methods

2.2.1. Experimental Design

Good experimentation requires the establishment of a precise goal and objective; an ill-defined experiment will often produce ambiguous results and fail to reach any conclusion. The simplest experimental design is one where screening is used to identify key factors affecting a measurable response (*see* **Note 2**). In our case the response to be maximized is soluble protein production. Utilizing the high-throughput platform described in **Chapter 10** enables the analysis of soluble protein produced in *E. coli* and insect cells. Analysis of this quantitative response allows the experimenter to identify and optimize conditions critical to production of soluble protein.

In order to express a protein, many factors need to be examined experimentally, as it is difficult to know a *priori* what will succeed. Performing one-factor-at-a-time experiments (**Fig. 2.1a**), especially when there are many potential important factors, raises the risk of locating a local maximum, thereby missing the actual best condition.

Also, experiments are best executed in an iterative manner so that information learned in one experiment can be applied to the next. Typically a sequence of experiments is used to meet a defined objective. The experiments include screening designs based on a fractional factorial (**Fig. 2.1b**) to identify signifi-

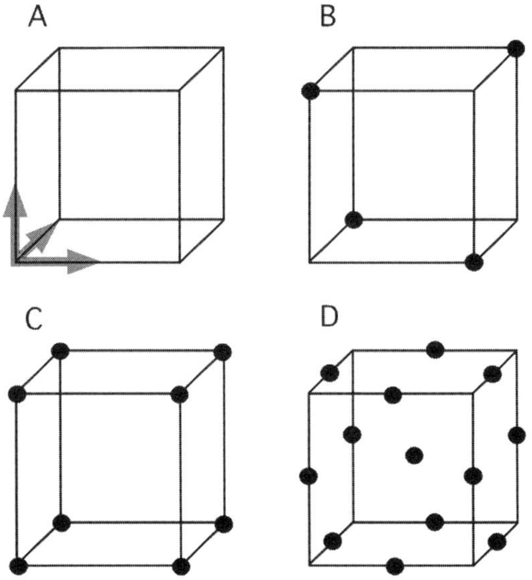

Fig. 2.1. Experimental design. (a) One factor at a time; (b) fractional factorial; (c) full factorial; (d) response surface model: Box-Behnken design for three factors.

cant factors, a full factorial (**Fig. 2.1c**) to identify interactions or response surface design (**Fig. 2.1d**) to fully characterize or model the effects, followed up with confirmation runs to verify the results (*see* **Note 3**).

2.2.2. Factors Affecting Expression

A factor is any variable associated with the product or process under experimental control. There are two different types of variables used in DOE: categorical and continuous. A categorical variable is a factor having only a discrete number of settings that have no intrinsic order, while a continuous variable can be assigned a numeric value. A number of factors affect recombinant protein expression including, but not limited to, construct length, vector, cell line, temperature, time, media, inducer concentration, and additives. The range of values used should be based on either literature precedent or previous experience expressing proteins (*see* **Note 4**). We will briefly discuss these factors and the approach we take to each expression system.

2.2.2.1. Construct

Once a target protein of interest is chosen, the first step is to design a number of constructs of varying length, as practical experience has shown that the exact construct limits can be critical to success. Alignments with homologous proteins that have been previously expressed can help limit the number of constructs, but it is unwise to choose only one. Limited proteolysis *(3)* or H/D exchange *(4)* of full-length protein can also be used to identify small, stable domains capable of being overexpressed and successfully used in structural studies downstream. Additional diversity can also be introduced into the experiment by exploring mutants and homologs of the target protein.

2.2.2.2. Expression System and Vector

The most commonly used source of recombinant expression is *E. coli*, but there are many other prokaryotic and eukaryotic systems available. We routinely use both *E. coli* and insect cells, having streamlined the process with a vector that can transform bacteria for direct expression using a bacteriophage T7 promoter and make baculovirus for insect cell expression using the polyhedrin promoter *(5)*. Typically, we produce a hexa-histidine (Hisx6)-tagged protein, but have explored other fusions options including glutathione-S-transferase (GST) and maltose-binding protein (MBP) where the literature suggests some advantage. The proliferation of commercially available vectors, especially those that facilitate rapid cloning like the Gateway™ destination expression vectors, has only added to the number of strategies that can be readily pursued.

2.2.2.3. Cell Line

For bacterial expression, choice of cell line can greatly affect the amounts of protein produced. There are a number of *E. coli* strains with genotypes engineered specifically to meet the needs of expressing recombinant proteins. While there is a wide variety of choice, the popularity of commercially available competent cells has reduced the number of cell lines more often used to just a few. Some of the most frequently used are derivatives of the BL21(DE3) cell line *(6)*, containing coresident plasmids to address specific protein expression issues, including toxicity, codon bias, and folding. For insect cell expression the choice of cell lines is smaller with Sf9, Sf21, and High-5 being the most commonly used. Despite the limited choice we have found, as in *E. coli*, proteins will have a distinct preference and it is worthwhile examining expression in as many cell lines as possible.

2.2.2.4. Temperature and Time

Temperature and time are frequently critical factors, especially since these two variables often interact. In bacteria, there are some proteins that benefit greatly from a slower, longer induction, which generally requires low temperature *(7)*. At high temperatures, bacterial cells will reach a maximum density and eventually run out of nutrients, at which point cell death will occur. If the protein of interest aggregates easily and cannot be overexpressed in a short time frame, then lowering the temperature is essential. We have expressed proteins anywhere from 15 to 37°C and 3–24 h. Insect cells are less tolerant of temperature variation, so we only examine expression at a single temperature (27°C). In both *E. coli* and insect cells the time of induction or infection, triggering the onset of expression, can also play a role in protein productivity. Induction of expression early or late in growth phase and its intensity directed either by IPTG *(8)* or multiplicity of infection *(9)* have been shown to influence protein levels.

2.2.2.5. Media

Specific media and additives have been shown to have an effect on expression *(10)*. We routinely use rich media (*see* **Note 5**) and serum-free media for bacterial and insect cell expression, respectively. The composition of the media, and whether or not it contains serum, can also have an effect on expression, though we do not vary this factor normally in our basic screens.

2.2.2.6. Additives

The inclusion of cofactors *(11)* and inhibitors *(12)* into the expression medium has also been shown to affect levels of recombinant protein expression. Additionally, the coexpression of partner proteins and chaperones can have a positive effect on the expression and solubility of certain proteins *(13)*.

2.2.3. Full Factorial Design

We have chosen one protein from our production portfolio to illustrate the various designs used in optimization. This process was applied to a fairly typical protein expression experiment: the expression of soluble HCV NS3 protease domains (NS3-prt) in *E. coli*. The objective was to identify the significant factors and interactions involved in maximizing soluble expression. Having this goal clearly established, the experimental design can then be chosen. Our choice of design for this problem was a full factorial. This selection was based on our previous experience producing this poorly expressed protease and the identification of a small number of factors, including genotype, capable of influencing its soluble production. The soluble expression of six different NS3-prt genotypes in total was examined using four factors: three continuous (temperature, time, IPTG concentration) and one categorical (cell line). Both nominal and discrete variables were examined at two levels, high (+) and low (–), resulting in a 2^4 full factorial design. A total of 16 conditions per construct were examined with each condition being tested in triplicate. A full factorial experiment containing all possible combinations of factors represents not only the most conservative approach, but also the most costly in terms of experimental resources. As mentioned before, the availability of DOE software with custom design capability greatly facilitates this process for the nonstatistician. The JMP DOE software will determine how large a sample size is needed to identify a significant effect (*see* **Note 6**), guard against uncontrolled (or unknown) variables during execution of the experiment through randomization (*see* **Note 7**), and introduce blocking (*see* **Note 8**) when appropriate.

The amount of protein expressed (the response) quantified by the Caliper LabChip 90 system was transferred into DOE analysis software (*see* **Note 9**). The expression data when shown graphically (**Fig. 2.2**) readily illustrate that the categorical factor BL21(DE3) pLysS has a negative effect on the levels of soluble protein expression. Subsequent multiple regression analysis (*see* **Note 10**) of the data generated by the most constructs in BL21(DE3) identified the significant factors conducive to soluble expression, lower temperature (22°C), and shorter induction period (3 h), while IPTG concentration was not significant over the range examined (*see* **Note 11**). Relationships between factors are readily exposed in an interaction plot, with nonparallel lines produced by the interactive plot of NS3-prt (1b) L13K expression demonstrating that the effect of temperature is highly dependent on time (**Fig. 2.3a**). The level of expression over the time of induction, which had previously appeared to have little significant effect on the level of soluble expression of NS3-prt (1b) L13K, diverges widely at higher values of temperature. The interaction of time with temperature tended to mask the effect of time as a main effect.

2.2.4. Fractional Factorial Design

Unlike the example we have just used, experiments are often initiated knowing very little about what factors influence the expression of a particular protein. In such situations the preference would be to examine as many factors as possible. A large screen-

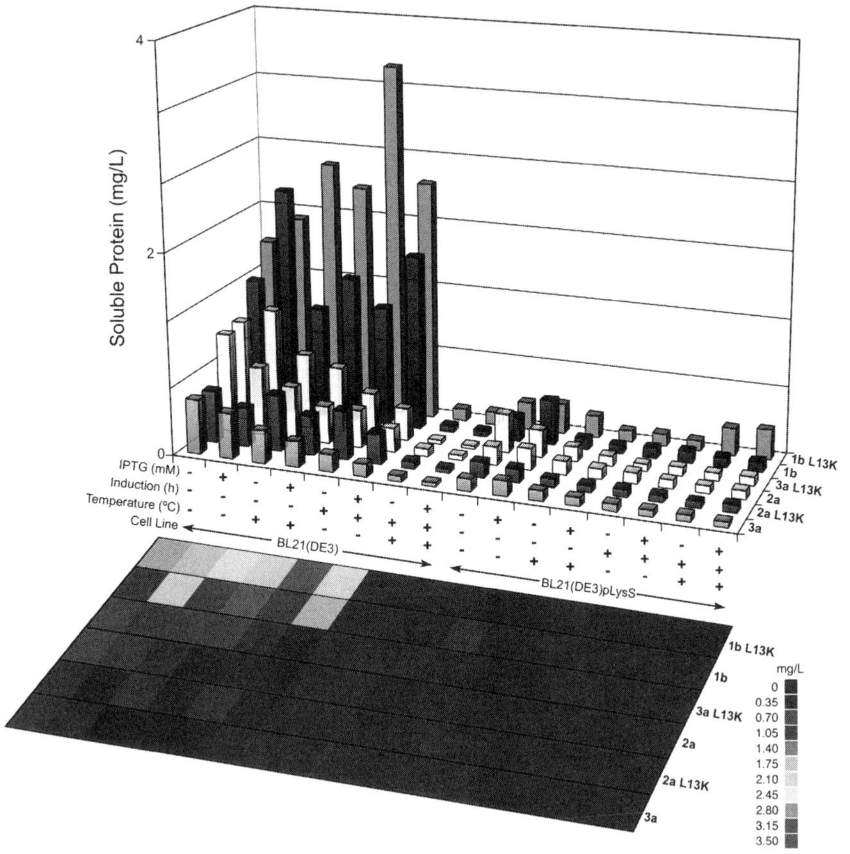

Fig. 2.2. Full factorial screening for effects on the soluble HCV NS3-prt expression in *E. coli*. Levels of expression are illustrated in a 3D Bar Chart and a Heat Map. Using a previously described high-throughput protein expression platform, each data point was obtained using the HT Protein Express 200 Chip run on the Caliper labchip 90 and measured in triplicate (*See Color Plate 1*).

ing approach is best accommodated using a fractional factorial experimental design, whereby the number of potential variables is reduced to a few effective ones. In this model, a partial combination of factors is capable of exploring the maximum number of variables, while requiring less experimentation, albeit at the cost of losing some information about possible interactions. Another consequence of using certain fractional factorial designs, particularly ones with low resolution, is effect aliasing (or confounding). This is where two or more variables have been changed at the same time in the same way resulting in their effects being aliased. This problem can be avoided using a 2-level full factorial or a higher resolution fractional.

2.2.5. Response Surface Designs

Once a process is close to optimum a response surface design can be used to fine-tune the conditions. Response surface designs are used to model the response of a curved surface to a range of continuous variables. The noninclusion of categorical variables is one limitation to response surface designs, and the reason that they are used in optimization and not the initial screening. A response surface model (RSM) provides a more complete understanding of the significant factors involved and is capable of identifying whether a minimum or maximum response exists within the model. There are two classical RSM designs, the Box-Behnken design and the central composite design (CCD). The Box-Behnken design (**Fig. 2.1d**) requires three factors and employs fewer data points than the CCD. Another important feature of the Box-Behnken design is that it has no points at the vertices of the cube as defined by the ranges of factors. This is sometimes useful when it is desirable to avoid these values due to engineering constraints. The cost of this characteristic is the higher uncertainty of predictions near the vertices compared to the CCD.

In the bacterial expression optimization described here a 3-factor Box-Behnken design was employed using 15 conditions. The cell line and significant factors identified in the initial screen were then applied to the customized RSM design. For instance, a protein with a strong preference for low temperature will be screened at lower temperatures in the RSM experiment. The design includes three center points, used to estimate the error of the process; each condition is run in triplicate to increase the accuracy. An RSM of NS3-prt (1b) L13K at three temperatures (21, 29, 37°C), times (3, 10.5, and 18 h), and IPTG concentrations (0.1, 0.55, and 1.0 mM) was produced using JMP software. The resultant RSM confirms the previous observation with expression peaking at high time and low temperature, and temperature being the most significant factor (**Fig. 2.3b**).

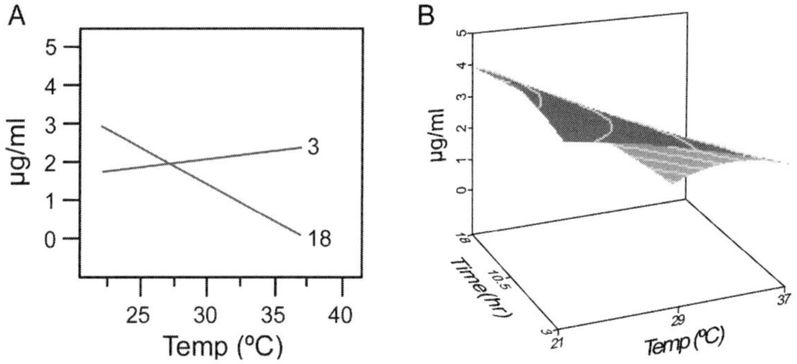

Fig. 2.3. (a) Interaction plot for temperature and time affects on the soluble expression of HCV NS3-prt (1b) L13K. The dependence of protein yield (y-axis) on temperature (x-axis) is plotted for two different times. The red line represents the 3-h data, while the purple line represents the 18-h data. (b) Response surface plot demonstrating expression as a function of time and temperature. The effect of time and temperature on protein yield at an optimal value of IPTG (0.78 mm) is depicted as a 3D surface using the statistical program JMP. The top surface is *colored purple* and the bottom is *colored gray*. Mapped to the surface in *blue* is the contour map of the same data (*See Color Plate 2*).

2.2.6. Validation

The final step in any DOE process is validation. Confirmation of conclusion(s) drawn by experimentation requires verification. Since the ultimate goal of the experiments we have described was to produce the maximum amount of soluble protein from liters of culture, the results of the optimization require verification at this greater volume. We have consistently found that optimal conditions determined by small-scale experimentation demonstrate excellent scalability in terms of soluble protein production. If there is any question as to the preferred condition, two conditions are chosen, grown side by side and compared. In the case of the NS3-prt (1b) L13K, our small-scale optimization results were confirmed by comparing two conditions head to head on a one-liter scale, where the optimized condition of 21°C and 18 h resulted in ~3.5 times more protein per liter and 2.5 times more protein per gram of cell paste when compared with production at the same temperature at 3 h. Once a process is validated, the experimenter can then reproducibly obtain the protein of interest, confident that the best yield is being obtained.

Finally, it is important to remember that DOE is merely a statistical tool, a means to an end. It does not guarantee success; it merely provides a framework for unraveling complex relationships between a response and multiple factors. Nor does it replace technical expertise or creativity in experimental work. When used correctly, DOE can be used to empower the role of investigator in the face of increasing automation.

2.3. Notes

1. Our preferred statistical program for use in experimental design is JMP 7.0 (SAS Institute Inc., Cary, NC, USA), but there are many other software packages are available. The following software contains DOE modules: Minitab (Minitab Inc, PA, USA), ECHIP (ECHIP Inc, DE, USA), and Stat-Ease (Stat-Ease Inc, MN, USA).
2. A single experiment can be defined as an experimental run. A design utilizes multiple runs directed toward meeting a single experimental objective.
3. One criticism of DOE is the potentially large number of runs at the onset of any investigation. Therefore many DOE texts recommend marshalling effort and not expending greater than 25% of resources on the initial screen, using the remainder for subsequent designs and the all-important validation step.

4. The change is response between two levels is termed the factor main effect. Since random error can easily obscure the main effect, if the levels are set too close together, the main effect can be estimated most precisely from extreme level settings of the factor.
5. We typically use brain heart infusion media (BHI) for *E. coli* growth, but have also explored other options, including autoinduction media formulated to support periods of cell growth leading to high densities at which point spontaneous induction of protein expression occurs from *lac* promoters, eliminating the need to monitor cell growth or induce with the addition of IPTG. For these reasons autoinduction medium, under its brand name Overnight Express™(Novagen, Madison, WI, USA), has been promoted as being ideally suited to high-throughput protein expression.
6. In order to design a meaningful experiment, an estimate of the response variable is required. Variability is expressed in terms of standard deviation, which is assumed to be constant over the range of response values encountered during experimentation. The spread of this variability will determine the size of the experiment and the number of runs required in the design.
7. Replication is used to dampen any uncontrolled variation (noise) that might occur, so that the variability associated with the phenomenon can be estimated. Replication requires more than simply resampling or taking additional measurements; the entire process must be repeated from start to finish. Where several samples are submitted from a given experiment, the response is generated as an average. The order in which the experiments are performed should also be randomized to avoid influences by uncontrolled variables such as material transfers, weighing error, and instrument readings. These changes, which often are time related, can significantly influence the response. If run order is not randomized, the analysis may indicate factor effects that are really due to uncontrolled variables that just so happen to change at the same time.
8. Blocking screens out noise caused by unknown sources of variation, such as raw materials, machine, or operator differences. By dividing experimental runs into homogeneous blocks and then arithmetically removing the differences, one increases the sensitivity of the DOE analysis. It is important not to block variables of potential interest.
9. Import raw expression data into JMP and convert to mg of protein per liter of culture. Multiply data by 0.033 and

0.024 for bacterial and insect cell expression, respectively (*see* **Chapter 10** for volumes of culture used).

10. In our example, we fitted the data using a multiple regression model since we were using only continuous predictors (time, temperature, and IPTG concentration) to explain a single continuous response (expression level). If one were using only categorical variables, one would fit an analysis of variance (ANOVA) model. In the case where both categorical and continuous predictors are used to fit a model for continuous response, it is called analysis of covariance (ANCOVA).

11. Look for significant factors and interactions by examining the *p*-values for each. The JMP program considers *p*-value <0.05 to be significant, but this is a matter of choice. Due to the variability in cell growth, we often use *p*-values of <0.01 as a cut-off, though one can choose to be more stringent.

References

1. Bousse L, Mouradian S, Minalla A, Yee H, Williams K, Dubrow R. (2001) Protein sizing on a microchip. Anal Chem 73(6):1207–12.
2. Box GEP, Hunter JS, Hunter WG. 2005 Statistics for Experimenters: Design, Innovation, and Discovery. 2nd ed. New York: Willey.
3. Chen GQ, Sun Y, Jin R, Gouaux E. (1998) Probing the d binding domain of the GluR2 receptor by proteolysis and deletion mutagenesis defines domain boundaries and yields a crystallizable construct. Protein Sci 7(12):2623–30.
4. Zhang Z, Smith DL. (1996) Thermal-induced unfolding domains in aldolase identified by amide hydrogen exchange and mass spectrometry. Protein Sci 5(7):1282–9.
5. Chambers SP, Austen DA, Fulghum JR, Kim WM. (2004) High-throughput screening for soluble recombinant expressed kinases in Escherichia coli and insect cells. Protein Expr Purif 36(1):40–7.
6. Studier FW, Moffatt BA. (1986) Use of bacteriophage T7 RNA polymerase to direct selective high-level expression of cloned genes. J Mol Biol 189(1):113–30.
7. Schein CH, Noteborn MHM. (1988) Formation of soluble recombinant proteins in *Escherichia coli* is favored by lower growth temperature. Biotechnology 6:291–4.
8. Bocanegra JA, Bejarano LA, Valdivia MM. (1997) Expression of the highly toxic centromere binding protein CENP-B in E. coli using the pET system in the absence of the inducer IPTG. Biotechniques 22(5): 798–800, 802.
9. Licari PB, Bailey J. (1991) Factors influencing recombinant protein yields in insect cell-baculovirus expression systems: multiplicity of infection and intracellular protein degradation. Biotechnol Bioeng 37:238–46.
10. Moore JT, Uppal A, Maley F, Maley GF. (1993) Overcoming inclusion body formation in a high-level expression system. Protein Expr Purif 4(2):160–3.
11. De Francesco R, Urbani A, Nardi MC, Tomei L, Steinkuhler C, Tramontano A. (1996) A zinc binding site in viral serine proteinases. Biochemistry 35(41):13282–7.
12. Kumagai A, Dunphy WG. (1996) Purification and molecular cloning of Plx1, a Cdc25-regulatory kinase from *Xenopus* egg extracts. Science 273(5280):1377–80.
13. Ying BW, Taguchi H, Kondo M, Ueda T. (2005) Co-translational involvement of the chaperonin GroEL in the folding of newly translated polypeptides. J Biol Chem 280(12):12035–40.

Chapter 3

Gateway Cloning for Protein Expression

Dominic Esposito, Leslie A. Garvey, and Chacko S. Chakiath

Summary

The rate-limiting step in protein production is usually the generation of an expression clone that is capable of producing the protein of interest in soluble form at high levels. Although cloning of genes for protein expression has been possible for some time, efficient generation of functional expression clones, particularly for human proteins, remains a serious bottleneck. Often, such proteins are hard to produce in heterologous systems because they fail to express, are expressed as insoluble aggregates, or cannot be purified by standard methods. In many cases, researchers are forced to return to the cloning stages to make a new construct with a different purification tag, or perhaps to express the protein in a different host altogether. This usually requires identifying new cloning schemes to move a gene from one vector to another, and frequently requires multistep, inefficient cloning processes, as well as lengthy verification and sequence analysis. Thus, most researchers view this as a linear pathway – make an expression clone, try it out, and if it fails, go back to the beginning and start over. Because of this, protein expression pipelines can be extremely expensive and time consuming.

The advent of recombinational cloning has dramatically changed the way protein expression can be handled. Rapid production of parallel expression clones is now possible at relatively low cost, opening up many possibilities for both low- and high-throughput protein expression, and increasing the flexibility of expression systems that researchers have available to them. While many different recombinational cloning systems are available, the one with the highest level of flexibility remains the Gateway system. Gateway cloning is rapid, robust, and highly amenable to high-throughput parallel generation of expression clones for protein production.

Key words: Recombinational cloning; Gateway; Cloning; Protein expression; HTP cloning; Site-specific recombination; *att* sites; Fusion proteins; Solubility tags

3.1. Introduction

The Gateway recombinational cloning system is a robust method for generating a wide variety of protein expression constructs for use in multiple host systems *(1)*. It also has the advantage of

Sharon A. Doyle (ed.), *Methods in Molecular Biology: High Throughput Protein Expression and Purification, vol. 498*

Book doi: 10.1007/978-1-59745-196-3

being readily scalable from a single construct cloning strategy to a high-throughput system for generating large numbers of parallel clones. In both cases, the more thorough the development of the initial cloning strategy is, the better the chances are for success downstream. We find that in many applications, increasing the work slightly at the initial cloning stage produces exponential increases in the downstream value of the clones generated. The choices which investigators need to make upfront are the focus of the discussion in the sections that follow. We have chosen to examine two possible scenarios that should benefit the largest number of researchers. First, there is the "low-throughput" scenario in which a single protein of interest is to be examined in either multiple host systems or with multiple fusion tags. Second, there is the "high-throughput" scenario, in which a large number of proteins (96 is used as a convenient example here) are expressed in the same host and under the same conditions. These two scenarios provide a range of methods that cover most possibilities for the use of Gateway cloning for protein production, and the similarities and differences in the methods for both will be highlighted later.

Gateway cloning is a multistep recombinational cloning method which eliminates the need for classical restriction enzyme and ligase (REaL) cloning for the transfer of genes between vectors. The parallel power of Gateway cloning involves the transfer of a gene of interest into many different expression vectors through simple recombination reactions which maintain protein reading frame (*see* **Fig. 3.1** for an overview). Two types of clones are generated using Gateway: Entry clones, which are transcriptionally silent "master" clones that are sequence verified, and Expression clones, which are the final protein production clones generated by recombination of the Entry clones. Because the Gateway site-specific recombination reaction does not involve PCR amplification, Expression clones do not need to be resequenced as long as their parent Entry clones have been sequence verified. For this reason, a large number of Expression clones can be generated easily from a single Entry clone without the need for additional amplification and sequencing. Gateway reactions are driven by recombination between sites called attachment sites (*att* sites), which come in four varieties, *attP*, *attB*, *attL*, and *attR*. All reactions are conservative, directional, and lead to the interconversion of these sites, the types of which can be used to differentiate the vectors. Entry clones contain *attL* sites, while Expression clones contain *attB* sites. The core recombination site in all these *att* sites is the same 21-bp DNA sequence which determines the directionality and specificity of the reaction. The protein-coding sequences of Gateway clones are always flanked by two slightly different *att* sites, identified with numbers as in *attB*1 and *attB*2. The *att*1 and *att*2 sites are unable to recombine with each other, and thus produce the unique order of the

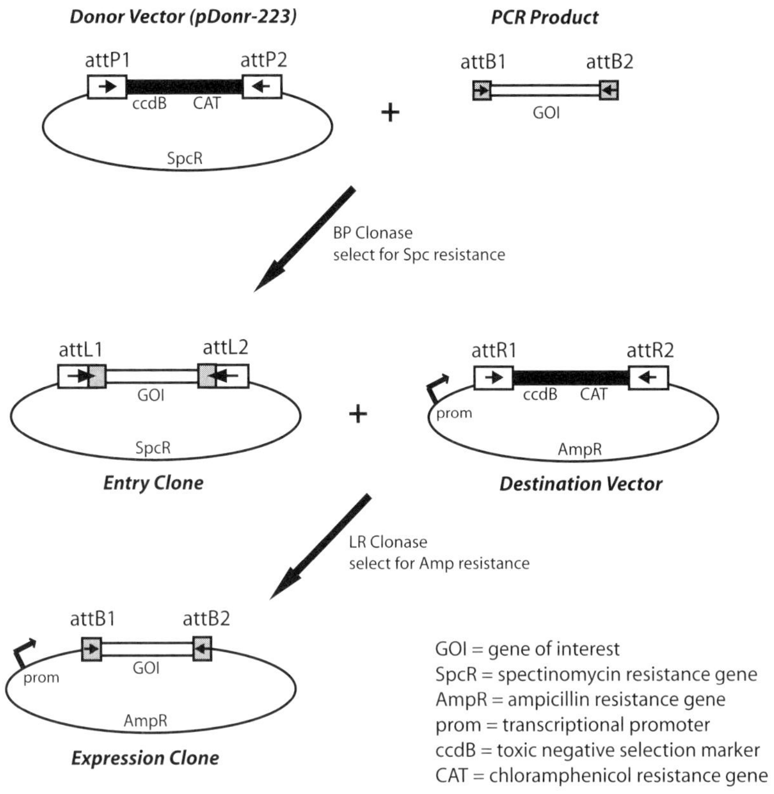

Fig. 3.1. General Gateway cloning scheme. This diagram presents a schematic representation of the standard Gateway reactions, identifying the different plasmid types found in the BP and LR reactions along with the relevant selection markers.

recombination reactions that maintains the reading frame and eliminates the need for directional screening.

Genes of interest can be inserted into Entry clones in several different ways. In order to provide most flexibility in the cloning strategy, we generally use a Gateway recombination reaction (called the BP reaction, *see* **Section 3.3.4**) to transfer genes into Entry clones. Additional entry routes are available, including traditional REaL cloning into commercially available Entry vectors, Topo cloning, or purchase of prepared Gateway ORFeome clones from suppliers such as Open Biosystems. In some cases, these methods are acceptable ways of obtaining Entry clones; however, the flexibility of adding epitope or purification tags is often lost in this process, making the recombination route much more useful. Once an Entry clone is constructed and sequence-verified, a second recombination reaction (called an LR reaction, *see* **Section 3.3.5**) is used to transfer the gene of interest from the Entry clone to the final Expression clone. The vectors used in the transfer are called Destination vectors (DVs). The DVs contain all the signals needed to make protein in the host of interest, including promoters, resistance markers, origins of plasmid replication, etc. In place of the gene of interest, they contain a Gateway "cassette" which encodes for a chloramphenicol resistance gene and a toxin called *ccdB*. These genes are flanked by *attR* recombination sites, and this whole cassette is transferred out during the LR reaction. After the LR reaction, the final Expression clone contains the whole backbone of the Destination vector, with the insert from the Entry clone in place of the Gateway cassette. A combination of positive selection using the marker on the Destination vector and negative selection to eliminate nonrecombinant vectors permits high efficiency generation of the Expression clone. Careful planning of the original Entry clone sequence also ensures that the reading frame of the cloned gene matches the reading frame of any tags in the Destination vector (*see* **Section 3.3.2**). Because each LR reaction can produce a separate Expression clone, a very large number of clones can theoretically be made in parallel from a single Entry clone, permitting Gateway to function in either low- or high-throughput modes depending on the choice of the investigator.

3.2. Materials

3.2.1. Determining Protein Context

1. TEV protease can be produced by published methods *(2)* using the pRK793 expression vector available from Addgene (http://www.addgene.org)
2. Gateway Destination Vector Conversion Kit (Invitrogen, Carlsbad, CA)

3.2.2. Oligonucleotide Design

1. Oligonucleotides can be ordered from numerous suppliers, such as Operon (Germantown, MD). We have found that they generally do not require HPLC or gel purification, and for Gateway reactions, the amount of oligonucleotide used is so small that a 50-nmol synthesis scale is more than sufficient.
2. Oligonucleotides for PCR amplification should be resuspended to a concentration of 10 μM in TE (10 mM Tris–HCl, pH 8.0, 0.1 mM EDTA).

3.2.3. PCR Amplification

1. 2× Phusion Master Mix HF (New England Biolabs, Beverley, MA)
2. DMSO: dimethyl sulfoxide (provided with Phusion Kit or available from Sigma, St. Louis, MO)
3. 96-Well PCR Plates (PGC Scientific, Frederick, MD)
4. QiaQuick PCR Purification Kit (Qiagen,Valencia, CA)
5. QiaQuick 96 PCR Purification Kit (Qiagen, Valencia, CA)

6. Ready-Load 1 kB Plus DNA Ladder (Invitrogen, Carlsbad, CA)
7. ReadyAgarose Gels (BioRad, Hercules, CA)
8. EGel-96 (Invitrogen, Carlsbad, CA)

3.2.4. BP Recombination

1. BP Clonase II kit (Invitrogen, Carlsbad, CA; comes with BP Clonase® II enzyme mix, 2 μg/mL Proteinase K solution).
2. FastPlasmid DNA Kit (Eppendorf, Hamburg, Germany).
3. DH5α chemically competent cells (Invitrogen, Carlsbad, CA). Store at −80°C, do not reuse open vials.
4. CG-Spec medium: Circlegrow medium (40 g/L, MP Biomedicals, Solon, OH), autoclave for 20 min, cool to 55°C, and add 50 μg/mL spectinomycin (Sigma, St. Louis, MO).
5. LB-Spec agar plates: LB-agar petri plates with 100 μg/mL spectinomycin (Teknova, Hollister, CA).
6. Falcon 2059 culture tubes (Fisher Scientific, Pittsburgh, PA).
7. 96-Well V-bottom plates (PGC Scientific, Frederick, MD).
8. Air-pore Sealing Tape (Qiagen, Valencia, CA).
9. Supercoiled DNA Ladder (Invitrogen, Carlsbad, CA). Store at 4°C.
10. Turbo 96 Plasmid Prep Kit (Qiagen, Valencia, CA).

3.2.5. LR Recombination

1. LR Clonase II enzyme mix (Invitrogen, Carlsbad, CA; comes with LR Clonase® II enzyme mix, 2 μg/mL Proteinase K solution.)
2. *E. coli ccdB* Survival competent cells (Invitrogen, Carlsbad, CA). Store at −80°C, do not reuse open vials.
3. *BsrG*I restriction enzyme (New England Biolabs, Beverley, MA).

3.2.6. Downstream Applications

1. GenElute Midiprep DNA Kit (Sigma, St. Louis, MO).
2. *E. coli* DH10Bac competent cells (Invitrogen, Carlsbad, CA). Store at −80°C, do not reuse open vials.
3. LB-KGTXI agar plates: LB-agar petri plates with 7 μg/mL gentamycin, 50 μg/mL kanamycin, 10 μg/mL tetracycline, 40 μg/mL IPTG, and 100 μg/mL Bluo-gal (Teknova, Hollister, CA).
4. CG-KG medium: Circlegrow medium (40 g/L, MP Biomedicals, Solon, OH), autoclave for 20 min, then add 50 μg/mL kanamycin and 7 μg/mL gentamycin (Sigma, St. Louis, MO).
5. LB-agar petri plates (Teknova, Hollister, CA).
6. Benzonase Cracking Buffer (BCB): 25 mM Tris–HCl pH 8.0, 5 mM $MgCl_2$. Store at room temperature.
7. Benzonase (EMD Biosciences, Darmstadt, Germany).

8. NuPAGE 4× LDS Sample Buffer (Invitrogen, Carlsbad, CA). Store at room temperature.
9. TCEP: Tris(2-carboxyethyl)phosphine hydrochloride (Pierce, Rockford, IL). Store at room temperature. Do not use for more than 3 months after opening.
10. MDG medium: a defined medium for growth of *E. coli* under conditions which will not induce T7 polymerase production. For detailed recipes, *see (3)*.
11. Plasmid Prep Buffer A: 25 mM Tris–HCl, pH 8.0, 10 mM EDTA, 0.9% D-glucose. Store at room temperature.
12. Plasmid Prep Buffer B: 0.20 M NaOH, 1.33% sodium dodecyl sulfate (SDS). Store at room temperature. SDS may precipitate at low temperatures – redissolve before use by heating at 37°C. Make fresh after 6 months, and be sure to cap immediately after use.
13. Plasmid Prep Buffer C: 7.5 M ammonium acetate (heat to 50°C to dissolve). Store at room temperature.
14. Plasmid Prep Resuspension Buffer: 10 mM Tris–HCl, pH 8.0, 0.1 mM EDTA, 0.1 μg/mL ribonuclease A (Sigma, St. Louis, MO). Store at room temperature for up to 6 months.
15. Benchmark Protein Ladder (Invitrogen, Carlsbad, CA).
16. Rosetta(DE3) competent cells (Novagen, La Jolla, CA). Other BL21(DE3) varieties can also be purchased as competent cells from Novagen.
17. Bac-to-Bac Baculovirus Expression Kit (Invitrogen, Carlsbad, CA).
18. Criterion SDS–PAGE gels (BioRad, Hercules, CA).
19. 10× TGS Buffer: 250 mM Tris base, 1.92 M glycine, 1% SDS (USB, Cleveland, OH).
20. GelStain: mix 175 mg Brilliant Blue G (Sigma, St. Louis, MO) into 10.5 mL acetic acid and add water to a final volume of 3.5 L. Stir for 15 min on a magnetic stirplate, and filter the solution over Whatman #1 filter paper. Some precipitation may be observed over time, and stain should not be stored for more than 3 months without refiltering.

3.3. Methods

3.3.1. Determining Protein Context

1. Prior to the start of any cloning project, a determination of the desired protein context must be made in order to maximize the downstream flexibility of the final expression clones. Proper choices at this stage can save time and money

later when expression may fail or be unacceptably low under certain conditions.

2. Gateway recombination replaces sequences between the *att* sites on a given plasmid with the sequences between the *att* sites on the other plasmid. For this reason, sequences such as fusion tags, translational start sites, etc. can be placed either in the Entry clone sequence between the *att* sites, or on the Destination Vector outside the *att* sites. Both will produce final Expression clones that make the protein of interest, but the difference in protein sequence may have experimental relevance.
3. For projects requiring the most flexibility, we suggest minimizing the amount of extra sequence present in the Entry clone. The simplest such construct would be an Entry clone containing only the gene of interest with a stop codon. Such a clone can be used to make aminoterminal fusions by subcloning to a Destination vector with an aminoterminal tag. Readthrough of the *attB*1 site would provide an additional 7–8 amino acids between the tag and the protein, but often this is not of major consequence for fusion proteins.
4. If near-native sequence is required, the simplest alteration to this scheme is to include a protease cleavage site in the Entry clone prior to the start of the gene of interest. Our preferred protease of choice is the Tobacco Etch Virus (TEV) protease, which very specifically cleaves a sequence ENLYFQ/X, where X is any amino acid except for proline *(2)*. The addition of this sequence to the Entry clone will permit native protein to be cleaved from any aminoterminal fusion generated by LR recombination (*see* **Note 1**).
5. In many cases, epitope or purification tags are necessary for steps in protein production. Often, multiple tags are necessary to permit multiple rounds of purification or detection. A common technique is to include a protease cleavage site at the start of the gene, and to add an epitope tag (FLAG, HA, His6) to the end of the gene prior to the stop codon. To avoid adding the 7–8 amino acids of the *attB*2 site to such a construct, it is best to add this tag inside the Entry clone sequence. Such small epitope tags are readily added in the PCR amplification step (*see* **Sections 3.3.2** and **3.3.3**).
6. Additional sequences may be required depending on the specific case. A common motif for prokaryotic expression is the addition of a Shine-Dalgarno translation initiation site *(4)*, which should be placed inside the Entry clone for maximal utility. Note, however, that such a sequence will eliminate the possibility of making aminoterminal fusions in bacteria, as all aminoterminal fusion Destination vectors carry their own Shine-Dalgarno sequence, which would lead to translation initiation at two sequences and a mixture of proteins.

Table 3.1
Common Entry clone sequence contexts

GOI-stop	Aminoterminal fusions
GOI-nostop	Aminoterminal and/or carboxyterminal fusions
Kozak-GOI-stop	Aminoterminal fusions or native eukaryotic expression
TEV-GOI-stop	Cleavable aminoterminal fusions
TEV-GOI-Tag	Cleavable am inoterminal fusions with carboxyterminal epitope/purification tag
SD-GOI-stop	Native expression in *E. coli*
Tag-GOI-stop	Aminoterminal tag inside the Entry clone

GOI gene of interest; Kozak, eukaryotic translation initiation sequence; TEV, TEV protease cleavage site; SD, bacterial translation initiation site; Tag, any small epitope or purification tag, such as His6 or FLAG

7. **Table 3.1** provides a list of common motifs for Entry clones which we and others have used. In most cases, these will suffice for wide flexibility. In some cases, additional sequences will be required that are too large for facile introduction into Entry clones. Such sequences (e.g., large fusion tags) can then be placed inside the Destination vectors instead (*see* **Note 2**).

3.3.2. Oligonucleotide Design

1. Entry into Gateway cloning using BP recombination requires a PCR amplification step to add the *attB* recombination signal sequences and any other desired sequences on to the gene of interest. Proper design of the oligonucleotides for this step ensures that the proper reading frame is generated in the final expression clone.
2. To clone most genes, 18–21 bp of gene-specific 5′ and 3′ sequences are used for primer annealing. For simple Entry clones which do not contain large amounts of additional features, the recombination signal sequences given in **Table 3.2** (*attB*1 at the 5′ end of the gene, and *attB*2 at the 3′ end) can be added directly to the gene-specific primers.
3. Longer PCR primers are required for more complicated tagging in Entry clones. Introduction of protease cleavage sites or epitope tags often leads to oligonucleotide lengths in excess of 60 bp. In our experience, such long oligonucleotides often are of reduced quality containing higher numbers of mutations or deletions. This requires more clones to be sequenced in order to avoid frameshift errors. To avoid this problem, a technique known as adapter PCR can be utilized.

4. Adapter PCR involves the use of multiple nested primers to add long 5′ or 3′ sequences to the gene of interest. First, a primer which contains the gene-specific portion and part or all of a tag sequence (such as a TEV protease site) is added to the PCR. After a few rounds of amplification, a second primer is added which contains the *attB* recombination signal and 12–18 bp of overlapping sequence with the first primer. A mixture of PCR products will be produced, but only the full-length product will have the *attB* recombination sites necessary for recombination to occur. **Table 3.2** shows some common adapter primer sequences and their corresponding gene-specific primers.

5. Adapter PCR can be used on both ends simultaneously by adding two different adapter primers. One can also nest multiple levels of adapter PCR to insert long 5′ or 3′ sequences if necessary

Table 3.2
Oligonucleotide sequences for Gateway sites and some common adapters

Standard Gateway primers	
*attB*1 (no Kozak)	5′-GGGGACA ACT TTG TAC AAA AAA GTT GGC – gsp
*attB*1 (Kozak-ATG)	5′-GGGGACA ACT TTG TAC AAA AAA GTT GGC ACC ATG – gsp
*attB*2 (stop)	5′-GGGGAC AAC TTT GTA CAA GAA AGT TGG *CTA* – gsp (reverse comp)
*attB*2 (stop)	5′-GGGGAC AAC TTT GTA CAA GAA AGT TGG – gsp (reverse comp)
Sequences for initial PCR of genes with additional sequences	
Amino-TEV	5′-GGC GAA AAC CTG TAC TTC CAA GGC – gsp
Amino-SD	5′-TTT AAC TTT AAG AAG GAG ATA TAT ACC ATG – gsp
Amino-His6	5′-ACC ATG TCA CAC CAT CAC CAT CAC CAT – gsp
Carboxy-His6	5′-GT TGG *CTA* ATG GTG ATG GTG ATG GTG ACC – gsp
Gateway adapter primers	
*attB*1-Tev adapter	5′-GGGGACAACTTTGTACAAAAAAGTTGGCGAAAAC CTGTACTTCCAAGGC
*attB*1-SD adapter	5′-GGGGACAACTTTGTACAAAAAAGTTGGCTCATT-TAACTTTAAGAAGG
*attB*1-aminoHis6 adapter	5′-GGGGACAACTTTGTACAAAAAAGTTGGCACCAT GTCACACCATCACCATCACCAT
*attB*2-carboxyHis6 adapter	5′-GGGGACAACTTTGTACAAGAAAGTT G*GCTA*ATGGTGATGGTGATGGTGACC

gsp gene-specific primer (should contain 18–21 bp of 5′ or 3′ of gene, in *attB*2 primers this sequence must be the reverse complement of the sense strand of the gene). *Underlined* sequences in the *att* sites identify the proper reading frame (*see* **step** 7 – **Section 3.3.2**). Stop codons in *attB*2 sequences are *italicized* (*see* **Note 3**). Often, the same adapter primers can be used for any gene which has a particular 5′ sequence (i.e., the TEV adapter listed in **Table 3.2**), thus minimizing the cost and number of oligos which need to be generated for a project.

6. In cases where multiple large genes are to be combined (e.g., a fusion of a protein of interest with a second protein of interest), or when large deletions are desired, overlap PCR can be used *(5)*. In this process, two separate PCR amplifications are carried out with 20–25 bp of overlapping sequence between the 3′ end of gene 1 and the 5′ end of gene 2. A third PCR is then carried out using the first two PCR products as templates along with flanking primers containing the *attB* sites. Again, the presence of the *attB* sites during only the last round of PCR ensures that no other side-products will be able to be cloned (*see* **Note 4**).
7. In order to maintain the proper reading frame for Gateway fusions (both amino and carboxy ends), care must be taken in the design of oligonucleotides. Key DNA sequences are present in both the *attB*1 and the *attB*2 sequences to guide reading frame alignment. As shown in **Table 3.2**, the *attB*1 reading frame is centered on the 2 lysine codons encoded by the 5′-AAA AAA sequence. For a gene to be in frame, it must be aligned properly with this sequence. The *attB*1 oligonucleotide sequence given in **Table 3.2** has an additional 3′ nucleotide not actually required for recombination in order to ensure the reading frame is intact. Likewise, the *attB*2 reading frame is centered on the 5′-TTT GTA sequence, and those codons must be in the same frame as the gene of interest (*see* **Note 5**).
8. In addition to oligonucleotides for amplification, oligonucleotides for sequence verification of Entry clones are also required. We generally order 1 primer for every 600 bp of sequence, and an additional primer on the reverse strand that is able to sequence back through the start of the gene. Typical primer lengths are 22–24 bp, and they can be selected manually or with the assistance of many common molecular biology computer programs. Standard Gateway sequencing primers can also be used to sequence into the gene of interest from both directions in the Entry clone (*see* **Note 6**).

3.3.3. PCR Amplification

3.3.3.1. Single Protein PCR Amplification

1. To a 200-μL thin-walled PCR tube, add 1 μL of each 10 μM oligonucleotide primer, 1.5 μL DMSO (*see* **Note 7**), 100–200 ng template DNA (*see* **Note 8**), and water to 25 μL final volume.
2. Add 25 μl 2× Phusion Master Mix HF, mix well, and carry out the PCR amplification using the following parameters: initial denaturation at 98°C for 30 s, 20 cycles of (98°C for 10 s, 55°C for 30 s, and 72°C for 30 s per kB of the expected

product), followed by a 10-min final elongation at 72°C, and cooling to 4°C (*see* **Note 9**).

3. For adapter PCR (*see* **step 5** – **Section 3.3.2**), after 5 cycles of amplification, pause the thermal cycler, and add 1 µL of 10 µM adapter primer(s) to the tubes. Continue cycling for an additional 15 cycles.
4. If multiple nested adapter primers are used, we suggest increasing the overall cycle time so that there are 5 cycles between adapter additions, and at least 15 additional cycles after the final adapter is added.
5. If only a small amount of template DNA is available, increase the number of overall cycles from 20 to 30. This will increase the likelihood of PCR errors, but may improve PCR product yield.
6. After cycling, load 5 µL of the PCR product on a 1% agarose gel to verify size by comparison to a linear DNA standard such as the ReadyLoad 1 kB DNA ladder (*see* **Note 10**).
7. Purify the PCR product using the QiaQuick PCR Purification Kit following the manufacturer's protocol and elute the DNA in 50 µL (*see* **Note 11**).

3.3.3.2. High-Throughput PCR Amplification

1. To each well of a 96-well PCR plate, add 0.5 µL of each 10 µM oligonucleotide primer, 1 µL DMSO (*see* **Note 7**), 50–100 ng template DNA (*see* **Note 8**), and water to 12.5 µL final volume. Addition of primers will be simplified if the oligonucleotides are also in a 96-well format, permitting addition by multichannel pipetting. A master mix of DMSO and water, as well as primers or template, can also be employed where applicable.
2. Add 12.5 µl 2× Phusion Master Mix HF per well, mix well, and carry out the PCR amplification using the following parameters: initial denaturation at 98°C for 30 s, 20 cycles of (98°C for 10 s, 55°C for 30 s, and 72°C for 30 s per kB of the expected product), followed by a 10-min final elongation at 72°C, and cooling to 4°C (*see* **Note 9**).
3. For adapter PCR (*see* **step 5** – **Section 3.3.2**), after 5 cycles of amplification, pause the thermal cycler, and add 0.5 µL of 10 µM adapter primer(s) to the tubes. Continue cycling for an additional 15 cycles.
4. If multiple nested adapter primers are used, we suggest increasing the overall cycle time so that there are 5 cycles between adapter additions, and at least 15 additional cycles after the final adapter is added.
5. After cycling, load 5 µL of the PCR product on a 1% agarose gel to verify size. It is recommended that a 96-well gel format be utilized at this step to reduce effort – options include the e-Gel 96 or various 96-well horizontal agarose gel platforms.

Verify size by comparison to a linear DNA standard such as the ReadyLoad 1 kB DNA ladder.

6. Purify the PCR products using the QiaQuick 96 PCR Purification Kit following the manufacturer's protocol and elute the DNA in 50 µL.

3.3.4. BP Recombination

3.3.4.1. Single Protein BP Recombination

1. Add the following reagents to a microcentrifuge tube in the order given (the total reaction volume should be 10 µL): 1–6 µL H_2O, *attB* flanked PCR fragment (15–150 ng, *see* **Note 12**), 150 ng pDonr-223 vector (*see* **Note 13**), 2 µL of BP Clonase II. A master mix can be used for all reagents except for BP Clonase II, which must be added last. Mix briefly by gentle vortexing.
2. Incubate the reaction mixture for at least 1 h at 30°C (*see* **Note 14**).
3. Add 1 µL of 2 mg/mL Proteinase K to inactivate the BP Clonase and incubate for 15 min at 37°C (*see* **Note 15**).
4. Add 1 µL of the BP reaction to a microcentrifuge tube containing 20 µL of chemically competent *E. coli* DH5 and incubate on ice for 5–10 min (*see* **Note 16**).
5. Heat shock the cells in 42°C water bath for 45 s and immediately add 80 µL of SOC medium. Shake the reaction for 1 h at 37°C.
6. Spread 50 µL of the transformation mix on LB-Spec agar plates and incubate overnight at 37°C. A good BP cloning result with a standard length (1 kB) gene should yield greater than 200 colonies per transformation (*see* **Note 17**).
7. Pick 2–4 separate Entry clone colonies into Falcon 2059 culture tubes containing 2 mL of CG-Spec media and grow overnight at 37°C with 200 rpm shaking.
8. Spin 1 mL of the culture in a microcentrifuge to pellet the cells, and isolate plasmid using the FastPlasmid kit, eluting the DNA in 75 µL of elution buffer (*see* **Note 18**).
9. Verify the size of the plasmid using agarose gel electrophoresis. Load 1 µL of the purified Entry clone DNA on a 1% agarose gel, and compare sizes to the Supercoiled DNA ladder.
10. Properly sized Entry clones should be sequence verified to ensure that no oligonucleotide or PCR-generated errors have been introduced.
11. Glycerol stocks of the *E. coli* strains containing Entry clones should be made by adding 250 µL sterile filtered 60% glycerol to 750 µL of culture. After mixing and incubation at room temperature for 5 min, these stocks can be frozen at –80°C and used to start new cultures if more Entry clone DNA is required in the future.

3.3.4.2. High-Throughput BP Recombination

1. Add the following reagents to a 96-well v-bottom plate in the order given (the total reaction volume should be 10 μL): 1–6 μL of H_2O, 50–100 ng *attB* flanked PCR fragment, 150 ng pDonr-223 vector, 2 μL of BP Clonase II. A master mix can be used for all reagents except for BP Clonase II, which must be added last.
2. Centrifuge the plate to mix the reaction thoroughly and incubate the reaction for at least 2 h at 30°C (*see* **Note 14**).
3. Add 1 μL of 2 mg/mL Proteinase K to inactivate the BP Clonase and incubate for 15 min at 37°C (*see* **Note 15**).
4. Add 1 μL of the BP reaction to a 96-well v-bottom plate containing 10 μL of chemically competent *E. coli* DH5α and incubate on ice for 5–10 min (*see* **Note 16**).
5. Heat shock the cells in a 42°C water bath for 45.0 s and immediately add 40 μL of SOC medium. Shake the reaction for 1 h at 37°C.
6. Spread the whole 50 μL of the transformation mix on LB-Spec agar plates and incubate overnight at 37°C.
7. Pick 1–2 separate Entry clone colonies into a 96-well deep well plate containing 1.5 mL of CG-Spec media per well, cover with Air-Pore sealing tape, and grow overnight at 37°C with 250 rpm shaking.
8. Isolate plasmid DNA using a commercially available high-throughput plasmid prep kit, such as the Qiagen Turbo 96 Kit (*see* **Note 19**).
9. Verify the size of the plasmids using agarose gel electrophoresis. Load 1 μL of the purified Entry clone DNA on a 1% agarose gel, and compare sizes to a supercoiled DNA ladder (Invitrogen, Carlsbad, CA).
10. Properly sized Entry clones should be sequence verified to ensure that no oligonucleotide or PCR-generated errors have been introduced.
11. Glycerol stocks of the *E. coli* strains containing Entry clones can be made by adding 50 μL sterile filtered 60% glycerol to the wells of a microtiter plate. Add 150 μL of Entry clone culture to each well, cover with aluminum sealing tape, and freeze at −80°C for long-term storage.

3.3.5. LR Recombination

3.3.5.1. Single Protein LR Recombination

1. Add the following reagents to a microcentrifuge tube in the order given (the total reaction volume should be 10 μL): 1–6 μL H_2O, Entry clone DNA (50–100 ng, *see* **Note 20**), Destination Vector DNA (150 ng, *see* **Note 21**), 2 μL of LR Clonase II.
2. Incubate the reaction mixture for at least 1 h at 30°C (*see* **Note 22**).

3. Add 1 μL of 2 mg/mL Proteinase K to inactivate the LR Clonase and incubate for 15 min at 37°C (*see* **Note 15**).
4. Add 1 μL of the LR reaction to a microcentrifuge tube containing 20 μL of chemically competent *E. coli* DH5α and incubate on ice for 5–10 min (*see* **Note 16**).
5. Heat shock the cells in 42°C water bath for 45 s and immediately add 80 μL of SOC medium. Shake the reaction for 1 h at 37°C.
6. Spread 50 μL of the transformation mix on LB agar plates containing the correct antibiotic (often ampicillin, but check the Destination vector literature) and incubate overnight at 37°C. A good LR cloning result should yield greater than 500 colonies per transformation.
7. Pick two separate Expression clone colonies into Falcon 2059 culture tubes containing 2 mL of CG media containing the necessary antibiotics and grow overnight at 37°C with 200 rpm shaking.
8. Spin 1 mL of the culture in a microcentrifuge to pellet the cells, and isolate plasmid using the FastPlasmid kit, eluting the DNA in 75 μL of elution buffer (*see* **Note 18**).
9. Verify the size of the plasmid using agarose gel electrophoresis. Load 1 μL of the purified Expression clone DNA on a 1% agarose gel, and compare sizes to the Supercoiled DNA ladder.
10. If additional confirmation of the Expression clone is required, restriction enzyme analysis can be performed (*see* **Note 23**).
11. Glycerol stocks of the *E. coli* strains containing Expression clones should be made by adding 250 μL sterile filtered 60% glycerol to 750 μL of culture. After mixing and incubation at room temperature for 5 min, these stocks can be frozen at –80°C and used to start new cultures if more Expression clone DNA is required in the future.

3.3.5.2. High-Throughput LR Cloning

1. Add the following reagents to a 96-well v-bottom plate in the order given (the total reaction volume should be 10 μL): 1–6 μL of H_2O, 50 ng Entry clone DNA, 150 ng Destination vector, 1 μL of LR Clonase II. A master mix can be used for all reagents except LR Clonase II, which must be added last.
2. Centrifuge the plate to mix the reaction thoroughly and incubate the reaction for at least 2 h at 30°C (*see* **Note 22**).
3. Add 1 μL of 2 mg/mL Proteinase K to inactivate the BP Clonase and incubate for 15 min at 37°C (*see* **Note 15**).
4. Add 1 μL of the LR reaction to a 96-well v-bottom plate containing 10 μL of chemically competent *E. coli* DH5α and incubate on ice for 5–10 min (*see* **Note 16**).

5. Heat shock the cells in a 42°C water bath for 45 s and immediately add 40 μL of SOC medium. Shake the plate for 1 h at 37°C.
6. Spread the whole 50 μL of each transformation mix on LB agar plates containing the correct antibiotic (often ampicillin, but check the Destination Vector literature) and incubate overnight at 37°C.
7. Pick one Expression clone colony from each plate into a 96-well deep well plate containing 1.5 mL of CG media plus antibiotics per well, cover with Air-Pore sealing tape, and grow overnight at 37°C with 250 rpm shaking.
8. Isolate plasmid DNA using a commercially available high-throughput plasmid prep kit, such as the Qiagen Turbo 96 Kit (*see* **Note 19**).
9. Verify the size of the plasmids using agarose gel electrophoresis. Load 1 μL of the purified Entry clone DNAs on a 1% agarose gel, and compare sizes to a supercoiled DNA ladder.
10. If additional confirmation of the Expression clone is required, restriction enzyme analysis can be performed (*see* **Note 23**).
11. Glycerol stocks of the *E. coli* strains containing Expression clones can be made by adding 50 μL sterile filtered 60% glycerol to the wells of a microtiter plate. Add 150 μL of Expression clone culture to each well, cover with aluminum sealing tape, and freeze at –80°C for long-term storage.

3.3.6. Downstream Applications

3.3.6.1. Preparation of Mammalian/Yeast Expression Clones

1. Generally for mammalian transfection or yeast transformation, large quantities of expression clone DNA are required. In these cases, a 50-mL culture of the expression clone should be grown using CG media with the proper selective antibiotics. Plasmid preparation from this culture can be accomplished using commercially available Midiprep DNA kits. We commonly use GenElute kits, which are rapid, inexpensive, and produce high-quality DNA (*see* **Note 24**).
2. For yeast transformation, expression clone DNA will sometimes require linearization with a restriction enzyme prior to use. This depends on the expression vector being used, and this information can usually be found in the literature that comes with the chosen vector.

3.3.6.2. Preparation of Baculovirus Expression Clones

1. Baculoviruses are used to express proteins in insect cells, which are often a more reliable host for the production of eukaryotic proteins. Numerous Gateway baculovirus vectors are available, and most use the Bac-to-Bac system for generation of recombinant baculoviruses (*6*).
2. Transform Baculovirus Expression clones into *E. coli* DH10Bac cells using the following protocol: mix 1 μL of Expression

clone DNA and 50 μL of competent cells in a Falcon 2059 tube, and incubate on ice for 5 min. Heat shock samples at 42°C for 45 s, and add 450 μL SOC medium to the tube. Incubate tubes at 37°C for 4 h with 200-rpm shaking.

3. After growth, plate 5 and 25 μL of the culture on LB-KGXTI agar plates and incubate the plates overnight at 37°C. Restreaking of positive clones may be required on the following day (*see* **Note 25**).
4. Pick positive white colonies into 2 mL LB-KG medium and grow overnight at 37°C with 200 rpm shaking. One milliliter of this culture can be used to prepare bacmid DNA using a standard alkaline lysis plasmid preparation procedure as follows (*see* **Note 26**).
5. Centrifuge 1 mL of culture in a microcentrifuge tube at maximum speed for 1 min, and aspirate the supernatant.
6. Add 150 μL Plasmid Prep buffer A and mix by pipetting.
7. Add 150 μL Plasmid Prep buffer B and mix by gentle inversion.
8. Add 150 μL Plasmid Prep buffer C and mix by gentle inversion. A white flocculent precipitate will form.
9. Centrifuge for 10 min at maximum speed, and transfer 400 μL sample to a new microcentrifuge tube. Centrifuge for an additional 5 min at maximum speed and transfer 350 μL to a new microcentrifuge tube.
10. Add 800 μL of 100% ethanol, mix by vortexing.
11. Centrifuge for 20 min at maximum speed, decant supernatant, recentrifuge briefly, and pipet out the remaining liquid.
12. Dry the tubes at 37°C for 5 min with the caps open to evaporate any remaining ethanol.
13. Resuspend in 50 μL Plasmid Resuspension buffer, incubate for 5 min at 37°C.
14. Centrifuge at maximum speed for 2 min and transfer the supernatant to a new tube, avoiding transfer of any solid pellet or particulates.
15. Generally, 5–10 μL of this material can be used for insect cell transfections. PCR or gel analysis can be performed to verify the bacmid DNA if desired (*see* **Note 27**).

3.3.6.3. Expression in Escherichia Coli

1. Expression clones should be transformed into an appropriate *E. coli* strain using chemical transformation (*see* **Section 3.3.4.1, steps 4–6**), and plated on LB-agar plates containing the proper selective antibiotics (*see* **Note 28**).
2. After overnight incubation at 37°C, pick a single colony using a sterile toothpick or pipet tip, and inoculate it into 2-mL liquid growth media containing the proper antibiotics

for selection. For clones with a T7 promoter, we suggest MDG media to avoid induction in the initial growth phase. For other promoters, standard LB or CG media can be substituted.

3. Incubate starter cultures for 12–16 h at 37°C with 200-rpm shaking.
4. Dilute starter cultures 1:100 into fresh growth media (CG or equivalent rich medium is preferred) containing the proper antibiotics for selection, and grow at 37°C with 200 rpm shaking until an OD_{600} of 0.4–0.6 is reached. For most clones, this will take between 2 and 3 h.
5. Induce cells for protein expression with the appropriate inducer for the particular vector – in most cases this will be IPTG at a final concentration of 0.5–1.0 mM. After induction, continue to grow the cells for either 3–4 h at 37°C, or alternately for 16–20 h at 18°C (*see* **Note 29**).
6. After induction, remove a sample of the culture for SDS–PAGE analysis of induced proteins. Generally, it is a good idea to remove a sample prior to induction as well in order to compare the uninduced and induced cells. 0.1 OD units (100 µL of a culture with an OD_{600} of 1.0) is usually appropriate for gel analysis.
7. Centrifuge the cells in a microcentrifuge at maximum speed, and aspirate the supernatant. Resuspend cells in 20 µL Benzonase cracking buffer (BCB), mix, and freeze on dry ice for at least 5 min.
8. Thaw the frozen cell pellet and add 1 unit of Benzonase, and incubate at 37°C for 10 min. After incubation, add 7.5 µL 4× SDS–PAGE loading buffer and 2 µL TCEP to each cell pellet.
9. Heat the samples at 70°C for 10 min prior to SDS–PAGE analysis.
10. Set up a Criterion SDS–PAGE gel of a proper percentage range for the proteins of interest (*see* **Note 30**).
11. Fill the upper and lower chambers of the gel apparatus with 1× TGS buffer.
12. Load 10 µl of sample in each well, along with 2 µl of Benchmark Protein Ladder as a molecular size standard.
13. Hook up the electrodes, and apply 200 V for 55 min.
14. At the end of the run, remove the gel from its cassette and place it in 100 mL of ultrapure water in a plastic tray. Heat for 1 min in a microwave oven at high power, and shake gently at room temperature for 3–5 min (*see* **Note 31**).
15. Pour off the water and add 70 mL of Gel Stain and microwave on high for 30–45 s or until the stain first starts to boil. Shake the gel gently at room temperature for 30 min (*see* **Note 32**).

16. Destain the gel by removing the stain, and washing 3× in ultrapure water. Add 200 mL ultrapure water and microwave for 2 min. Add 2–3 KimWipes to the tray, and gently shake for 15–30 min at room temperature (*see* **Note 33**).

3.4. Notes

1. The use of TEV protease is preferred by many laboratories for several reasons. First, TEV protease is relatively easy to produce in the lab, thus reducing the costs of purchasing more expensive reagents (e.g., thrombin). Second, TEV sites are rarely found in proteins, increasing the specificity of cleavage. Third, TEV is a highly efficient enzyme and has relatively mild optimal cleavage conditions. Finally, the flexibility at the final amino acid position means that often one can make a truly native aminoterminus after cleavage *(7)*.
2. An additional advantage of the Gateway system is that it is trivial to convert any expression vector (commercial or your favorite lab vector) to a Gateway Destination Vector. The Gateway Destination Vector Conversion Kit contains a Gateway cassette with positive and negative selection markers, and this can be ligated into any blunt-ended restriction site in your vector of interest. The positive selection marker (CAT) enables easy selection of the final Destination vector, and there are three different reading frame cassettes available to fit any vector. The whole process takes no more than 2–3 days to complete, and the final vector can be used in a Gateway LR reaction.
3. Adapter PCR with more than two nested adapters tends to be highly inefficient. If more than two adapters are required on the same end of the DNA, we suggest considering an alternative method of generating the construct. If not, a larger amount of PCR product may be necessary in the Gateway BP reaction due to the small amount of PCR product that contains the full-length product with *att* sites.
4. Overlap PCRs and PCRs with multiple adapters tend to create more side-products called primer-dimers. These are small DNA molecules that result from improper annealing of primers, and which can sometimes contain 1 or 2 *attB* sites. Due to their small size, primer-dimers are highly efficient in the BP recombination reaction and may produce false positive Entry clones.
5. It is a common problem for Gateway cloners to confuse the *attB2* reading frame due to the use of the reverse complement

sequence and its strong similarity to the *attB*1 sequence. We strongly recommend some kind of in silico modeling of Gateway reactions prior to primer design. Several commercial molecular biology software programs can carry out virtual Gateway cloning, including VectorNTI (Invitrogen, Carlsbad, CA). Other programs in common use can often be tricked into carrying out Gateway reactions by pretending that Gateway *att* sites are restriction enzymes. In either case, a quick check of the reading frame is always a worthwhile endeavor.

6. Due to the presence of inverted *attL* repeats in the Entry clones, sequencing of small inserts (<500 bp) can be difficult. For this purpose, one can use special blocking oligonucleotides to eliminate the problem *(8)*, or simply sequence with gene-specific primers. The latter is often easier for single protein cloning projects, while the former is more efficient for high-throughput projects.
7. The addition of 3% DMSO to the PCR can help to minimize the effects of very GC-rich primers or template. Often it is not required, but we have seen no detrimental effect from including it in most PCRs. If templates are very AT rich, it is suggested that the DMSO be left out.
8. The use of large amounts of template DNA helps to dramatically reduce PCR errors by forcing the use of original template molecules for subsequent PCR cycles rather than PCR products which may contain errors. If limited template is available, this amount can be reduced by 10- to 20-fold, but errors will likely increase.
9. Phusion polymerase has become the standard PCR reagent in our lab due to its robust activity and extremely high fidelity. Other polymerases (KOD, Pfu) can also be used, but we would recommend using only high-fidelity enzymes to limit PCR mutations, particularly with long genes. The suggested conditions are optimized for use on BioRad or Applied Biosystems PCR machines; some optimization may be required if PCR machines with slower ramp times are used.
10. The appearance of a properly sized band on an agarose gel does not guarantee the success of the adapter PCR. In most cases, the extra length added by the adapter will not be long enough to distinguish the full-length product from a product of only the first set of primers. For this reason, if failures are observed in BP reactions after adapter PCR, it may be worthwhile to split the PCR into two separate reactions to ensure that the adapter PCR is actually working.
11. Column purification of PCR products is only successful for products >150 bp in length. For smaller products, a PEG

precipitation can be carried out as detailed in the Gateway product manuals. Failure to purify the PCR products will generally lead to a large amount of primer-dimers, small fragments caused by primer misannealing, which will clone very efficiently in the BP reaction. In extreme cases, gel purification may be necessary to eliminate these products completely.

12. Generally, the more PCR product used in the reaction, the higher the efficiency will be. Particularly with long PCR products (>5 kB), the higher end of the concentration range should be used. Be aware that with adapter PCRs, the effective concentration of PCR product with both *attB* sites may be lower than the actual concentration.
13. PDonr-223 is one example of a Donor vector – the *attP*-containing vector which becomes the backbone of the Entry clone after the BP reaction. pDonr-223 is a spectinomycin-resistant version of pDonr-221, the standard Gateway Donor vector. We suggest using pDonr-223 rather than pDonr-221 because the kanamycin resistance marker on many cDNA templates can interfere with the kanamycin selection of pDonr-221. Additional Donor vectors are available from Invitrogen (Carlsbad, CA). Note that Donor vectors must be propagated in *E. coli ccdB* Survival or another strain which is resistant to the *ccdB* toxin.
14. To increase the recombination efficiency of the BP reaction an overnight incubation (not longer than 18 h) at 25°C should be performed. A fivefold increase in colonies can be expected, though in some cases, background will also be increased. Also, longer incubations are recommended for PCR products >5 kB.
15. Failure to Proteinase K treat the BP reaction will result in dramatically reduced colony counts due to the inability of the DNA to transform while coated with Clonase proteins.
16. DH5α is a recommended *E. coli* strain for Gateway reactions. However, it can be substituted with any other *recA endA* strain (such as TOP10 or DH10B) if necessary. Be sure that any strain used does not have the F′ episome as it contains the *ccd*A gene which will detoxify the *ccdB* gene resulting in failure of the negative selection. For good BP results, be sure that the competent cells have a transformation efficiency of at least 1×10^8 cfu/μg. Electrocompetent cells can also be used instead of chemically competent cells; however, the only advantage would be in the case of a very low efficiency BP reaction (such as with a very long gene) – usually the number of colonies obtained with standard chemically competent cells is more than sufficient.

17. For high-throughput screening and plating, a regular petri plate (100 × 15 mm) can be divided into three or four quadrants to plate multiple 50-μL transformation mixtures at the same time. The cells can be spread using plastic L-spreaders or inoculating loops.
18. Many commercial kits are available for generating plasmid DNA from *E. coli*. We prefer the FastPlasmid kit for routine plasmid preps due to its high speed and consistent results. FastPlasmid can only be used for DNA generated in $EndA^+$ hosts, as it does not remove nucleases which could affect downstream processes. A standard alkaline lysis plasmid prep will also work – *see* **Section 3.3.6.2, steps 5–14** for a protocol.
19. There are many commercial kits for high-throughput plasmid preparations, but the Qiagen Turbo Plasmid Prep Kit is fast, cheap, and produces DNA of a reasonable quality for downstream sequencing and subcloning reactions.
20. Addition of extra Entry clone may improve efficiency, but this will likely also increase the chance of cotransformation of Expression clone DNA and Entry clone DNA into the same cell. We suggest a maximum of 50–100 ng Entry clone to avoid cotransformation problems. Increases in efficiency, if needed, can be achieved by increasing the amount of Destination vector.
21. Destination vectors can be added as either supercoiled plasmids, or as plasmids that have been linearized within the Gateway cassette. Contrary to the manufacturer's claim, using linearized Destination vector will actually improve LR efficiency 2- to 5-fold. However, due to the high efficiency of the LR reaction, we do not find the extra effort required to prepare and purify linearized Destination vector to be worthwhile.
22. To increase the recombination efficiency of the LR reaction an overnight incubation (not longer than 18 h) at 25°C should be performed. However, LR reactions are generally efficient enough that a 1-h incubation at 30°C produces more than sufficient numbers of colonies, and a reduced reaction time may even be acceptable.
23. Gateway reactions are usually so efficient and accurate that further confirmation of Expression clones is not necessary. However, if desired, the Gateway *attB* sites can be cleaved with the restriction enzyme *Bsr*GI, which will cut out your gene of interest (if it has no additional sites) and allow verification of insert size. Alternately, other restriction sites can be employed.
24. For some applications, DNA for transfection into mammalian cells must be generated with very low levels of bacterial

endotoxin. For these purposes, plasmid prep kits specifically designed for endotoxin removal should be used.

25. In a DH10Bac transformation, two types of colonies will be observed – blue colonies are nonrecombinant, while white colonies contain the proper recombinant baculovirus. Some white colonies may contain small amounts of nonrecombinant baculovirus which can affect downstream success. In order to ensure low background, if time is available white colonies should be restreaked onto fresh LB-KGXTI plates and grown for an additional day at 37°C. Colonies which produce streaks with any blue color at all should be discarded. You should also expect to find that the blue colonies are smaller in size – for this reason, the largest colonies should be picked where possible.
26. DH10Bac cells harbor a single copy plasmid containing the entire baculovirus genome. This "bacmid" DNA cannot be purified by many commercially available kits due to its size and low copy number; however, a classical alkaline lysis preparation will produce enough material for subsequent insect cell transfections.
27. Gel analysis of bacmid DNA is difficult due to the size of the bacmid, interference from leftover plasmid DNA, and low concentration. We prefer PCR verification of bacmids using the suggestions in the manufacturer's Bac-to-Bac kit. The utility and process of the PCR verification will vary depending on your choice of vectors.
28. Choice of expression strains is dependent partially on the type of expression vector being used. In most cases, vectors contain the T7 promoter, and thus require a strain that makes the T7 RNA Polymerase. Common choices are BL21(DE3) or its derivatives. A frequent choice in our laboratory is Rosetta(DE3), which is a derivative of BL21 which contains a plasmid that encodes tRNA genes for rare codons often found in mammalian genes.
29. Lower temperature induction can dramatically improve soluble protein expression for many proteins *(9)*. If an 18°C shaker is not available, induction at 25°C or even 30°C often gives improvement of soluble protein expression. A comparison of solubility at 37°C and lower temperatures may be valuable.
30. Criterion gels offer a wide range of polyacrylamide concentrations to optimize separation of proteins of various molecular weights. We prefer the wide-range 4–12 and 10–20% gels for proteins in the 40–150 and 10–80 kDa ranges, respectively. Criterion offers 26-well gels that have high resolution and fast run times; however, other SDS gel systems can be substituted, and voltages and run times may vary by manufacturer.

31. The heating in water is an essential step to remove SDS from the gel prior to staining. Elimination of this step will result in significantly less sensitive staining, as the dye will not bind well to regions of the gel coated with detergent. Contrary to many published protocols, fixation of the gel is not required, and a short soaking in water will not adversely affect resolution of your protein samples.
32. This staining protocol produces the same sensitivity as traditional Coommasie Blue staining without the need for methanol and in considerably less time. It is equivalent to the commercially available "safe" staining reagents such as Invitrogen's SimplyBlue or Pierce's GelCode Blue in sensitivity and speed, but is considerably cheaper.
33. We find that KimWipes are an economical destaining agent, soaking up much of the dye and accelerating the destaining process significantly. One must be careful not to allow gels to destain for too long or protein bands will begin to destain. If time is a problem, gels should be stained as mentioned, and then diluted with 100 mL water and let sit overnight. Destaining can then be carried out the next day. We do not recommend leaving the gels in destain overnight.

Acknowledgments

The authors would like to thank Kelly Esposito and Dr. William Gillette for assistance in protocol development. Some Gateway adapter PCR protocols are derived from original work by Dr. David Cheo and Dr. Jim Hartley. This work has been funded in whole or in part with federal funds from the National Cancer Institute, National Institutes of Health, under contract N01-CO-12400. The content of this publication does not necessarily reflect the views or policies of the Department of Health and Human Services, nor does mention of trade names, commercial products, or organizations imply endorsement by the U.S. Government.

References

1. Hartley, J. L., Temple, G. F., and Brasch, M. A. (2000) DNA cloning using in vitro site-specific recombination. Genome Res. 10, 1788–1795.
2. Kapust, R. B., Tozser, J., Fox, J. D., Anderson, D. E., Cherry, S., Copeland, T. D., and Waugh, D. S. (2001) Tobacco etch virus protease: mechanism of autolysis and rational design of stable mutatnts with wild-type catalytic efficiency. Protein Eng. 14, 993–1000.
3. Studier, F. W. (2005) Protein production by auto-induction in high density shaking cultures. Protein Expr. Purif. 41, 207–234.

4. Shine, J. and Dalgarno, L. (1975) Determinant of cistron specificity in bacterial ribosomes. Nature 254, 34–38.
5. Horton, R. M., Cai, Z. L., Ho, S. N., and Pease, L. R. (1990) Gene splicing by overlap extension: tailor-made genes using the polymerase chain reaction. Biotechniques 8, 528–535.
6. Luckow, V. A., Lee, S. C., Barry, G. F., and Olins, P. O. (1993) Efficient generation of infectious recombinant baculoviruses by site-specific transposon-mediated insertion of foreign genes into a baculovirus genome propagated in *Escherichia coli*. J. Virol. 67, 4566–4579.
7. Kapust, R. B., Tozser, J., Copeland, T. D., and Waugh, D. S. (2002) The P1′ specificity of tobacco etch virus protease. Biochem. Biophy. Res. Commun. 294, 949–955.
8. Esposito, D., Gillette, W. K., and Hartley, J. L. (2003) Blocking oligonucleotides improve sequencing through inverted repeats. Biotechniques 35, 914–920.
9. Hammarstrom, M., Hellgren, N., van den Berg, S., Berglund, H., and Hard, T. (2002) Rapid screening for improved solubility of small human proteins produced as fusion proteins in *Escherichia coli*. Protein Sci. 11, 313–321.

Chapter 4

Flexi Vector Cloning

Paul G. Blommel, Peter A. Martin, Kory D. Seder, Russell L. Wrobel, and Brian G. Fox

Summary

A protocol for ligation-dependent cloning using the Flexi Vector method in a 96-well format is described. The complete protocol includes PCR amplification of the desired gene to append Flexi Vector cloning sequences, restriction digestion of the PCR products, ligation of the digested PCR products into a similarly digested acceptor vector, transformation and growth of host cells, analysis of the transformed clones, and storage of a sequence-verified clone. The protocol also includes transfer of the sequence-verified clones into another Flexi Vector plasmid backbone. Smaller numbers of cloning reactions can be undertaken by appropriate scaling of the indicated reaction volumes.

Key words: Flexi vector; Gene cloning; High throughput; Expression vectors; Proteomics; Genomics

4.1. Introduction

The availability of sequenced genomes has stimulated investigations into the best high-throughput methods to obtain the encoded proteins and enzymes. Structural genomics *(1–3)*, functional proteomics *(4)*, drug discovery *(5–7)*, agricultural research *(8)*, environmental studies *(9)*, and many other topics of current research in protein biochemistry and enzymology benefit from these efforts.

An essential prerequisite is the establishment of reliable and reproducible protocols for high-throughput cloning. Current methods include recombinational *(10–12)*, ligation-independent *(13–16)*, and ligation-dependent cloning *(17)*. Because of the complexity of protein expression and folding in heterologous hosts, methods to efficiently transfer cloned and sequence-verified genes to many different expression contexts are desirable.

Sharon A. Doyle (ed.), *Methods in Molecular Biology: High Throughput Protein Expression and Purification, vol. 498*

Book doi: 10.1007/978-1-59745-196-3

Flexi Vector cloning is a ligation-dependent method facilitated by selection for the replacement of a toxic gene insert in an acceptor vector *(17)*. Genome-scale restriction mapping has shown that the combination of SgfI and PmeI restriction sites used by the Flexi Vector method allows cloning of 98.9% of all human genes, 98.9% of mouse, 98.8% of rat, 98.5% of *C. elegans*, 97.8% of zebra fish, 97.6% of *Arabidopsis*, and 97.0% of yeast genes, suggesting broad overall utility for use with eukaryotes. This protocol covers Flexi Vector cloning of genes from cDNA directly into an expression vector and the subsequent high-fidelity transfer of the sequence-verified coding region into alternate expression vectors.

4.2. Materials

4.2.1. Flexi Vector Plasmids

Flexi Vector plasmids for bacterial, cell-free, and mammalian cell expression are available from Promega Corporation (Madison, WI). The University of Wisconsin Center for Eukaryotic Structural Genomics (CESG) will be depositing Flexi Vector plasmids and genes cloned by the Flexi Vector method into the Materials Repository of the Protein Structure Initiative at the Harvard Institute of Proteomics (http://plasmid.hms.harvard.edu). Among these are the *Escherichia coli* expression vectors pVP56K and pVP68K, and the wheat germ cell-free expression vector pEU-His-FV.

Figure 4.1 shows a map of the plasmid pVP56K, which is used to create a His8-MPB-target fusion protein. The target protein can be liberated from the N-terminal portion of the fusion by treatment with tobacco etch virus (TEV) protease ***(12, 18)***. The Bar-CAT cassette, bounded by SgfI and PmeI restriction sites, consists of the lethal barnase gene to select against the parental plasmid during cloning and the chloramphenicol acetyltransferase gene to select for the presence of the cassette during construction and propagation of the vector. Plasmids containing the lethal barnase gene must be propagated in a barnase-resistant strain (e.g., *Escherichia coli* BR610, which is available through Technical Services, Promega Corporation).

4.2.2. Target Genes

Target cDNA originally cloned by the Mammalian Gene Collection (http://mgc.nci.nih.gov/) can be purchased from Open Biosystems (http://www.openbiosystems.com/), Invitrogen (https://www.invitrogen.com/), or American Type Culture Collection (ATCC, http://www.atcc.org/catalog/molecular/index.cfm). Other sources of eukaryotic cDNAs that the CESG has used are the Kazusa DNA Research Institute (http://www.kazusa.or.jp/eng/index.html), and the Arabidopsis Biological Resource Center (ABRC, http://www.biosci.ohio-state.edu/pcmb/Facilities/

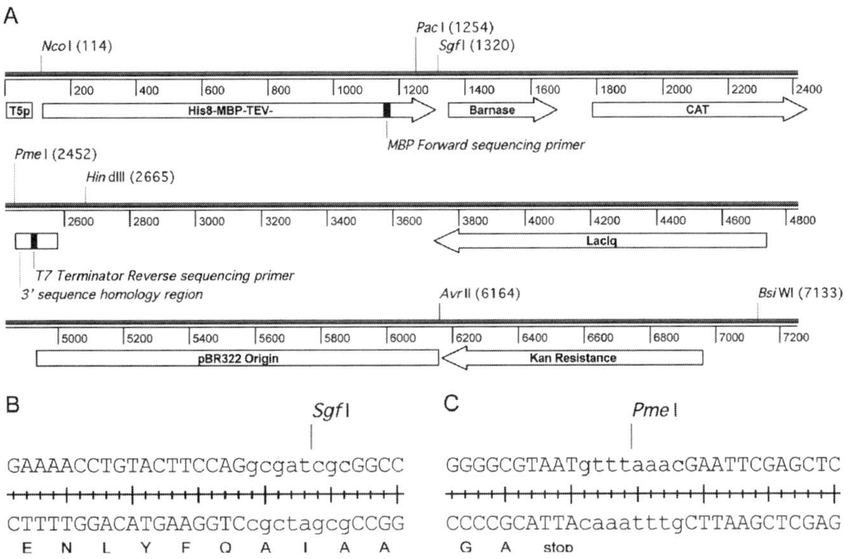

Fig. 4.1. *Escherichia coli* expression vector pvp56k. (a) Linear map showing key features of the vector. (b) Sequence in the region near to the sgfi site. The nucleotide and encoded protein sequence of a portion of the linker between His8-MBP and the target is shown. The TEV protease site is ENLYFQA, where proteolysis occurs between the Q and A residues. After expression of the fusion protein, an N-terminal AIA-target is released by treatment with TEV protease. The identity of the next residues in the target is determined by the PCR primer design. (c) Sequence in the region near to the pmeI site, including the stop codon of the target gene.

abrc/abrchome.htm). Flexi Vector cloning can also be applied to cDNA libraries or genomic DNA prepared from natural organisms or tissues. Genes already cloned by the Flexi Vector method are available from Origene (Rockville, MD) and the Kazusa DNA Institute.

4.2.3. Flexi Vector Reagents

The SgfI/PmeI 10X Enzyme Blend and Buffer (Product No. R1852), high concentration T4 DNA ligase (M1794), Magnesil PCR cleanup kits (A923A), Magnebot II magnetic bead separation block (V8351), and DNA molecular weight markers (PR-67531) are from Promega. (*see* **Note 1**).

4.2.4. PCR Reagents

ORF specific primers (25-nmol synthesis with standard desalting) can be obtained from IDT (Coralville, IA). The dNTP mix (10 mM of each nucleotide, U1515) is from Promega. An MJ DNA Engine, DYAD, Peltier Thermal Cycler (MJ Research, Waltham, MA) can be used. HotMasterTaq DNA Polymerase (0032 002.676) is from Eppendorf (Hamburg, Germany). PCR plates (T-3069-B) are from ISC Bioexpress (Kaysville, UT). Adhesive covers for PCR plates (4306311) are from Applied Biosystems (Foster City, CA). The protocol was originally developed using YieldAcc HotStart DNA polymerase (600336) from Stratagene (La Jolla, CA) and can be substituted by Pfu Ultra II Hotstart DNA polymerase

(600672). PCR plates are centrifuged in an Allegra 6R centrifuge with a GH3.8 rotor (Beckman Coulter, Fullerton, CA).

4.2.5. Bacterial Cell Culture Materials

Select96 competent cells (L3300) are from Promega. Deep-well growth blocks (19579) are from Qiagen (Valencia, CA). CircleGrow medium (3000-132-118268) is from Q-BIOgene (Morgan Irvine, CA). Secure Seal sterile tape (05-500-33) is from Fisher Scientific. AeraSeal gas permeable sealing tape (T-2421-50) is from ISC Bioexpress (Kaysville, UT). The microplate shaker (12620-926) is from VWR (West Chester, PA). Growth blocks are centrifuged in a Beckman Allegra 6R centrifuge with a GH3.8 rotor or in a Beckman Avanti J30-I with a JS5.9 rotor.

4.2.6. Plasmid Preparation

ColiRollers plating beads (71013-3) are from EMD Biosciences (San Diego, CA). Agar plates contain Luria Bertani medium plus 0.5% (w/v) glucose and 50 μg/mL of kanamycin. CircleGrow medium is from Q-BIOgene and is supplemented to contain 50 μg/mL of kanamycin. QiaVac96 vacuum plasmid preparation materials are from Qiagen.

4.2.7. DNA Analysis and Sequence Verification

Optical spectroscopy is used to assess DNA concentration and purity *(19)*. In this work, measurements are made with a SpectroMax Plus Model 01269 spectrophotometer (Molecular Devices, Sunnyvale, CA). Samples are measured in UV Star plastic 384-well plates (T-3118-1) from ISC Bioexpress.

PCR is used for qualitative insert size mapping and for DNA sequencing. The vector-directed forward and reverse primers are 5′-GATGTCCGCTTTCTGGTATGC-3′ (MBP Forward sequencing primer, *black* rectangle starting at 1,155 bp, **Fig. 4.1a**) and 5′-GCTAGTTATTGCTCAGCGG-3′, (T7 Terminator sequencing primer, *black* rectangle starting at 2,501 bp, **Fig. 4.1a**), respectively. A 2.5X PCR Mastermix (FP-22-004-10) from Fisher and the 2% E-gel 96 system (G7008-02) from Invitrogen (Carlsbad, CA) are used for insert size determination. Big Dye Version 3.1 sequencing reagents are from Applied Biosystems. DNA sequencing can be performed at the University of Wisconsin Biotechnology Center.

4.3. Methods

Standard molecular cloning techniques are used *(20)*. A comparison of Flexi Vector and Gateway cloning methods has been published *(17)*. Promega also provides detailed instructions for Flexi Vector cloning *(21)*.

The complete protocol consists of PCR amplification of the desired gene to append Flexi Vector cloning sequences, restriction digestion of the PCR products, ligation of the digested PCR products into a similarly digested acceptor vector, transformation and growth of host cells, analysis of the transformed clones, and storage of a sequence-verified clone. The protocol also includes transfer of the sequence-verified clone into another Flexi Vector plasmid backbone.

The following protocol is for cloning in a 96-well format. Smaller numbers of cloning reactions can be undertaken by appropriate scaling of the indicated reaction volumes. This protocol describes production of plasmid constructs that yield an N-terminal fusion to the expressed protein, as illustrated in **Fig. 4.1**. A section is provided on modifications that yield alternative N-terminal constructs, and thus serve to illustrate how expression vector and primer design can be used to provide useful variations of expression constructs.

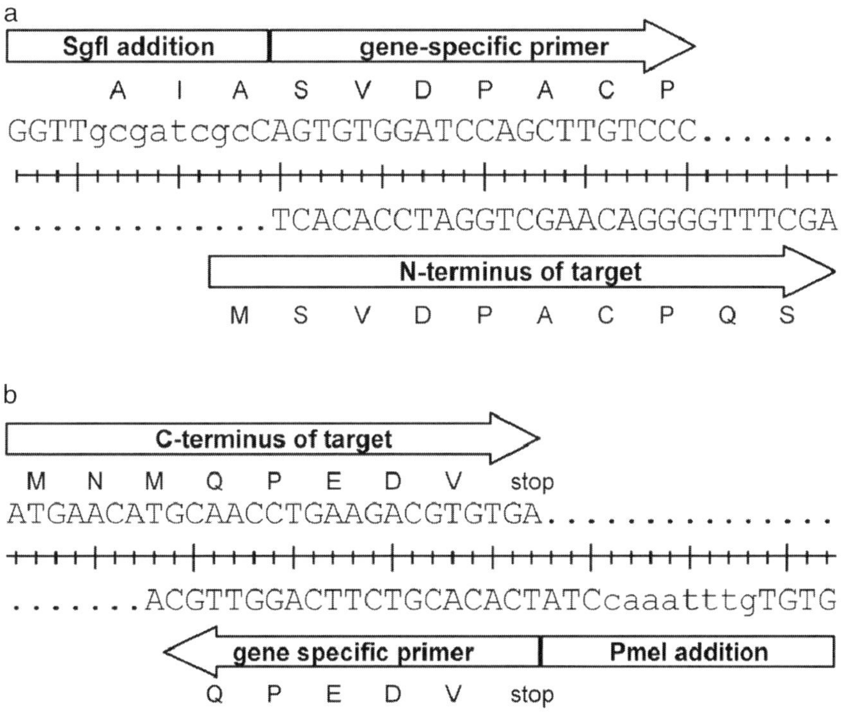

Fig. 4.2. An example of 5′ coding and 3′ reverse complementary strand primers created for Flexi Vector cloning. (a) The 5′ primer consists of an exact match of the desired gene-specific sequence and an additional sequence encoding an sgfi site (34 nucleotides). (b) The 3′ complement reverse primer consists of an exact match of the gene-specific sequence including the stop codon, a primer-encoded stop codon and an additional sequence adding a pmel site (33 nucleotides).

4.3.1. Attachment of Flexi Vector Cloning Sequences

4.3.1.1. PCR Primers

In the Flexi Vector cloning approach described in this section, target genes are amplified using a single-step PCR. **Figure 4.2** shows an example of primers designed to clone a structural genomics target, human stem cell Nanog protein (NM_024865), into pVP56K. In general, the forward and reverse primers are 28–36 nucleotides in length. The gene-specific portion includes 14–23 nucleotides that exactly match the target gene beginning at the second codon. Whenever possible, the gene-specific primers end with a C or G nucleotide to enhance DNA polymerase initiation. The invariant sequence 5′-GGTTgcgatcgcC-3′(including an SgfI site, *lower* case) is added to the 5′ end of the forward primer. The reverse primer consists of the invariant sequence 5′-GTGTgtttaaacCTA (including a PmeI site, *lower* case) followed by the reverse complement of the 3′ gene-specific 14–23 nt including the stop codon. The additional nucleotides are added to the 5′end of these sequences to promote restriction nuclease digestion of the PCR products. For this example, the synthesis of primers containing a total of 68 nucleotides is required.

4.3.1.2. PCR Amplification

The following steps are used to PCR amplify the desired gene and append the sequences required for cloning (*see* **Note 2**).

(1) Create a PCR Primers plate by combining forward and reverse primers for each target gene to 10 μM each. Label this plate and save it at 4°C.

(2) Create the Flexi-PCR Master Mix consisting of 2.23 mL of water, 225 μL of either 10X YieldAce buffer or 10X Pfu Ultra II Buffer, 55 μL of dNTPs (10 μM each), and 23 μL of YieldAce or Pfu Ultra II Hotstart polymerase.

(3) Aliquot 23.5 μL of Flexi-PCR Master Mix to each well of an ISC PCR plate. The remaining Flexi-PCR Master Mix can be saved at 4°C for possible follow-up use.

(4) Add 1 μL of the mixture from the PCR Primers plate to each well of the ISC PCR plate.

(5) Add 1 μL of plasmid cDNA for each gene to be cloned. If the clones are provided as glycerol stocks, use a multichannel pipette to add a stab of the frozen culture to the ISC PCR plate.

(6) Centrifuge the plate briefly in an Allegra 6R centrifuge and 6H3.B rotor to get liquid to the bottom of the wells and then cover the plate with sealing tape.

(7) Put the plate in the thermocycler and cycle using the following parameters for reactions using YieldAce polymerase: (1) 95°C for 5.00 min; (2) 94°C for 30 s; (3) 50°C for 30 s; (4) 72°C, 1.00 min/kb; (5) repeat **steps 2–4** for 4 more times; (6) 94°C for 30 s; (7) 55°C for 30 s; (8) 72°C for 1.00 min/kb plus 10 s per cycle; (9) repeat **steps 6–8** for 24 more times; (10) 72°C for 30.00 min; and (11) 4°C and hold. For amplification of

~1 kb genes, this PCR takes ~3 h to complete. Use the following conditions for PfuUltraII polymerase: (1) 95°C for 3.00 min; (2) 95°C for 20 s; (3) 50°C for 20 s; (4) 72°C, 15 s/kb; (5) repeat steps 2–4 for 4 more times; (6) 95°C for 20 s; (7) 55°C for 20 s; (8) 72°C for 15 s/kb; (9) repeat steps 6–8 for 24 more times; (10) 72°C for 3 min; and (11) 4°C and hold.

(8) Analyze the completed Flexi-PCR reactions on a 2% E-gel 96 by loading 15 μL of the gel running buffer plus 5 μL of the reaction samples. Load 5 μL of PCR molecular weight markers with 15 μL of the gel running buffer.

(9) Retry the PCR for any genes that fail to amplify.

(10) Create a master PCR plate of all successfully amplified genes, label the plate, and begin the restriction digestion step or store the plate at –20°C until needed.

4.3.2. Restriction Digestion of PCR Products

The following steps are used to digest the acceptor vector and PCR products with SgfI and PmeI prior to the ligation (*see* **Note 3**).

4.3.2.1. Digestion Reaction

(1) Create the Acceptor Vector Digest Master Mix consisting of 158.3 μL of sterile, deionized water, 44.0 μL of 5X Flexi-Digest Buffer, 2.20 μL of 10X SgfI/PmeI Enzyme Blend, and 13.5 μL of Acceptor Vector (e.g., purified pVP56K at a concentration of 150 ng/μL). Mix the solution well, as the enzyme blend is dense and tends to settle. Substitution of individual preparations of SgfI and PmeI will require extensive optimization beyond the scope of this protocol.

(2) Place the Acceptor Vector digest reaction in the thermocycler and cycle using the following parameters: (1) 37°C for 40.00 min; (2) 65°C for 20.00 min; and (3) hold at 4°C until needed.

(3) Create the PCR-Digest Master Mix consisting of 638 μL of sterile, deionized water, 220 μL 5X of Flexi-Digest Buffer, and 22 μL of 10X SgfI/PmeI Enzyme Blend.

(4) Add 8.0 μL of the PCR-Digest Master Mix to each well of an ISC PCR plate.

(5) Add 2.0 μL of the Flexi-PCR obtained from procedure 3.1 to each well.

(6) Place the PCR-Digest reaction in a 37°C incubator for 40 min and then move to 4°C.

4.3.2.2. Cleanup

The restriction digests are purified using the Wizard Magnesil PCR Cleanup system and a Magnebot II plate. An important part of the cleanup is to thoroughly dry the sample after the final wash step to evaporate all residual ethanol.

(1) Add 10 μL of well-mixed Magnesil Yellow to each well of the PCR digest plate.

(2) Mix the 20 μL of solution 4 times, incubate for 45 s, and then mix four more times. If any bubbles are present, briefly centrifuge the plate.

(3) Place the PCR digest plate on the Magnebot II magnetic stand, wait 30 s for the magnetic beads to adhere to the right side of the plate, and remove and discard the liquid. The PCR product is now bound to the Magnesil Yellow beads.

(4) Remove the plate from the Magnebot II, and add 20 μL of Magnesil Wash Solution to each well. Mix each well 4 times, wait 60 s, and then mix 4 more times.

(5) Place the plate on the Magnebot II, wait 30 s for the magnetic beads to adhere to the right side of the plate, and remove and discard the liquid.

(6) Wash the beads two more times using 30 μL of 80% ethanol as described in **steps 4** and **5**.

(7) Place the PCR digest plate on a 42°C heating block for 10 min or until all of resin has dried.

(8) To elute the DNA, add 10 μL of water, mix well, and wait 60 s.

(9) Place the plate on the Magnebot II, remove the Magnesil particles, and save the supernatant for use in the ligation reaction.

4.3.3. Ligation of PCR Products into an Acceptor Vector

Ligation reactions are performed in a 96-well PCR plate using the restriction-digested and purified PCR products and Acceptor vector prepared in **step 3.2**. The following steps are used.

(1) Create the Ligation Master Mix containing 225 μL of sterile, deionized water, 110 μL of 10X T4 Ligase Buffer, and 50 μL of T4 DNA Ligase HC.

(2) Add 5.0 μL of cleaned-up PCR product digest, 2.0 μL of Acceptor vector digest, and 3.5 μL of Ligation Master Mix to each well of a new PCR plate.

(3) Store the clearly labeled plate of leftover cleaned-up PCR product at −20°C.

(4) Incubate the reaction in a thermocycler at 25°C for 3 h. Proceed to the transformation step (**Section 4.3.4**) or the reaction can be left overnight at 4°C.

4.3.4. Transformation and Growth of Host Cells

The material from the ligation reaction is used to transform Select96 competent cells by the following steps.

(1) Thaw four strips of Select96 cells on ice. Distribute 15 μL into each well of a prechilled PCR plate. Begin warming SOC medium in a 37°C incubator.

(2) Add 1 μL of the ligation reaction to the Select96 cells, cover the PCR plate and incubate at on ice for 20 min.

(3) Heat shock at 42°C for 30 s on a heat block.

(4) Chill on ice for 1 min.

(5) Add 90 μL of prewarmed SOC medium.

(6) Cover the plate and incubate at 37°C for 1 h with no shaking.

(7) Label Luria Bertani agar plates containing 0.5% (w/v) glucose and 50 μg/mL of kanamycin with the corresponding plate position numbers (A1-H12 for a 96-well plate).

(8) Add 5–10 sterile ColiRoller glass beads to each plate.

(9) Apply the entire volume of the transformation reaction onto the glass beads.

(10) Shake the plates horizontally for 15 s and then dump the glass beads off the plate into a beaker of 80% ethanol. The beads can be washed with 1% (v/v) nitric acid, rinsed extensively with deionized water, and then sterilized and dried for reuse.

(11) Incubate the plates overnight at 37°C.

4.3.5. Analysis of Plasmid DNA

A colony PCR amplification step provides a qualitative check for the presence of the target gene before more labor-intensive plasmid preparation, quantification by optical spectroscopy, and DNA sequence verification. (*see* **Note 4**).

4.3.5.1. Colony PCR Screening of Transformants

Two colonies are selected from the transformation plate and used for colony PCR screening. The same colonies are also used to prepare an inoculum for subsequent plasmid isolation. The preparation of the colony replicates and the colony PCR screening are accomplished as follows.

(1) Prepare 50 mL of CircleGrow medium containing 50 μg/mL of kanamycin.

(2) Pour the medium into a sterile multichannel reagent reservoir.

(3) Use a multichannel pipette to aliquot 200 μL of the CircleGrow medium into each well of two deep-well growth blocks. Label the blocks "Screening Block 1" and "Screening Block 2."

(4) Pipette 10 μL of water into each well of two PCR plates. Label the PCR plates "PCR Screen Plate 1" and "PCR Screen Plate 2."

(5) Pick a colony off the transformation plate with a pipette tip and dab the tip into the water in a well of the PCR plate labeled "PCR Screen Plate 1" and then eject the same pipette tip into the deep-well growth block labeled "Screening Block 1." Repeat the procedure with a second colony into "PCR Screen Plate 2" and "Screening Block 2."

(6) Cover Screening Blocks 1 and 2 with AeraSeal breathable sealing tape.

(7) Place the Screening Blocks 1 and 2 on an orbital shaker at 37°C and aerate vigorously (800 rpm on the VWR microplate shaker) for ~16 h.

(8) Create the Colony PCR Screen Master Mix containing 880 μL of sterile, deionized water, 2.20 mL of 2.5X Eppendorf Hot Master Mix, and 110 μL of MBP Forward sequencing primer (10 μM).

(9) Add 14.5 μL of Colony PCR Master Mix to each well of PCR Screen Plate 1 and PCR Screen Plate 2.

(10) Add 0.5 μL of the gene-specific reverse primers (10 μM) used in **step 3.1** (e.g., the primer designed as implied in **Fig. 4.2b**) to each well of PCR Screen Plate 1 and PCR Screen Plate 2.

(11) Centrifuge the PCR plate to eliminate air bubbles.

(12) Put the two PCR Screen plates in a thermocycler and use the following parameters: (1) 95°C for 5.00 min; (2) 94°C for 30 s; (3) 50°C for 30 s; (4) 72°C for 1.00 min/kb; (5) repeat **steps 2–4** for 19 more times; (6) 72°C for 10.00 min; (7) 4°C and hold. For amplification of ~1 kb genes, this PCR takes ~90 min to complete.

(13) Analyze the completed colony PCRs on a 2% E-gel 96 by loading 15 μL of the gel running buffer plus 5 μL of the reaction sample. Load 5 μL of PCR molecular weight markers plus 15 μL of the gel running buffer.

4.3.5.2 Growth of Individual Isolates

The following steps are used to obtain a 1-mL culture of individual isolates from the transformation plate that are found by colony PCR to have an insert of appropriate size for the gene cloned.

(1) Prepare 200 mL of CircleGrow medium containing 50 μg/mL of kanamycin.

(2) Aliquot 1 mL of CircleGrow medium per well of a Qiagen flat-bottom growth block.

(3) Inoculate the growth block with 5 μL of a culture in the Screening Block 1 or 2 that was identified to have an insert of appropriate size by the colony PCR analysis. Fill any empty wells of the growth block with culture medium or water to help assure the vacuum miniprep procedure used later.

(4) Cover the growth block with AeraSeal Plate Sealers.

(5) Shake the growth block overnight at 37°C orbital plate shaker set at the maximum value (800 rpm on the VWR microplate shaker) for ~16 h.

(6) Store the growth block at 4°C for later use.

4.3.5.3. QiaPrep 96 Turbo Plasmid DNA Purification

This section uses reagents from the QiaPrep 96 kit from Qiagen. Several steps described later are modified from the manufacturer's protocol. These modifications are essential to prepare plasmid DNA of sufficient quantity and purity for subsequent use in Flexi Vector reactions.

(1) Add RNase A to buffer P1 (Qiagen). P1 buffer is stored at 4°C.

(2) Check to make sure that the buffer P2 (Qiagen) has not precipitated during storage. If it has, warm it to 37°C until all precipitate has been redissolved.

(3) The growth block obtained in **Section 4.3.5** is centrifuged either at 2,100 × *g* for 30 min in an Allegra 6R centrifuge with a GH3.8 rotor or at 5,000 × *g* for 15 min in an Avanti J30-I with a JS5.9 rotor.

(4) Discard the supernatant.

(5) *Resuspension*: Add 250 µL of buffer P1 to each well of the growth block. Seal the block with tape and vortex thoroughly to resuspend the cells. Ensure that no cell clumps remain.

(6) *Alkaline lysis*: Add 250 µL of buffer P2 to each sample. Use a clean, dry paper towel to dry the top of the growth block. Seal the growth block tightly with aluminum tape seal and gently invert the Blocks 4–6 times to mix. Incubate the growth block at room temperature for no more than 5 min.

(7) *Neutralization*: Add 350 µL of buffer N3 (Qiagen) to each sample. Dry the top of the growth block and tightly seal the block with a new sheet of aluminum tape. Gently invert the Blocks 4–6 times. To avoid localized precipitation, mix the samples gently but thoroughly immediately after addition of buffer N3. The solution will become cloudy.

(8) Place a Turbofilter plate (white) on the top of a QiaPrep Plate (blue) seated together on top of an empty growth block. Apply 850 µL of neutralized lysate from previous step to the top most Turbofilter plate. Centrifuge at 3,000 × *g* for 5 min to filter the lysate and bind the plasmid to the membrane of QiaPrep Plate. Discard the supernatant captured in the growth block.

(9) Place a QiaPrep plate (blue) on the top of the vacuum manifold. The white plastic reservoir should be placed beneath the plate to collect the flow-through waste.

(10) Wash the QiaPrep plate with 0.5 mL/well of Buffer PB.

(11) Wash the QiaPrep Plate with 0.7 mL/well of Buffer PE.

(12) Prepare to centrifuge the QiaPrep plate by placing the QiaPrep plate on top of a used Qiagen 96-well elution plate. Prepare an accurate counterbalance with another elution plate and the used Turbofilter plate.

(13) Spin the balanced plates at 5,000 × *g* for 5 min on the Avanti J30-I centrifuge with a J30-I rotor. The membrane will be dry after this step. Discard the flow-through liquid.

(14) *Elution*: To elute the plasmid DNA, ensure that the QiaPrep plate is in place over a clean elution plate and then add 100 µL of buffer TE (Qiagen) to the center of the filter well. Let the plate stand for 1 min.

(15) Elute the dissolved plasmid DNA by spinning at 6,000 × *g* for 5 min in the Avanti J30-I centrifuge with a JS5.9 rotor.

4.3.5.4. Determination of DNA Concentration and Purity

The concentration and purity of plasmid preparations used for Flexi Vector cloning must be determined. This can be accomplished using UV-visible spectroscopy *(19)*. The minimum useful concentration of plasmid DNA for subsequent work is 25 ng/µL, and the ratio of A_{260}/A_{280} must be between 1.8 and 2.0. Ratios outside this range indicate contamination that will interfere with subsequent sections.

(1) Add 95 µL of water to each well that will be used in a UV Star 384-well plate.

(2) Insert the plate into the spectrophotometer and obtain a reference setting using water as the blank.

(3) Using the multichannel pipette, add 5 µL of the purified plasmid sample to wells (dilution 1:20). Mix carefully to avoid creating air bubbles.

(4) Insert the plate into the spectrophotometer and read the absorbance values at 260 and 280 nm.

(5) Calculate the plasmid concentration (ng/µL) by the following formula: (A_{260} of the sample – A_{260} of the blank) × 1,000. Alternatively, the plasmid concentration can be calculated by the following approach. Measure the A_{260} value of a 1:20 dilution of a 100 ng/µL plasmid DNA standard. Multiply A_{260} measured for the unknown sample by (100/A_{260} of the diluted standard). This calculation also gives a concentration estimate in ng/µL. The plasmid concentration must be greater than or equal to 25 ng/µL. Typical plasmid concentrations from this procedure are ~100 ng/µL in a typical volume of ~100 µL.

(6) Calculate the ratio of A_{260}/A_{280}. This value should be between 1.8 and 2.0. Values that deviate from this range have a suspect concentration estimate and also likely contain contaminants that will interfere with subsequent Flexi Vector transfer reactions.

4.3.5.5. Sequence Analysis of Plasmid DNA

Previous work with genes from eukaryotes has revealed the necessity for DNA sequence verification before extensive downstream studies of protein expression are undertaken *(12)*.

(1) Prepare the Forward Sequencing Master Mix containing 660 µL of sterile, deionized water, 165 µL of 2.5X Buffer 3.1, 110 µL Big Dye v3.1, and 27.5 µL of MBP Forward sequencing primer (10 µM).
(2) Prepare the Reverse Sequencing Master Mix containing 660 µL of sterile, deionized water, 165 µL of 2.5X Buffer 3.1, 110 µL of Big Dye v3.1, and 27.5 µL of T7 Terminator sequencing primer (10 µM).
(3) Aliquot 8.75 µL of each master mix into two separate PCR plates.
(4) Aliquot 1.25 µL of purified plasmid DNA into both the forward and the reverse reactions.
(5) Spin down both plates in the Allegra 6R centrifuge with 6H3.B rotor.
(6) Put the PCR plates into a thermocycler and cycle using the following parameters: (1) 95°C for 3.00 min; (2) 94°C for 10 s; (3) 58°C for 4.00 min; (4) repeat **steps 2–4** for 50 more times; (5) 72°C for 10.00 min; (6) hold at 4°C. This PCR takes ~5 h to complete.
(7) When the PCR is complete, the materials are suitable for submission to a DNA sequencing facility for automated sequence analysis.

4.3.6. Creation of Glycerol Stocks

Glycerol stocks of sequence-verified clones are prepared in the following manner.

(1) Add 20 µL of 80% sterile glycerol to a new PCR plate.
(2) Add 80 µL of culture containing the sequence-verified clones obtained in **Section 4.3.5.2** to create one plate.
(3) Mix the culture and the glycerol stock well with the pipette.
(4) Add an appropriate barcode or other labeling to the PCR plate.
(5) Cover the PCR plate with foil tape and store the plate at −80°C.

4.3.7. Flexi Vector Transfer Reaction

Cloned genes are moved between different Flexi Vectors by restriction digestion and ligation. The donor and acceptor plasmids must encode resistance to different antibiotics in order to permit positive selection of an acceptor plasmid that has accepted an insert and negative selection of the unchanged donor plasmid. The lethal barnase gene will provide selection against the acceptor plasmid that has not been digested. For 96-well operation, the following steps accomplish the transfer reaction. For smaller number of reactions, the volumes should be scaled to avoid waste of reagents. It is essential to have the highest quality plasmid DNA preparations for these transfer reactions. Plasmid contaminated with *E. coli* genomic DNA will yield false positive colonies, and plasmid contaminated with residuals from the plasmid preparation will have lower efficiency of gene transfer (*see* **Note 4**).

(1) Create the Flexi Transfer Master Mix from 305 μL of deionized, sterile water, 110 μL 5X Flexi-Digest Buffer, 5.5 μL of 10X SgfI/PmeI Enzyme Blend, and 22.0 μL of purified Acceptor Vector (nominal DNA concentration of 150 ng/μL).

(2) Add 4.0 μL of Flexi Transfer Master Mix to each well of a 96-well ISC PCR plate stored on ice.

(3) Add 1.0 μL of donor vector (nominal DNA concentration of 30 ng/μL) to each well of the 96-well plate, cover the plate with an adhesive cover, and centrifuge the plate for 1 min in the Allegra 6R centrifuge and 6H3.8 rotor.

(4) Incubate the plate for 40 min at 37°C in the thermocycler.

(5) Incubate the plate for 20 min at 65°C in the thermocycler to inactivate the restriction enzymes.

(6) Create the Ligation Transfer Master Mix from 440 μL deionized, sterile water, 110.0 μL of 10X Ligase Buffer, and 55 μL of T4 DNA Ligase HC.

(7) Add 5.5 μL of Ligation Transfer Master Mix to each well of the plate containing the heat-inactivated acceptor vector digests. Mix the contents of the plate thoroughly, cover the plate with an adhesive cover, and centrifuge the plate for 1 min in an Allegra 6R centrifuge and 6H3.8 rotor.

(8) Incubate the plate for 1 h at 25°C in the thermocycler to complete the ligation reaction.

(9) The ligation reaction can be stored at −20°C until needed for transformation as described in **Section 4.3.4**.

(10) Sequence verification is not routinely necessary after Flexi Vector transfers *(17)*.

4.3.8. Alternative Constructs for Flexi Vector

The following sections provide information on approaches to develop custom plasmids with Flexi Vector capabilities and to provide primer design examples for the production of other types of expression constructs and fusion proteins.

4.3.8.1. Creation of an Antibiotic Resistance Cassette

The vectors recently developed at CESG reserve the AvrII and Bsi WI restriction sites to define an antibiotic resistance cassette (**Fig. 4.1**). By use of these restriction sites, the kanamycin resistance gene and promoter can be swapped with either the ampicillin resistance gene and promoter or other antibiotic resistance genes and promoters. For workers interested in creating new Flexi Vector backbones, these sites should be created by PCR mutagenesis before the Flexi Vector barnase cassette is introduced. Other antibiotic resistance genes and promoters can be introduced into this site after similar PCR mutagenesis, digestion, and ligation. Plasmids containing the lethal barnase gene must be propagated in a barnase-resistant strain (e.g., *Escherichia coli* BR610, which is available through Technical Services, Promega Corporation).

4.3.8.2. Design of 3′ Sequence in Flexi Vector Plasmids

Self-ligation of the vector backbone through the SgfI and PmeI sites can be reduced by including a region of sequence identity adjacent to either the 3′ or 5′ end of the Flexi Vector cloning cassette *(17)*. This region of identity acts to inhibit replication by forming of an extensive DNA palindrome when two vectors with substantial sequence identity ligate to each other *(22)*. Therefore, inclusion of a region of either 3′ or 5′ sequence identity of ~100 bp or longer should be included in the design when different vector backbones are customized for use with the Flexi Vector system. One example is the transfer pairing of CESG plasmids having a kanamycin resistance marker (originally a Qiagen pQE80 backbone) with pEU plasmids having an ampicillin resistance marker (Cell-Free Sciences, Yokohama, Japan).

The pVP56K vector shown in **Fig. 4.1** includes 131 bp of the DNA sequence 3′ from the PmeI site of pF1K (Promega) as the 3′ homology region. This fragment can either be cloned by PCR from pF1K or moved from CESG vectors to other compatible vectors as a separate piece obtained by PmeI and HindIII digestion. The 3′ homology region can also be included with the BarCAT cassette by restriction digestion of CESG vectors with SgfI and HindIII.

4.3.8.3. Two-Step PCR for Fusion Protein Expression

pVP56K encodes the tobacco etch virus protease recognition site in a 5′ position relative to the SgfI site used for Flexi Vector cloning (**Fig. 4.1**). The protein sequence of this site is ENLYFQA, where proteolysis occurs between Q and A. Thus, when TEV protease is used to proteolyse the His8-MBP-target fusion protein produced from pVP56K, an AIA-target protein is liberated. In some circumstances, this modified N-terminal may be undesirable.

Figure 4.3 shows a variation of the vector backbone, pVP68K, and a primer design that allows liberation of S-target after TEV protease processing of the His8-MBP-target fusion protein. Since a significant fraction of natural proteins have a serine as the second residue, CESG primers used for high-throughput cloning encode this residue. However, TEV protease is relatively tolerant of substitution of other residues at the P1 position *(23)*, so a native N terminus (after bacterial N-terminal Met processing) can be engineered through primer design in many cases.

This approach requires a two-step PCR procedure similar to that we have previously used for Gateway cloning *(24)*. In the example shown in **Fig. 4.3b**, the first PCR forward primer contains 14 gene-specific nucleotides (**Fig. 4.3b**). An invariant

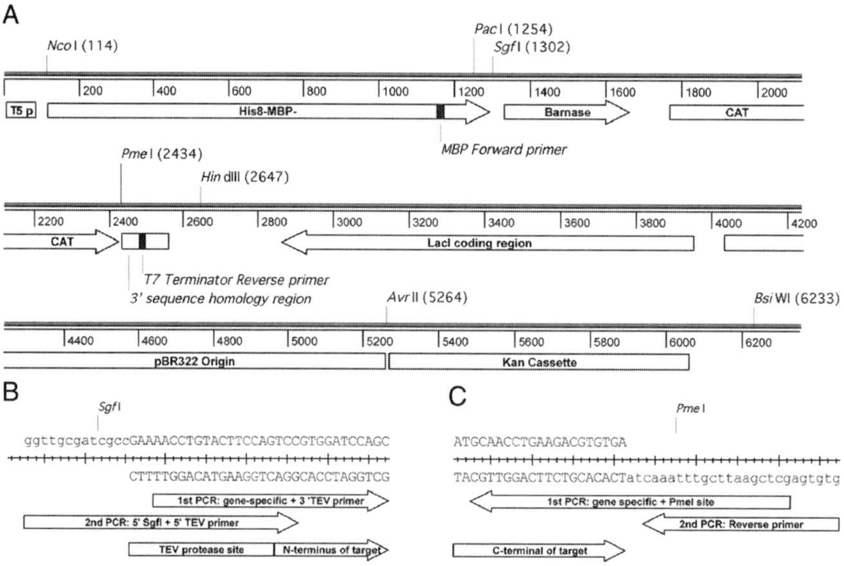

Fig. 4.3. *Escherichia coli* expression vector pvp68k. (a) Linear map showing key elements in the vector. (b) Sequence in the 5′ region near to the sgfi site. The first round forward PCR primer includes a gene-specific sequence and a portion of the TEV protease site (29 nucleotides). The second round forward PCR primer is a universal sequence that completes the TEV protease site and appends a 5′ sgfi sequence (34 nucleotides). The first round 3′ reverse complement primer includes a gene-specific portion, an additional stop codon and the pmeI site and a redundant ecori site (39 nucleotides). The second round reverse complement PCR primer is a universal sequence that duplicates the pmeI and EcoRI sites (24 nucleotides). After expression of the His8-MBP-target fusion protein, an N-terminal S-target can be released by treatment with TEV protease.

sequence is added to the first PCR forward primer to encode a portion (15 nucleotides) of the TEV protease site. The design of the first step 3′ reverse complement primer (**Fig. 4.3c**) is the similar to that described in **Fig. 4.2** and an example is shown in **Fig. 4.3c**. The first step PCR is as described in **Section 4.3.1.2** using 0.2 μM of the appropriately designed forward and reverse primers. For the second PCR, a universal forward primer (39 nucleotides) is used to add the nucleotides required to complete the TEV site and add the SgfI site, and a universal reverse PCR primer is used to duplicate the PmeI site and add additional nucleotides. For the second step PCR, one-fifth of the first-step reaction is added into a new PCR using 0.2 μM of the universal forward primer and reverse primers. The second-step reaction is then completed as described in **Section 4.3.1.2**. For this example, the synthesis of primers containing a total of 126 nucleotides is required.

4.3.8.4. Native N Terminus

The example of **Fig. 4.4** is compatible with use of pFK1 (Promega). Primer design places the native start codon immediately downstream from the SgfI site and ~35 residues downstream from the promoter. This example also places the SgfI site

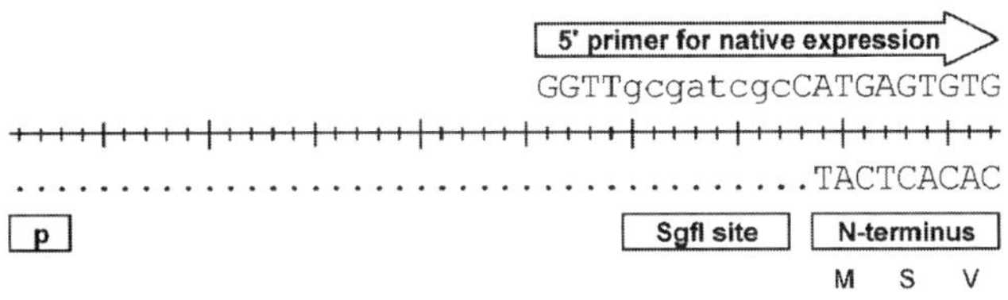

Fig. 4.4. A primer design example for expression of a native protein using Flexi Vector cloning. The promoter region is located upstream of the sgfi site and the start codon encoded by the 5′ primer.

so that the consensus-10-region is retained. For newly engineered plasmids, the nucleotide sequence between the promoter and the desired start codon should not encode alternative start codons in the other translation frames. The design of the 3′ reverse complement primer is the same as described in **Fig. 4.2**. For this example, the synthesis of primers containing a total of 58 nucleotides is required.

4.3.8.5. C-Terminal Fusion Proteins

C-terminal fusion proteins can be produced by Flexi Vector cloning into vectors such as pFC7K (C-terminal HQ tag) or pFC8K (C-terminal HaloTag). This is accomplished by digestion of the C-terminal fusion acceptor vector by SgfI and the alternative blunt end restriction enzyme Eco IRCI and digestion of an existing Flexi Vector clone with SgfI and PmeI. Ligation of the insert will place the C terminus of the target in frame with the fusion protein encoded by the acceptor vector, but will destroy the PmeI site in the ligated product. Vectors created by this approach can no longer be used for transfer to another Flexi Vector, so this approach should not be used as the first step in assembling a family of Flexi Vector expression contexts.

4.4. Notes

1. Use of the Promega blend of SgfI and PmeI is simpler than trying to optimize a mixture of separately purchased SgfI and PmeI.
2. The majority of errors in cloned genes occur in sequence associated with primers. It is advisable to us high-quality primers to minimize the number mutations introduced in these regions, which are critical for successful cloning and gene transfer. It may be necessary to sequence several clones in order to find those without errors.
3. The timing of **step 6** is important as overdigestion can lead to lowered efficiency of cloning. Also, do not heat inactivate the PCR digest because the residual DNA polymerase

present can exhibit significant endonuclease activity during the temperature rise with the consequent removal of nucleotides before and after the restriction enzymes have digested the ends of the PCR products.

4. In **Sections 4.3.5** and **4.3.7**, the purity of the DNA preparations used is essential for efficient digestion and ligation during initial cloning and subsequent transfer reactions. It is advisable to prepare large stocks of the plasmids and verify their purity and function. Care in the preparation of the inserts must be taken to avoid transfer of guanidinium, detergents, or solvents from the plasmid and DNA preparations into the restriction digestion reactions, as these enzymes can be inactivated by these contaminants.

Acknowledgments

Protein Structure Initiative Grant 1U54 GM074901 (J.L. Markley, PI; G.N. Phillips, Jr. and B.G. Fox, Co-Investigators) and a Sponsored Research Agreement from Promega Corporation (B.G. Fox, PI) generously supported this research. The authors enthusiastically acknowledge the efforts all other coworkers of the University of Wisconsin Center for Eukaryotic Structural Genomics for their work in establishing our complete pipeline effort and thank Dr. Mike Slater (Promega) for many useful scientific discussions.

References

1. Zhang, C. et al. (2003) Overview of structural genomics: from structure to function. *Curr Opin Chem Biol* 7, 28–32
2. Terwilliger, T.C. (2000) Structural genomics in North America. *Nat Struct Biol* 7 **Suppl**, 935–939
3. Chandonia, J.M. et al. (2006) The impact of structural genomics: expectations and outcomes. *Science* **311**, 347–351
4. Becker, K.F. et al. (2006) Clinical proteomics: new trends for protein microarrays. *Curr Med Chem* **13**, 1831–1837
5. Lundstrom, K. (2006) Structural genomics: the ultimate approach for rational drug design. *Mol Biotechnol* **34**, 205–212
6. Schnappinger, D. (2007) Genomics of host-pathogen interactions. *Prog Drug Res* **64**, 311, 313–343
7. Su, Z. et al. (2007) Emerging bacterial enzyme targets. *Curr Opin Investig Drugs* **8**, 140–149
8. Green, R.D. et al. (2007) Identifying the future needs for long-term USDA efforts in agricultural animal genomics. *Int J Biol Sci* **3**, 185–191
9. Venter, J.C. et al. (2004) Environmental genome shotgun sequencing of the Sargasso Sea. *Science* **304**, 66–74
10. Witt, A.E. et al. (2006) Functional proteomics approach to investigate the biological activities of cDNAs implicated in breast cancer. *J Proteome Res* **5**, 599–610
11. Chapple, S.D. et al. (2006) Multiplexed expression and screening for recombinant protein production in mammalian cells. *BMC Biotechnol* **6**, 49

12. Thao, S. et al. (2004) Results from high-throughput DNA cloning of *Arabidopsis thaliana* target genes by site-specific recombination. *J Struct Funct Genomics* **5**, 267–276
13. Alzari, P.M. et al. (2006) Implementation of semi-automated cloning and prokaryotic expression screening: the impact of SPINE. *Acta Crystallogr D Biol Crystallogr* **62**, 1103–1113
14. Stols, L. et al. (2002) A new vector for high-throughput, ligation-independent cloning encoding a tobacco etch virus protease cleavage site. *Protein Expr Purif* **25**, 8–15
15. de Jong, R.N. et al. (2007) Enzyme free cloning for high throughput gene cloning and expression. *J Struct Funct Genomics*
16. Doyle, S.A. (2005) High-throughput cloning for proteomics research. *Methods Mol Biol* **310**, 107–113
17. Blommel, P.G. et al. (2006) High efficiency single step production of expression plasmids from cDNA clones using the Flexi Vector cloning system. *Protein Expr Purif* **47**, 562–570
18. Jeon, W.B. et al. (2005) High-throughput purification and quality assurance of *Arabidopsis thaliana* proteins for eukaryotic structural genomics. *J Struct Funct Genomics* **6**, 143–147
19. Gallagher, S.R. et al. (2006) In: Current Protocols in Molecular Biology, Vol. Appendix 3D A.3D.1-A.3D.21. Wiley Interscience: Hoboken, NJ
20. Sambrook, J., Russell, D.W. (2001) In: Molecular Cloning: A Laboratory Manual, Vol. 3, 3rd ed. 15.44–15.48. Cold Spring Harbor Laboratory Press: Cold Spring Harbor, NY
21. Part# TM254. (2005) Promega Corporation: Madison, WI
22. Yoshimura, H. et al. (1986) Biological characteristics of palindromic DNA (ii). *J Gen Appl Microbiol* **32**, 393–404
23. Kapust, R.B. et al. (2002) The P1′ specificity of tobacco etch virus protease. *Biochem Biophys Res Commun* **294**, 949–955
24. Thao, S. et al. (2004) Results from high-throughput DNA cloning of *Arabidopsis thaliana* target genes using site-specific recombination. *J Struct Funct Genomics* **5**, 267–276

Chapter 5

The Precise Engineering of Expression Vectors Using High-Throughput In-Fusion™ PCR Cloning

Nick S. Berrow, David Alderton, and Raymond J. Owens

Summary

In this chapter, protocols for the construction of expression vectors using In-Fusion™ PCR cloning are presented. The method enables vector and insert DNA sequences to be seamlessly joined in a ligation-independent reaction. This property of the In-Fusion process has been exploited in the design of a suite of multi-host compatible vectors for the expression of proteins with precisely engineered His-tags. Vector preparation, PCR amplification of the sequence to be cloned and the procedure for inserting the PCR product into the vector by In-Fusion™ are described.

Key words: In-Fusion; PCR cloning; High throughput

5.1. Introduction

The first step in the high-throughput production of proteins is the construction of vectors for the expression of the target proteins. Conventionally, this involves the manipulation of DNA using restriction enzymes and DNA ligase to combine sequences for expression into the appropriate plasmid vector. The process is relatively time consuming, for example, involving gel electrophoresis to purify the component DNA fragments and is limited by the availability of unique restriction enzyme sites for cloning. To overcome these limitations and hence increase the efficiency of vector construction for high-throughput applications, a number of ligation-independent methods have been developed, for example Gateway™ cloning based on λ phage site-specific recombination *(1, 2)* and Ligation-Independent PCR cloning (LIC) which involves hybridization of single-stranded ends produced

Sharon A. Doyle (ed.), *Methods in Molecular Biology: High Throughput Protein Expression and Purification, vol. 498*

Book doi: 10.1007/978-1-59745-196-3

by treatment of both linearized vector and insert with T4 polymerase *(3, 4)*. Although highly efficient, both these methods do not allow the precise fusion of vector and insert sequences. In the case of the Gateway™ system, the addition of the *att* recombination sites to the 5′ and 3′ ends of the cloned sequence means that extra amino acids are incorporated into the expressed protein. In the LIC method the sequences flanking the insert can only be composed of three of the four bases, since the fourth base is required as a 'lock' for stopping, at a specified point, the single-strand production by the 3′ to 5′ processing activity of T4 polymerase. Recently, an alternative method has been developed and commercialized by Clontech as In-Fusion™ PCR cloning (http://bioinfo.clontech.com/infusion/). The system is based on an enzyme with proof-reading exonuclease activity that catalyses the joining of DNA duplexes via exposure of complementary single-stranded sequences. Consequently, vectors and inserts can be precisely joined in an entirely sequence-independent manner. Further, it has been shown that the In-Fusion reaction is efficient over a wide insert DNA concentration range and suitable for the cloning of large PCR products (3–11 kb) *(5)*. We have combined In-Fusion™ PCR cloning with customized multi-promoter plasmids to produce a suite of expression vectors (pOPIN series) that enable the precise engineering of (His_6-) tagged constructs with no unwanted vector sequences added to the expressed protein. The use of a multiple host-enabled vector permits rapid screening for expression in both *E. coli* and eukaryotic hosts (HEK293T cells and insect cells, e.g. *Sf*9 cells) *(6)*. In this chapter, the protocols for using In-Fusion™ PCR cloning as part of an HTP protein production pipeline will be described.

5.2. Materials

5.2.1. Primers and Vectors

1. The 3′ and 5′ regions of homology required for In-Fusion™ cloning are generated by adding approximately 15-bp extensions to both forward and reverse PCR primers. The sequences of these extensions should match precisely the 5′ and 3′ ends of the recipient vector exposed by linearization of the vector at the position into which the PCR product is to be inserted. Typically, oligonucleotide primers are approximately 35-bp long (including the extension and gene-specific region), and purification of the primers is not necessary.
2. Any vector can be used with the In-Fusion™ cloning method. A unique restriction enzyme site(s) is required at the point of cloning for linearization of the vector by enzymatic cleavage. Alternatively the need for a restriction site(s) can be avoided

by producing the linearized vector by inverse PCR, though depending upon the size of the starting vector this risks introducing unwanted PCR errors into the vector backbone *(7)*. The pOPIN suite of vectors developed in the OPPF are designed so that linearization requires cutting with two restriction enzymes releasing a beta-galactosidase expression cassette (**Fig. 5.1**). Therefore blue-white selection is used to screen out colonies transformed with non-linearized vector following the In-Fusion™ cloning reaction (*see* **Section 5.3.4**). The pOPIN vectors and the extensions required for In-Fusion™ cloning into them are listed in **Table 5.1**. The vectors are available from the corresponding author on request.

5.2.2. Enzymes and Buffers

1. In-Fusion™ enzyme is available lyophilized with the buffer components in microtube format (8 or 96) which conveniently can be stored at room temperature. The enzyme and buffers can also be purchased separately in liquid form (Clontech, Oxford, UK).
2. There are a number of high-fidelity PCR polymerases which are suitable for amplification of DNAs, for example, the KOD Hi-Fi™ and KOD Hotstart™ polymerases (Novagen, Nottingham, UK). Most manufacturers supply the reaction buffer including dNTPs with the enzymes.

5.2.3. Gel Electrophoresis and Purification of DNA

1. Tris borate EDTA (TBE) running buffer (×10): 108 g Tris base, 55 g Boric acid, 9.3 g EDTA dissolved in 1 L water; store at room temperature.
2. DNA gel loading dye: 0.25% (w/v) Bromophenol Blue in 30% (v/v) glycerol/TE: store at room temperature.
3. Visualization of DNA: SYBRSafe™ (InVitrogen, Paisley, UK).
4. AMPure magnetic beads and SPRIPlate 96R magnet (Agencourt Biosciences, Beverley, MA, USA).
5. QIAquick and QIAprep kits (Qiagen, Crawley, UK); Wizard® kit (Promega, Madison Wisconsin, USA).
6. QIAVac 96 vacuum manifold (or similar).
7. GS96 medium (Qbiogene, Morgan Irvine, CA, USA).
8. 5-Bromo-4-chloro-3-indolyl-β-D-galactoside (X-gal) and Isopropyl-β-D-thio-galactopyranoside (IPTG), (Melford Laboratories Ltd, Ipswich, UK).

5.2.4. Cloning-Grade E. Coli

Chemically competent cells with an efficiency of at least 10^8 cfu/μg circular plasmid DNA are required for In-Fusion™ cloning, for example, TAM1 cells from ActivMotif (Rixensart, Belgium) OmniMax2 cells (InVitrogen, Paisley, UK) and Fusion-Blue™ Competent Cells (Clontech).

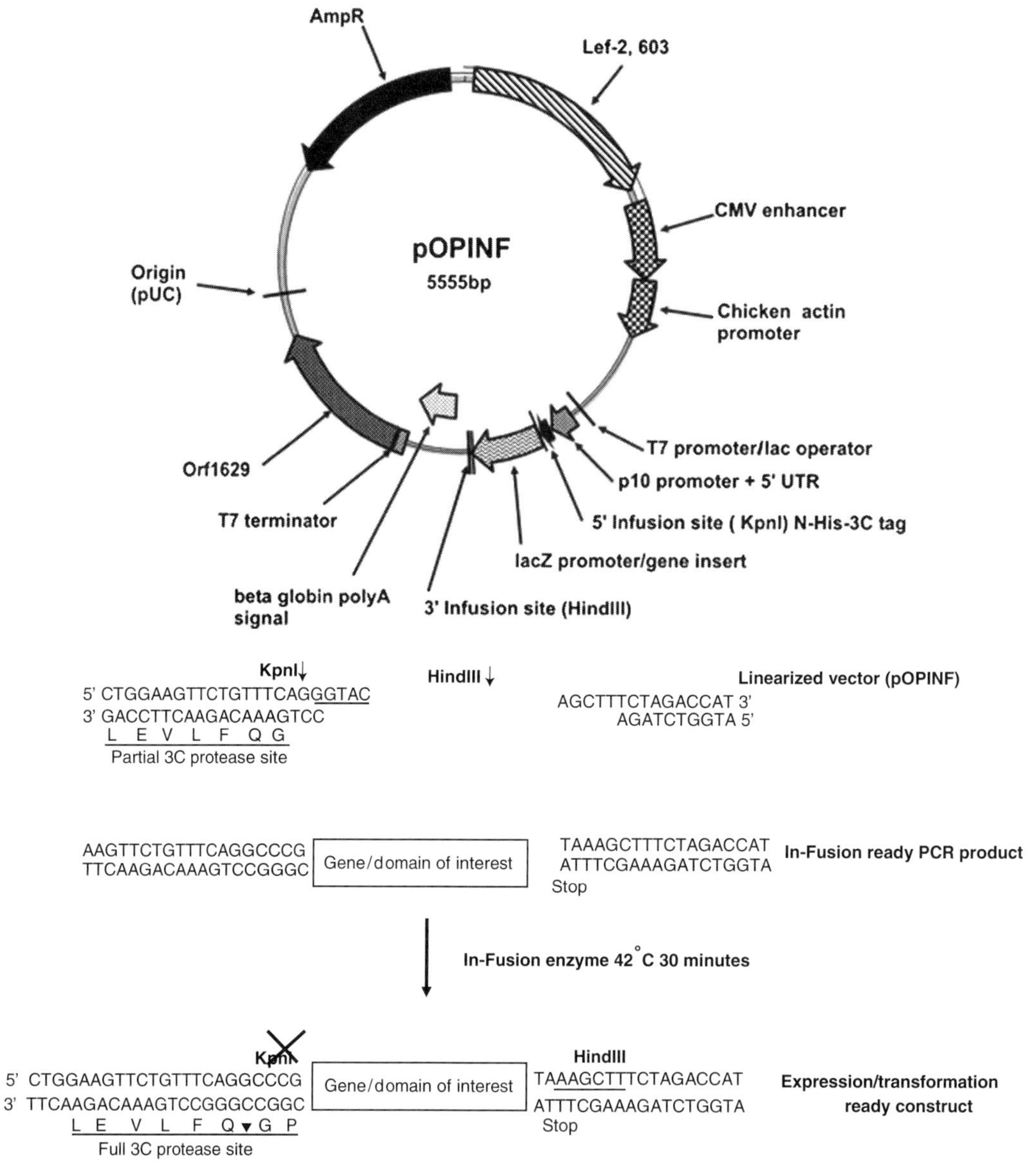

Fig. 5.1. Schematic representation of the pOPINF vector showing key features of the plasmid (a) and details of the In-Fusion cloning region (b) (a) vector map showing key features of pOPINF as follows: pUC origin for high-copy replication in E. coli and ampicillin-resistance marker (AmpR), T7 promoter/lac operator for high-level transcription of insert in E. coli containing the λ (DE3) prophage and T7 transcription terminator, CMV Enhancer and Chicken β-Actin Promoter for transcription of gene insert in mammalian cell lines, a β-globin polyA signal is included to enhance transcript stability, the 'flanking' baculoviral ORFs Lef-2603 and 1629 for recombination into the baculovirus genome and the p10 promoter/5′UTR for expression in insect cell lines, N-His-3C tag for simple affinity purification of gene product from any host cell. The positions of the 5′(KpnI) and 3′(HindIII) In-Fusion sites are indicated. (b) Schematic of an In-Fusion reaction into pOPINF. The two 'free' ends of the linearized vector (top) are shown with a partial 3C Protease cleavage site as the 5′ In-Fusion site and the standard 3′ In-Fusion site. The PCR-amplified insert (centre) flanked by the In-Fusion extensions. During the reaction the In-Fusion enzyme removes the 3′ overhang left by the KpnI digest of the vector, the 5′ overhang generated by the HindIII digestion of the vector is not removed by the In-Fusion enzyme. The transformation-ready reaction product (lower) is shown with the re-constituted 3C Protease cleavage site and translation stop codon now flanking the inserted gene in the circularized plasmid. N.B. the In-Fusion enzyme has no ligase activity but generates single strands on both PCR product and vector to the extent of the homology between the two.

Table 5.1
Summary of In-Fusion™ site sequences and characteristics of the pOPIN vectors where ↓ represents the 3C protease cleavage site and ↑ represents the cleavage sites for either eukaryotic signal peptidase or the specific SUMO protease. Underlined sequences represent sites where translation initiation and stop codons are present in In-Fusion primer extensions. Vectors marked with dagger symbols use the same primer extensions.

Vector	Fusion tag	Parent vector/antibiotic resistance	Restriction sites for linearization of the vector	Forward primer extension	Reverse primer extension	Approximate increase in size of PCR product with T7 primer (bp)
pOPINA	...KHHHHHH tag	pET28a/Kanamycin	NcoI and DraI	AGGAGATATACCATG	GTGGTGGTGGT-GTTT	110
pOPINB	MGSSHHHHHHSSGLEVL-FQ↓GP ... tag	pET28a/Kanamycin	KpnI and HindIII	AAGTTCTGTTTCAG-GGCCCG‡	ATGGTCTA-GAAAGCTTTA‡	130
pOPINC	...KHHHHHH tag	pTriEx4/Ampicillin	NcoI and PmeI	AGGAGATATACCATG†	GTGATGGTGAT-GTTT†	200
pOPIND	MAHHHHHHSSGLEVL-FQ↓GP... tag	pTriEx4/Ampicillin	KpnI and HindIII	AAGTTCTGTTTCAG-GGCCCG‡	ATGGTCTA-GAAAGCTTTA‡	225
pOPINE	...KHHHHHH tag	pTriEx2/Ampicillin	NcoI and PmeI	AGGAGATATACCATG‡	GTGATGGTGAT-GTTT†	170
pOPINF	MAHHHHHHSSGLEVL-FQ↓GP... tag	pTriEx2/Ampicillin	KpnI and HindIII	AAGTTCTGTTTCAG-GGCCCG‡	ATGGTCTA-GAAAGCTTTA‡	225
pOPING	MGILPSPGMPALLSLVSLLSVLL MGCVA↑ETG... cleavable secretion leader and.KHHH-HHH tags	pTriEx2/Ampicillin	KpnI and PmeI	GCGTAGCTGAAACCGGC	GTGATGGTGAT-GTTT	260
pOPINH	MGILPSPGMPALLSLVSLLSVLL MGCVA↓ETMAHHHHHHS SGLEVLFQ↓GP...... cleavable secretion leader and cleavable N-his tag	pTriEx2/Ampicillin	KpnI and HindIII	AAGTTCTGTTTCAG-GGCCCG‡	ATGGTCTA-GAAAGCTTTA	315
pOPINI	MAHHHHHHSSG... tag	pTriEx2/Ampicillin	KpnI and HindIII	ACCATCACAGCAGCGGC	ATGGTCTA-GAAAGCTTTA	200

(continued)

Table 5.1
(continued)

Vector	Fusion tag	Parent vector/antibiotic resistance	Restriction sites for linearization of the vector	Forward primer extension	Reverse primer extension	Approximate increase in size of PCR product with T7 primer (bp)
pOPINJ	MAHHHHHHSSG-*GST*-LEVLFQ↓GP… tag	pTriEx2/Ampicillin	KpnI and HindIII	AAGTTCTGTTTCAG-GGCCCG‡	ATGGTCTA-GAAAGCTTTA‡	890
pOPINK	MAHHHHHHSSG-*GST*-LEVLFQ↓GP… tag	pET28a/Kanamycin	KpnI and HindIII	AAGTTCTGTTTCAG-GGCCCG‡	ATGGTCTA-GAAAGCTTTA‡	790
pOPINM	MAHHHHHHSSG-*MBP*-LEVLFQ↓GP… tag	pTriEx2/Ampicillin	KpnI and HindIII	AAGTTCTGTTTCAG-GGCCCG‡	ATGGTCTA-GAAAGCTTTA‡	1,330
pOPINS	MGSSHHHHHH-*SUMO*↓… tag	pET28a/Kanamycin	KpnI and HindIII	GCGAACAGATCGGTGGT	ATGGTCTA-GAAAGCTTTA	400

5.2.5. Plastic-Ware in Multi-Well Format

1. Thermo-Fast 96 v-well skirted PCR plates sealed with a clear film (e.g. ABGene, Epsom, UK).
2. 96 deep-well plate sealed with gas-permeable film (e.g. ABGene).
3. 96 v-well microtitre plates (e.g. Greiner Bio-One, Frickenhausen, Germany) and foil seals (e.g. ABGene).
4. 24-Well tissue culture plates with lids (e.g. Corning, Lowell, MA, USA).

5.3. Methods

5.3.1. Preparation of Vector

1. Digest vector DNA prepared using QIAPrep (Qiagen) spin column (or similar DNA purification method) with appropriate restriction enzyme(s) using standard conditions. For example, incubate 5 μg pOPINF plasmid DNA with 50 Units KpnI and 50 Units of HindIII, in a total reaction volume of 100 μl for 2 h at 37°C.
2. Purify linearized vector by preparative gel electrophoresis (*see* **Section 5.3.3.2**) or by a spin column (e.g. QIAquick PCR purification kit). If the plasmid DNA has been prepared without the use of an alcohol precipitation step the spin column method is sufficient, otherwise gel purification is recommended.

5.3.2. HTP PCR Amplification

1. Dilute the primers (100-μM stocks) 1–10 with either sterile UHQ water or a buffer such as EB (10 mM Tris–HCl, pH 8.0) prior to use (*see* **Note 1**).
2. The standard amplification reaction for In-Fusion™ cloning comprises 30 pmol each primer (final concentration of 0.6 μM each), 50–100 ng of template, dNTPs (final concentration of 200 μM each), 1 mM $MgCl_2$ (final concentration) and 1 U of KOD HiFi™ polymerase in a final volume of 50 μl (*see* **Note 2**).
3. Perform the thermal cycling in a 96 v-well PCR microtitre plate using the following parameters (*see* **Note 3**):

   Step 1: 94°C 2 min (not necessary for KOD HiFi)

   Step 2: 94°C 30 s

   Step 3: 60°C 30 s

   Step 4: 68°C 2 min

   Step 5: Go to **step 2** and repeat cycle 29 times

   Step 6: 72°C 2 min

   Step 7: 4°C Hold

- When thermal cycling is complete add 10 µl of DNA gel loading buffer to each well of the PCR plate. Analyse 6 µl aliquots of each product (after mixing by pipette) on a 1.25% TBE agarose gel.

5.3.3. Purification of PCR Products

Providing the PCR products are of good quality (i.e. few multiple bands and 'smeared' products) then the AMPure™ magnetic beads can be used for purification avoiding the need to run preparative agarose gels (**Section 5.3.3.1**). However, if a significant number of the products contain multiple bands, then gel purification and extraction is advisable. The PCR products may be separated by agarose gel electrophoresis on any suitable gel apparatus. A minimum run length of 30 mm is recommended, and the process is simplified if the gel wells are spaced such that they can be used in conjunction with a multi-channel pipette. The wells must also accommodate a sample volume of at least 50 µl/well. The use of SYBRSafe stain in the gel and a Blue Light illuminator (e.g. Clare Chemicals DR88 Dark Reader) is recommended to prevent UV damage to the PCR products during visualization and excision. DNA is extracted using a modified QIAquick® 96 PCR (**Section 5.3.3.2**).

5.3.3.1. Ampure Magnetic Bead

1. If required add approximately 5 U DpnI enzyme to each PCR and incubate at 37°C for 1 h prior to the clean-up step (see Note 4).
2. Add 90 µl of the AMPure™ resin to each 50 µl PCR mix thoroughly by pipette mixing 10 times or orbital shaking for 30 s. This step binds PCR products 100 bp and larger to the magnetic beads. The colour of the mixture should appear homogenous after mixing. Incubate the mixed samples for 3–5 min at room temperature to ensure maximum binding of PCR products to the resin.
3. Place the reaction plate onto a SPRIPlate 96R magnet for 5–10 min to separate beads from solution. The separation time is dependent on the volume of the reaction. Wait for the solution to clear before proceeding to the next step.
4. Aspirate the cleared solution from the reaction plate and discard. This step must be performed while the reaction plate is located on the SPRIPlate 96 magnet and can be carried out with any suitable multi-channel pipette. Do not disturb the ring of separated magnetic beads.
5. Dispense 200 µL of (freshly prepared) 70% ethanol to each well of the reaction plate and incubate for 30 s at room temperature. Remove the ethanol by aspiration and discard. Repeat wash step a further two times. It is important to perform these steps with the reaction plate situated on a SPRIPlate 96R. Do not disturb the separated magnetic beads.

Be sure to remove all the ethanol from the bottom of the well as it may contain residual contaminants. The ethanol can also be discarded by inverting the plate to decant off the liquid, but this must be done while the plate is situated on the SPRIPlate 96R.

6. Air dry the plate for 10–20 min on a bench to allow complete evaporation of residual ethanol. If the samples are to be used immediately, proceed to step 7 for elution. If the samples are not to be used immediately, the dried plate may be sealed and stored at 4 or –20°C.
7. To elute the DNA from the resin, add 40 µL of elution buffer (EB: 10 mM Tris–HCl, pH 8.0) to each well of the reaction plate, seal and either vortex for 30 s or pipette mix 10 times (see Note 5).

5.3.3.2. Gel Purification and Extraction of PCR Products

1. Excise the electrophoretically separated PCR products from the agarose gel and place each gel slice into an individual well of a 96 deep-well block. Trim the slices if necessary so that there is no more than 400 mg of agarose in each well. Add 3 volumes of Buffer QG (QIAquick kit) to 1 volume of gel (i.e. 100 µl/100 mg) to each well and cover the block with a plate seal.
2. Incubate the block at 50°C for 10 min (or until gel is completely dissolved) in an oven or shallow water bath. Mix by inverting the block every 2–3 min during the incubation to help dissolve the gel.
3. Assemble the QIAvac 96 vacuum manifold (Fig. 5.2) as follows:

   – Place waste tray inside the QIAvac base, and put the top plate squarely over the base.

   – Position the QIAquick™ 96 plate securely into the QIAvac top plate.
4. Attach the QIAvac 96 to a vacuum source (15.20 in. of Hg or the equivalent).
5. Once the gel slices have dissolved completely, remove the seal from the block and check that the colour of the mixture in each well is yellow (similar to Buffer QG without dissolved agarose). If, after solubilization of the agarose, the binding mixture appears orange or violet add 10 µl of 3 M sodium acetate, pH 5.0, to the respective samples, seal the block and mix by inversion.
6. Add 1 gel volume of isopropanol to each of the samples, for example, if the agarose gel slice weighs 200 mg, add 200 µl isopropanol (*see* **Note 6**). Cover the block with a plate seal and mix by inverting 6–8 times.

7. Apply maximum volume of 1 ml sample to the wells of the QIAquick™ 96 Plate and switch on vacuum source. After the samples in all wells have passed through the plate, switch off vacuum source. Repeat if the volume of the sample is greater than 1 ml until all the dissolved gel has been applied to the binding plate.
8. Wash each well of the QIAquick™ 96 plate twice by adding 1 ml of Buffer PE to the well and apply vacuum.
9. After washing apply *maximum* vacuum for an additional 10 min to dry the membrane of the binding plate.
10. Switch off the vacuum source and slowly ventilate the QIAvac 96 manifold. Lift the top plate from the base keeping the QIAquick Plate and the top plate together and vigorously rap the top plate onto a stack of absorbent paper until no further liquid comes out. Blot the nozzles of the QIAquick Plate with clean absorbent paper, reassemble the manifold and apply *maximum* vacuum for an additional 10 min to dry the membrane (*see* **Note** 7).
11. To elute directly into a standard height microtitre plate replace waste tray with the inverted plate holder (provided with QIAvac 96) and place a 96-well plate directly onto the rack. Place the top plate with QIAquick™ 96 back onto the base.
12. To elute, add 50 µl of Buffer EB to the centre of each well of the QIAquick 96 Plate, allow to stand for 1 min, and switch on vacuum source for 5 min. Once finished, switch off vacuum source and ventilate QIAvac 96 slowly (*see* **Note 8**).
13. Take 5-µl samples from the elution plate to check recovery of PCR products by Agarose gel electrophoresis, and either seal the plate for storage at –20°C or use immediately in In-Fusion reactions.

5.3.4. In-Fusion Reaction and HTP Transformation of E. Coli

1. Take 10–100 ng of purified insert and 100 ng of linearized and purified vector (these are convenient to prepare and store at 100–200 ng/µl) in a total volume of 10 µl of either EB/H2O (**Note 9**).
2. Add this to a well of the dry-down In-Fusion™ plate. Mix contents briefly by pipetting up and down, taking care that the lyophilized enzyme/buffer pellet is resuspended. Cover the In-Fusion plate with a self-adhesive foil plate seal.
3. Incubate the plate for 30 min at 42°C in either a thermal cycler or water bath.
4. Dilute immediately with 40 µl TE and either transform into E. coli straight away or freeze the reaction for use later. Five microlitres of the diluted reaction should give tens to hundreds of colonies per well of a 24-well plate.

5. Thaw competent *E. coli* on ice, add 50 μl of cells to 5 μl of the diluted In-Fusion reaction and incubate on ice for 30 min.
6. Heat shock the cells for 30 s at 42°C and return the cells to ice for 2 min.
7. Add 450 μl of GS96 media supplemented with glycerol (0.05%, v/v) or Luria Broth (LB) per tube. The use of GS96 here allows the cells to recover without shaking, and this enables a concentrated aliquot of cells to be pipetted from the bottom of the tube for plating.
8. Transfer to 37°C incubator (shaking is not required if GS96 media is used) and incubate for 1 h.
9. Plate on LB Agar supplemented with the appropriate antibiotic for the vector, X-Gal and IPTG. Plates are prepared by the addition of 1 ml of molten LB agar (plus appropriate supplements) to each well of the 24-well plates. Plate 10 μl of cells/well, shake plates laterally/orbitally by hand to distribute the culture and allow at least 10–15 min for the plates to dry before inverting.
10. Following overnight incubation at 37°C, wells should contain predominantly white colonies. Any blue colonies are derived from inefficiently linearized parental plasmid and are non-recombinant. Picking two colonies should be sufficient to obtain a cloned PCR product (*see* **Note 10**).

5.3.5. Colony Picking, Culture, Preparation of Glycerol Stock

1. Prepare 96 deep-well blocks by addition of 1.5 ml of GS96 (plus glycerol) supplemented with the appropriate antibiotic for the pOPIN vector used.
2. Using 200/300-μl pipette tips pick individual white colonies into each well, leaving the tips in the deep well plate to keep track of which wells have been picked into.
3. When picking is complete, remove tips (tips can be removed eight at a time using a multi-channel pipette) and seal plates with gas-permeable adhesive seals.
4. Shake the filled plates at 200–225 rpm at 37°C overnight; microplate holders for standard shakers are available (e.g. single-layer plate holders or plate 'stackers' from New Brunswick Scientific, St. Albans, UK).
5. Make a glycerol stock of all the cultures by transferring 100 μl from each well to a microtitre plate containing 100 μl of filter sterililized LB/30% (v/v) glycerol, seal and store at –80°C (*see* **Note 11**).
6. Replace the gas-permeable seal on each plate with a solid seal and harvest the cells by centrifugation at 5,000 × *g* for 15 min (the Beckman JS5.3 rotor for the Beckman *Avanti* centrifuge is ideal for this).

7. Decant the media to waste by inverting the plate and then rest the plate upside down on a wad of absorbent tissue to remove residual media (make sure that the pellets are tightly stuck to the blocks). The cell pellets may be stored at –20°C until required or used immediately for extracting plasmid DNA.

5.3.6. Plasmid Preparation (Wizard® Protocol)

1. Resuspend each cell pellet by adding 250 µl of Cell Resuspension Solution. This may be done by pipetting with a multi-channel pipette or on a microtitre plate shaker for 30–60 s until a uniform cell suspension is achieved.
2. Add 250 µl of Cell Lysis Solution to each sample. Seal and mix by inversion 3–4 times or 30 s on microtitre plate shaker, incubate for 3 min at room temperature but do not incubate for longer than 5 min (*see* **Note 12**).
3. During the incubation, prepare the vacuum manifold with the Binding plate (unmarked plate) in the QIAvac base (on plate holder) and the Lysate Clearing plate (with blue spot) in the upper plate holder of the manifold (**Fig. 5.2**).
4. Add 350 µl of the Neutralization Solution to each sample. Mixing is not necessary.

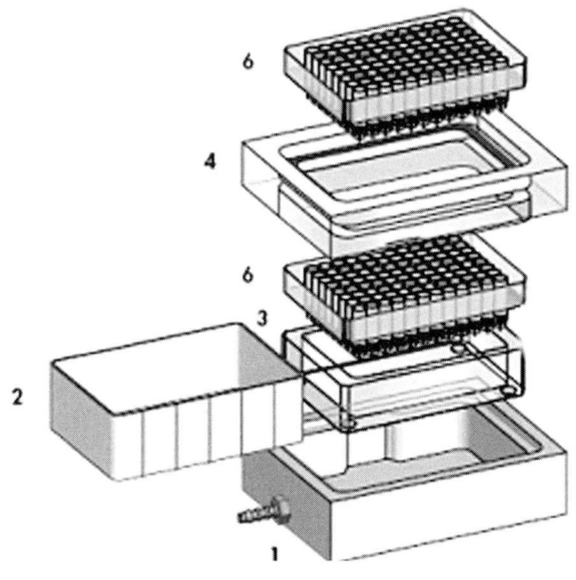

Fig. 5.2. Components of the qiavac 96-exploded view-reproduced from the qiavac handbook, QIAGEN Ltd (UK permission currently being sought from QIAGEN, UK). (1) QIAvac base: holds either the waste tray or the lower plate holder; (2) Waste tray; (3) Lower Plate holder (shown with 96-well plate); (4) QIAvac 96 top plate with aperture for 96-well filter plate; (5) Microtube rack: elution into standard height microtitre plate is described in text. If using racked 1-ml tubes in this format place directly into QIAvac base for elution; (6) 96-well filter plate, e.g. QIAquick™ 96 plate or Wizard® Lysate and Binding plates.

5. Transfer the bacterial lysates to the Lysate Clearing Plate assembled on the Vacuum Manifold. Allow 1 min for the filtration discs to wet uniformly, and then apply a vacuum to the manifold (15.20 in. of Hg or the equivalent) using a vacuum pump fitted with a control valve. Allow 3–5 min under vacuum for the lysates to pass through the Lysate Clearing Plate.
6. Release the vacuum, discard the Lysate Clearing Plate and move the Binding Plate from the Lower Qiavac plate holder to the QIAvac top plate. Place the white Qiavac waste tray in the lower chamber.
7. Add 500 µl of the Neutralization Solution to each well of the Binding Plate in the QIAvac top plate.
8. Apply a vacuum for 1 min, then turn off the pump. Ensure all the wash solution has passed through the Binding Plate.
9. Add 1.0 ml of Wash Solution containing ethanol to each well of the Binding Plate. Apply a vacuum for 1 min.
10. Turn off the pump and repeat the wash procedure (**step 9**). After the wells have been emptied, continue for an additional 10 min under vacuum to allow the binding matrix to dry.
11. Remove the Binding Plate from the Qiavac and blot by tapping onto a clean paper towel to remove residual ethanol. If you find residual ethanol in your mini-preps then you can augment this drying step by wrapping the plate in tissue and incubating at 37°C for 10–15 min.
12. Place a V-bottomed well microplate, or PCR plate, in the Qiavac base on the inverted plate holder. Return the Binding Plate to the QIAvac top plate, ensuring that the Binding Plate tips are centred over the Elution Plate wells and both plates are in the same orientation.
13. Add 100 µl of Nuclease-Free Water/EB (10 mM Tris–HCl, pH 8.0, to the centre of each well of the Binding Plate and incubate 1 min at room temperature.
14. Apply a vacuum for 2–3 min as previously described. Ensure that the entire elution buffer has passed through the plate.
15. Release the vacuum and remove the Binding Plate and Qiavac Upper plate holder. Carefully remove the Elution Plate from the QIAvac base seal the plate and store at 4°C or –20°C. Eluate volumes may vary but are generally 60–70 µl.

5.3.7. Verification of Constructs

1. PCR screen the plasmid mini-preps using the PCR protocol described in **Section 5.3.2** replacing the gene-specific forward primers with a T7 forward primer (5′ TAATACGACTCACTATAGGG 3′). Twenty-five microlitres reactions can be used for screening.
2. Amplify and analyse the products as described in **Section 5.3.2**. The T7 forward primer is present in all the pOPIN vectors; the

increases in PCR size with this primer relative to the original gene-specific forward primer are shown in **Table 5.1**.

3. PCR-verified pOPIN vectors can be used directly for expression screening. The DNA prepared by the procedure described in **Section 5.3.6** is of sufficient quantity and quality for transforming the appropriate *E. coli* expression strains, transiently transfecting mammalian cells (e.g. HEK 293T cells; see Nettleship et al. this volume for protocols) and for constructing recombinant baculoviruses. In this way expression of the cloned gene can be evaluated in multiple hosts in parallel.

5.4. Notes

1. Do not use buffers containing chelating agents such as EDTA as these will inhibit the Mg^{2+}-dependent activity of the DNA polymerases to be used.
2. Making a 'master mix' of all the common reagents is convenient and reduces the possibility of pipetting errors.
3. We have obtained the best results, in terms of target coverage and product quality with KOD HiFi™ in Buffer 2 with a 60 °C annealing temperature. If reactions produce multiple or smeared bands then consider using KOD Hot Start™ as this may reduce non-specific product formation. These parameters represent good starting conditions for testing and can usually amplify products up to 2 kbp in size. Optimization with specific primer pairs and templates may still be necessary.
4. If the PCR template has the same antibiotic resistance as your target pOPIN vector you must DpnI treat your PCR to digest away any template DNA. The DpnI enzyme is active in most PCR buffers and therefore can simply be added to each reaction. If large numbers of samples are to be processed simultaneously then you may consider making a large 'master mix' of buffer and DpnI such that approximately 5 μl (containing 0.5–1.0 units) of this mix can be added to each reaction.
5. When setting up downstream reactions, pipette the DNA from the plate while it is situated on the SPRIPlate96R. This will prevent bead carry over (however, the beads do not inhibit In-Fusion™ reactions). For long-term freezer storage, transferring AMPure™ purified samples into a new plate away from the magnetic particles is recommended.

6. This step increases the yield of DNA fragments <500 bp and >4 kb; for DNA fragments from 500 bp to 4 kb addition of isopropanol has no effect on the yield.
7. This step removes residual Buffer PE which may be present around the outlet nozzles and collars of the QIAquick Plate. Residual ethanol, from Buffer PE, may inhibit the subsequent In-Fusion reactions.
8. It is important to ensure that the elution buffer is dispensed directly onto the centre of QIAquick membrane for complete elution of bound DNA.
9. The In-Fusion reaction volume may be reduced by splitting the contents of the In-Fusion™ enzyme well into two or more wells. Adjust dilution volume in **step 4** accordingly. Total reaction volumes of 2.5 µl have been reported, but we have only tried down to 5 µl to date.
10. In our experience, picking two clones gives an average cloning efficiency for cloning 96 PCR products in parallel of approximately 90%; picking a further two clones can improve this to approximately 95%.
11. Glycerol stocks are a very convenient and stable archive format and can save time if plasmids require re-prepping at later dates.
12. Over-incubation during the alkaline lysis step can lead to nicking of the plasmid DNA.

Acknowledgments

The Oxford Protein Production Facility is supported by grants from the Medical Research Council, UK, the Biotechnology and Biological Sciences Research Council, UK and Vizier (European Commission FP6 contract: LSHG-CT-2004-511960).

References

1. Walhout, A.J., Temple, G.F., Brasch, M.A., Hartley, J.L., Lorson, M.A., van den Heuvel, S. and Vidal, M. (2000) GATEWAY recombinational cloning: application to the cloning of large numbers of open reading frames or ORFeomes. *Methods Enzymol*, **328**, 575–592.
2. Hartley, J.L., Temple, G.F. and Brasch, M.A. (2000) DNA cloning using in vitro site-specific recombination. *Genome Res*, **10**, 1788–1795.
3. Aslandis, C.D., P.J. (1990) Ligation-independent cloning of PCR products (LIC-PCR). *Nucleic Acids Res*, **18**, 6069–6074.

4. Haun, R.S., Serventi, I.M. and Moss, J. (1992) Rapid, reliable ligation-independent cloning of PCR products using modified plasmid vectors. *Biotechniques*, **13**, 515–518.
5. Marsischky, G. and LaBaer, J. (2004) Many paths to many clones: a comparative look at high-throughput cloning methods. *Genome Res*, **14**, 2020–2028.
6. Berrow, N.S., Alderton, D., Sainsbury, S., Nettleship, J., Assenberg, R., Rahman, N., Stuart, D.I. and Owens, R.J. (2007) A versatile ligation-independent cloning method suitable for high-throughput expression screening applications. *Nucleic Acids Res*, **35**, e45.
7. Benoit, R.M., Wilhelm, R.N., Scherer-Becker, D. and Ostermeier, C. (2006) An improved method for fast, robust, and seamless integration of DNA fragments into multiple plasmids. *Protein Expr Purif*, **45**, 66–71.

Chapter 6

The Polymerase Incomplete Primer Extension (PIPE) Method Applied to High-Throughput Cloning and Site-Directed Mutagenesis

Heath E. Klock and Scott A. Lesley

Summary

Significant innovations in molecular biology methods have vastly improved the speed and efficiency of traditional restriction site and ligase-based cloning strategies. "Enzyme-free" methods eliminate the need to incorporate constrained sequences or modify Polymerase Chain Reaction (PCR)-generated DNA fragment ends. The Polymerase Incomplete Primer Extension (PIPE) method further condenses cloning and mutagenesis to a very simple two-step protocol with complete design flexibility not possible using related strategies. With this protocol, all major cloning operations are achieved by transforming competent cells with PCR products immediately following amplification. Normal PCRs generate mixtures of incomplete extension products. Using simple primer design rules and PCR, short, overlapping sequences are introduced at the ends of these incomplete extension mixtures which allow complementary strands to anneal and produce hybrid vector/insert combinations. These hybrids are directly transformed into recipient cells without any post-PCR enzymatic manipulations. We have found this method to be very easy and fast as compared to other available methods while retaining high efficiencies. Using this approach, we have cloned thousands of genes in parallel using a minimum of effort. The method is robust and amenable to automation as only a few, simple processing steps are needed.

Key words: Cloning; Ligase independent; Enzyme free; Site-directed mutagenesis; PIPE; Incomplete primer extension

6.1. Introduction

Contemporary cloning strategies outline mainly iterative protocols based on ligase-independent methods *(1–13)*. Most of these methods require specific sequences for successful cloning. Recombinatorial cloning is only one example where specific sequences must be incorporated and can encode extra, unwanted residues into expressed proteins *(9)*. "Enzyme-free" cloning alleviates sequence

Sharon A. Doyle (ed.), *Methods in Molecular Biology: High Throughput Protein Expression and Purification, vol. 498*

Book doi: 10.1007/978-1-59745-196-3

requirements through a series of PCR steps and product treatments *(10)*. Likewise, the PIPE method also eliminates sequence constraints, and it also reduces cloning and site mutagenesis to a single PCR step and transformation. These combined innovations make the PIPE method very fast, cost effective, and highly efficient. The following protocol includes all the wet lab steps, from making competent cells to submitting samples for sequencing, necessary for successful cloning and mutagenesis. Although the various examples presented here are shown with the expression vector pSpeedET (in-house), the PIPE method can be used with other vectors.

6.2. Materials

6.2.1. Preparation of Competent Cells

1. Milli-Q water (*see* **Note 1**).
2. LB Broth: 25 g of Difco™ LB Broth (Miller) per 1 L of deionized (DI) water and autoclaved for 30 min at 121°C. The media is autoclaved in 2,000 ml Kimax® Baffled Culture Flasks (ThermoFisher Scientific, Waltham, MA).
3. Sterile, disposable Corning Erlenmeyer polycarbonate flasks (500 ml) (available through ThermoFisher Scientific).
4. $MgCl_2$ Solution: 100 mM $MgCl_2$ in Milli-Q water, then sterile filtered. Stored at 4°C.
5. $CaCl_2$ + Glycerol Solution: 100 mM $CaCl_2$ and 15% glycerol in Milli-Q water, then sterile filtered. Stored at 4°C.
6. Microcentrifuge tubes (2 ml).

6.2.2. Preparation of Selective LB Agar Plates

1. Sterile square BioAssay trays with 48-well dividers (Genetix, Boston, MA).
2. LB Agar: 40 g of Difco™ LB Agar (Miller) per 1 L of DI water and autoclaved for 30 min at 121°C.
3. Antibiotics (working concentration): Ampicillin (100 μg/ml), chloramphenicol (20 μg/ml), kanamycin (30 μg/ml).

6.2.3. PCRs for Amplifying Cloning Vectors and Inserts or Generating Site-Directed Mutants

1. Template DNA (20–100 pg per PCR amplification).
   i. For cloning vector amplifications (*V-PIPE*), use a recipient expression plasmid such as pSpeedET (*see* **Fig. 6.1a**).
   ii. For insert amplifications (*I-PIPE*), use genomic DNA, cDNA, PCR product, or miniprepped DNA from a previously generated clone (*see* **Fig. 6.1b**).
   iii. For mutagenic amplifications (*M-PIPE*), use the miniprepped DNA from a previously generated clone (*see* **Fig. 6.1c**).

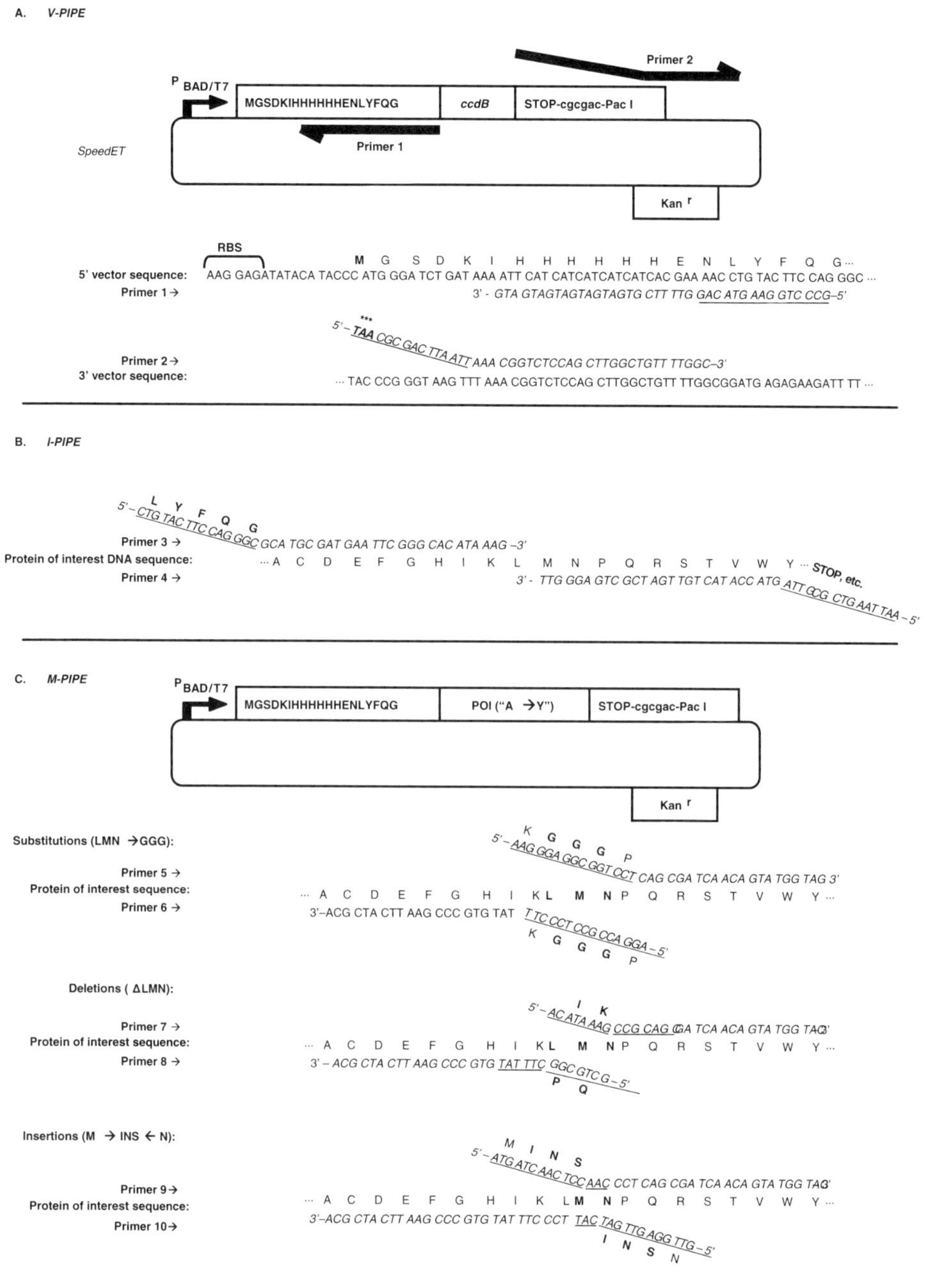

Fig. 6.1. Oligonucleotide design for the three common PCR amplifications used in the PIPE method. The 15 base complementary overlaps are shown as the *underlined* portion of each primer sequence. (**a**) V-PIPE (vector PCR). Primers 1 and 2 are examples of primers that could be used to PCR amplify a vector, such as speedet, in a way suitable for annealing to inserts amplified in (**b**). (**b**) I-PIPE (insert PCR). Primers 3 and 4 are examples of primers that could be used to PCR amplify inserts from various templates in a way suitable for annealing to the vector amplified in (**a**). PIPE cloning works by intermolecular annealing across the two annealing sites of the (**a**) and (**b**) PCRs. (**c**) M-PIPE (mutagenic PCR). Primers 5 and 6 represent primers which could be used to create substitution mutants. Primers 7 and 8 represent primers which could be used to create deletion mutants. Primers 9 and 10 represent primers which could be used to create insertion mutants. PIPE mutagenesis works by intramolecular annealing across the single site of a (**c**) PCR.

2. Oligonucleotide primers: Ordered at 50 μM (Integrated DNA Technologies, Coralville, IA).
   a) *V-PIPE* primers (*see* **Fig. 6.1a**).
   b) *I-PIPE* primers (*see* **Fig. 6.1b**).
   c) *M-PIPE* primers (*see* **Fig. 6.1c**).
3. *Pfu*Turbo® DNA polymerase: 2.5 U/μl (Stratagene, La Jolla, CA) (*see* **Note 2**).
4. Cloned *Pfu* DNA Polymerase Reaction Buffer (10×): 200 mM Tris–HCl (pH 8.8), 20 mM $MgSO_4$, 100 mM KCl, 100 mM $(NH_4)_2SO_4$, 1% Triton® X-100, 1 mg/ml nuclease-free BSA (Stratagene, La Jolla, CA).
5. 10 mM dNTP mix (contains all four dNTPs at 10 mM each).
6. Milli-Q water.
7. PIPE *Pfu* Master Mix (1×): 5 μl of 10× Cloned *Pfu* DNA Polymerase Reaction Buffer, 2.5 units of *Pfu*Turbo® DNA polymerase, 1 μl of 10 mM dNTPs to a volume of 35 μl. (Template DNA (1–5 ng) and forward and reverse primers are added separately and the reaction is brought up to a final volume of 50 μl.).
8. Thermocycler: MJ Research PTC-200 (Bio-Rad Laboratories, Hercules, CA) or similar.

6.2.4. Transforming Competent Cells and Plating Transformed Cells

1. Competent cells (from **Section 6.2.1**).
2. Water bath (42°C).
3. 96-Well microtiter plates: Falcon® U-bottom, polystyrene plates (Becton Dickinson, Franklin Lakes, NJ).
4. Selective LB agar plates (from **Section 6.2.2**).
5. Sterile glass beads, 5-mm diameter, autoclaved.
6. Two to five sterile glass beads are dispensed into each of the 48 wells of the selective LB agar BioAssay tray for spreading cultures.

6.2.5. Screening Colonies by Diagnostic PCR (dPCR)

1. LB Broth (*see* **step 2 – Section 6.2.1**).
2. 96-Well microtiter plates: Costar® flat-bottom, polystyrene plates.
3. 96-Well PCR plates: Costar® Thermowell®, polypropylene plates.
4. Thermocycler: MJ Research PTC-200 (Bio-Rad Laboratories, Hercules, CA) or similar.
5. *Taq* Reaction Buffer (10×): 100 mM Tris–HCl (pH 8.8), 500 mM KCl, 15 mM $MgCl_2$, 0.01% (w/v) gelatin.
6. *Taq* DNA Polymerase (*see* **Note 3**).

7. dNTPs (10 mM) (*see* **step 5** – **Section 6.2.3**).
8. Forward sequencing primer (50 μM).
9. Reverse sequencing primer *or* an insert-specific reverse primer (50 μM) (*see* **Note 4**).
10. Cells cultured from isolated colony picks.
11. Milli-Q water.
12. *Taq* Master Mix (for 1 reaction): 5 μl 10× *Taq* Reaction Buffer, 1 μl 10 mM dNTPs, 1 μl *Taq* DNA Polymerase, 0.5 μl pBAD forward primer, 0.5 μl pBAD reverse or universal insert-specific reverse primer and 39 μl Milli-Q water.
13. *Taq* Master Mix (for 96 reactions): 510 μl 10× *Taq* Reaction Buffer, 102 μl 10 mM dNTPs, 102 μl *Taq* DNA Polymerase, 51 μl pBAD forward primer, 51 μl pBAD reverse (for dPCR) or a universal insert-specific reverse primer (for SYBR-PCR) and 3,978 μl Milli-Q water (*see* **Note 5**).

6.2.6. SYBR-PCR Assay (for Optional Use with the Insert-Specific Reverse Primer Amplification)

1. 96-Well microtiter plates: Costar® flat-bottom, polystyrene plates.
2. SYBR-PCR products.
3. Quench Buffer: 10 mM EDTA.
4. Dilution Buffer: 50 mM Tris–HCl (pH 8.0).
5. SYBR Buffer: 40 μl of 10,000× SYBR Green I Dye (Sigma) is diluted in 20 ml of Dilution Buffer to make a 20× solution (*see* **Note 6**).
6. Microtiter Plate Fluorescence Reader: Spectramax Gemini XS with SoftMax Pro 4.6 software or similar.

6.2.7. Preparation of dPCR Products for Sequencing

1. PCR product (*see* **Section 6.2.5**).
2. Exo/SAP Solution *(14)*: 0.5 μl Exonuclease I, 0.5 μl Shrimp Alkaline Phosphatase, 4.0 μl Milli-Q water (enzymes available through USB, Cleveland, OH) (*see* **Note 7**).
3. Thermocycler: MJ Research PTC-200 (Bio-Rad Laboratories, Hercules, CA) or similar.

6.2.8. Glycerol Stock Archival of Putative Clones

1. 96-Well microtiter plates: Costar® flat-bottom, polystyrene plates.
2. Glycerol (80% (v/v)): 1 kg of 100% glycerol (ρ = 1.25 g/cm^3) is diluted to 1 L with Milli-Q water, then autoclave.
3. Cultures (*see* **step 5** – **Section 6.3.5**).
4. Aluminum foil lids: Biomek® Seal and Sample (#538619, Beckman Coulter, Fullerton, CA) or similar.
5. Freezer (–80°C).

6.3. Methods

The PIPE method is based on the observation that, contrary to popular assumption, normal PCR amplifications result in mixtures of products which are not fully double stranded *(15)*. The 5′ ends of such products are left variably unpaired by incomplete 5′ → 3′ primer extension caused by sequence-specific stalling and changes in the reaction equilibrium (less dNTPs available, more template copies to synthesize) in the final cycles of PCR. These unpaired 5′ ends on the PCR products are the same 5′ ends on the synthetic amplification primers. Therefore, a simple oligonucleotide design rule can control the sequences of these ends in a way that promotes easy cloning and mutagenesis.

The first 15 bases on the 5′ ends of the primers are designed to be directionally complementary such that the resultant PCR fragment(s) can anneal as desired and become viable plasmids upon transformation. In basic PIPE cloning, the vector is linearized by *V-PIPE* PCR amplification and contains two distinct 5′ ends (**Fig. 6.1a**). The inserts are *I-PIPE* PCR amplified with primers which contain 5′sequence complementary to the two distinct ends of the amplified vector (**Fig. 6.1b**). In this manner, annealing occurs directionally and creates a viable plasmid. In basic PIPE mutagenesis (*M-PIPE*), the entire plasmid is amplified (**Fig. 6.1c**). The two primers used are designed to create the mutation (a substitution, deletion, or insertion) and to be complementary to each other so that the linearized PCR product can self-anneal to recreate a viable, mutant plasmid. The following protocol describes PIPE cloning and mutagenesis in a 96-well plate.

6.3.1. Preparation of Competent Cells

1. Start a 10-ml overnight culture of cells in LB Broth at 37°C for a 1-L batch.
2. On the following morning, seed 1 L of LB Broth with the 10-ml overnight culture.
3. Incubate the culture at 37°C while shaking at 250 rpm until the optical density measured at 600 nm (OD_{600}) reaches 0.4–0.6. (The remainder of this preparation is done on ice or in the cold room.).
4. Split the 1-L culture into 2 × 500 ml sterile, disposable Corning Erlenmeyer polycarbonate flasks and pellet the cells by chilled centrifugation at 2,500 × *g* for 20 min (*see* **Note 8**).
5. Decant and discard the LB Broth.
6. Add $MgCl_2$ Solution at one-tenth the original volume (100 ml) to the pellet.

7. Resuspend the pellet by gentle swirling the $MgCl_2$ Solution over the pellet. Pipette up and down to break up dislodged pellet, if necessary (*see* **Note 8**).
8. Incubate the resuspended cells on ice for 30 min.
9. Pellet the cells by chilled centrifugation at 2,500 × *g* for 20 min.
10. Decant and discard the supernatant.
11. Add $CaCl_2$ + Glycerol Solution at one-fiftieth the original volume (20 ml) to the cell pellet.
12. Resuspend the pellet by gentle swirling the $CaCl_2$ + Glycerol Solution over the pellet. Pipette up and down to break up dislodged pellet, if necessary.
13. Divide the competent cells into 2 ml aliquots using 2-ml microcentrifuge tubes.
14. Flash freeze the aliquots in liquid nitrogen.
15. Store aliquots at −80°C.

6.3.2. Preparation of Selective LB Agar Plates

1. LB Agar plates: 200 ml of liquefied LB Agar is supplemented with an appropriate antibiotic at 55°C and then poured into the BioAssay tray with the 48-well divider removed from the tray (*see* **Note 9**).
2. After the LB agar has been poured, the 48-well divider is placed back into the tray and partially submerged into the LB agar creating 48 separate squares.
3. The LB agar is allowed to solidify at room temperature.
4. The plates are then dried overnight by propping the lids slightly open to allow sufficient gas exchange while minimizing potential contamination.

6.3.3. PCRs for Amplifying Cloning Vectors and Inserts or Generating Site-Directed Mutants

1. Dilute the 50-μM oligonucleotide primer stocks to 10 μM with Milli-Q water.
2. Set up the PIPE PCRs.
   a) (*V-PIPE*) Tranfer 5 μl of both 10 μM forward and reverse primers (*see* **step 2a – Section 6.2.3**) into PCR tubes and then add 5 μl of template (*see* **step 1i – Section 6.2.3**) and 35 μl of PIPE *Pfu* Master Mix (*see* **Notes 10** and **11**).
   b) (*I-PIPE*) Tranfer 5 μl of both 10 μM forward and reverse primers (*see* **step 2b – Section 6.2.3**) into a PCR plate and then add 5 μl of template (*see* **step 1ii – Section 6.2.3**) and 35 μl of PIPE *Pfu* Master Mix (*see* **Note 10**).
   c) (*M-PIPE*) Tranfer 5 μl of both 10 μM forward and reverse primers (*see* **step 2c – Section 6.2.3**) into a PCR plate and then add 5 μl of template (*see* **step 1iii – Section

6.2.3) and 35 μl of PIPE *Pfu* Master Mix (*see* **Notes 12** and **13**).

3. Thermocycler conditions for PIPE amplifications.
   a) (*V-PIPE*) 95°C for 2 min, then 25 cycles of 95°C for 30 s, 55°C for 45 s and 68°C for 14 min, and finally a 4°C hold.
   b) (*I-PIPE*) 95°C for 2 min, then 25 cycles of 95°C for 30 s, 55°C for 45 s and 68°C for 3 min, and finally a 4°C hold (*see* **Note 14**).
   c) (*M-PIPE*) 95°C for 2 min, then 25 cycles of 95°C for 30 s, 55°C for 45 s and 68°C for 14 min, and finally a 4°C hold.
4. Confirm successful amplifications by gel electrophoresis.

6.3.4. Transforming Competent Cells and Plating Transformed Cells

1. Thaw 2 ml aliquot of competent cells (*see* **step 15** – **Section 6.3.1**) on ice for 10–15 min.
2. Chill a 96-well microtiter plate on ice for 10–15 min.
3. Transfer 2 μl of each of the PCRs into wells of a prechilled microtiter plate.
   – When cloning, first mix 2 μl from the V-PIPE and 2 μl I-PIPE reactions together (*see* **Note 15**).
   – For mutagenesis, 2 μl from the M-PIPE reaction can be used directly (*see* **Note 16**).
4. Dispense 20 μl of competent cells into each well. Pipette up and down ONCE to ensure DNA has mixed with the cells.
5. Incubate the DNA-cell mixture on ice for 15 min.
6. Heat shock the cells by floating the microtiter plate in a 42°C water bath for 45 s (*see* **Note 17**).
7. Immediately return the microtiter plate to ice.
8. Dispense 100 μl of LB Broth (no antibiotic) into each well.
9. Recover the transformed cells by incubating at 37°C while shaking at 250 rpm for 1 h.
10. Dispense 100 or 40 μl of the recovered cells into the respective wells of the selective LB agar trays with glass beads.
11. Shake (by hand) the trays enough to move the glass beads and evenly distribute the cells across the entire well.
12. Invert the tray to drop the glass beads off of the LB agar and onto the lid.
13. Remove the glass beads from the lid.
14. Incubate the inverted trays overnight (12–16 h) in a stationary 37°C incubator to grow the bacterial colonies (*see* **Note 18**).

6.3.5. Screening Colonies by Diagnostic PCR (dPCR) or SYBR-PCR

1. Dispense 200 µl of LB Broth with appropriate antibiotic into the wells of flat-bottom 96-well plate.
2. Using aseptic technique, pick and transfer 1–4 isolated colonies per transformation (*see* **step 14** – **Section 6.3.4**) into unique wells of the microtiter plate.
3. Incubate the plate at 37°C while shaking at 250 rpm for at least 3 h (up to overnight).
4. Transfer 3-µl samples from each culture into a 96-well PCR plate.
5. Put the remainder of the cultures (~197 µl/well) back into the shaking incubator to continue growth for future glycerol stock archival.
6. Add 47 µl of *Taq* Master Mix to each well containing cells in the PCR plate (*see* **Note 19**).
7. Place the PCR plate into a thermocycler.
8. Amplify the DNA fragments using the following cycling conditions: 95°C for 2 min followed by 30 cycles of 95°C for 30 s, 55°C for 45 s and 72°C for 3 min, then finally a 4°C incubation.

6.3.6. SYBR Assay (for Optional Use with the Universal Insert-Specific Reverse Primer Amplification)

1. Dispense 50 µl of Quench Buffer in a flat-bottom 96-well plate.
2. Transfer 5 µl of each SYBR-PCR product (*see* **step 8** – **Section 6.3.5**) into the wells with Quench Buffer (*see* **Note 20**).
3. Dispense 150 µl of SYBR Buffer into each well.
4. Prepare an unamplified sample for a negative control.
5. Measure fluorescence of each sample using a microtiter plate fluorescence reader. Excitation: 485 nm. Emission: 525 nm. Auto-cutoff enabled.
6. SYBR results are determined on a relative scale with the positive wells having at least fourfold higher fluorescence than the negatives or the control.

6.3.7. Preparation of dPCR Products for Sequencing

1. Dispense 5 µl of Exo/SAP Solution[15] directly into the dPCR positive wells (determined by gel analysis, SYBR assay, or simply assumed to be positive).
2. Incubate the Exo/SAP reaction at 37°C for 30 min, then 75°C for 15 min.
3. Submit these samples directly for sequencing.

6.3.8. Glycerol Stock Archival of Putative Clones

1. Add 50 µl of 80% (v/v) glycerol (20% final) to 150 µl of each culture (*see* **step 5** – **Section 6.3.5**) in a 96-well microtiter plate.
2. Mix the glycerol into the culture by pipetting up and down.

3. Seal the plate using the aluminum foil lid.
4. Store the plate in a –80°C freezer.

6.4. Notes

1. Milli-Q water is purified to a resistivity of 18.2 MΩ cm and contains total organics at less than five parts per billion using the Milli-Q Synthesis System (Millipore, Billerica, MA).
2. Thermostable DNA polymerases from *Pyrococcus furiosus* (*Pfu*), *Thermococcus kodakaraensis* (KOD), and *Thermus aquaticus* (*Taq*) as well as Phusion™ DNA Polymerase have all been used successfully. However, the majority of our experience and, therefore, success has come from using *Pfu*Turbo DNA Polymerase. We also observe spurious mutations using *Taq* DNA Polymerase which have not been observed using the proofreading enzymes.
3. *Taq* DNA Polymerase works very well for amplifying DNA directly from cell cultures. The ability to PCR from these cells grown from isolated colonies eliminates the need to miniprep DNA for sequencing. Amplifying DNA from cell cultures using *Pfu*Turbo DNA Polymerase has NOT worked well for us.
4. The insert-specific reverse primer is designed such that successful PCR amplification is dependent on the insert annealing to the vector to form a viable colony. Conditional PCR can be used in series with a fluorescence assay (SYBR) to determine insert-containing plasmids without running a DNA gel.
5. This master mix is actually made at 102× to account for volume losses in the multiple transfer steps.
6. Although SYBR dyes are reported to be far less carcinogenic than ethidium bromide, they still bind tightly to DNA. Appropriate care should be observed in handling and disposal.
7. Alternate Exo/SAP protocols may use 4 μl of buffer instead of 4 μl of water, but our sequencing results have not suffered.
8. Use centrifugation vessels which maximize the surface area across which the cells are pelleted. This makes it much easier to resuspend the cells without excessive stress to the cells.
9. Subjecting antibiotic(s) to the LB agar at temperatures greater than 55°C for extended periods of time can decrease relative effectiveness on susceptible cells.
10. Our vector pSpeedET contains the *ccdB* gene which is toxic to our expression strain (HK100). This creates a negative

selection against vector template DNA. Strains carrying the *ccdA* antidote on the F′ episome are not susceptible to this negative selection. In the V-PIPE PCR, the entire plasmid is amplified except for the *ccdB* cassette.

11. For large cloning projects, amplifying the vector separately from the inserts is logistically best. It is possible, however, to amplify both the vector and the insert (from the same or different templates) in the same reaction (IV-PIPE). This is done by adding both templates and all four primers to the same 50 μl PCR. Each primer is added at half the original concentration so that the total primer concentration in the reaction remains fixed at 2 μM.
12. Mutagenesis is not limited to a single substitution, deletion, or insertion per reaction. We have made up to five substitutions using a single primer pair when the sites are all in close proximity. We have made four codon substitutions using four primer pairs in the same reaction when the sites were disbursed. We have made N- and C-terminal insertions and deletions separately and simultaneously. In all cases, the total primer concentration in the reactions is 2 μM.
13. In most cases, the M-PIPE templates encode the same antibiotic resistance as the desired mutant plasmids will have. We have found that ~1 ng (20 pg/reaction) is typically low enough to eliminate background transformants. In the cases where PCR amplification is only successful using higher amounts of template, *DpnI* may be used to digest the template DNA to reduce background.
14. Use the thermocycling conditions for V-PIPE PCR when trying to amplify the vector and insert in the same reaction.
15. The PCR products are used directly out of the thermocycler. There are no post-PCR treatments to the PCR products unless the background is determined to be too high (*DpnI* treatment, **Note 13**). If the vector and the insert were amplified together (IV-PIPE), then 2 μl from that reaction is sufficient.
16. The PCR products are used directly out of the thermocycler. There are no post-PCR treatments to the PCR products unless the background is determined to be too high (*DpnI* treatment).
17. Efficient heat shock requires direct water to well contact. Be careful to avoid introducing air pockets between the plate and water.
18. If satellite colonies are present, incubate the trays overnight at 30°C instead of 37°C.
19. For SYBR-PCR, the universal insert-specific reverse primer is designed to anneal to any amplified DNA insert. PCR amplification is conditional on the insert fragment annealing to the

vector in the correct orientation. Therefore, PCR is successful on insert-containing plasmids but not with background transformants such as vector only or PCR template contamination.

20. This assay is not helpful when using two vector-specific primers since amplification will occur with or without the insert. Although the assay is very quick in identifying insert-containing plasmids (compared to agarose gel), the sizes of the PCR products cannot be determined. In our experience, the SYBR assay and PCR products of the correct size have at least an 80% correlation.

Acknowledgments

This work was supported, in part, by NIH grant PSI U54 GM074898. We thank Eric Koesema for his experiments in confirming the proof of principle and testing the initial applications of the PIPE method. We also thank Mark Knuth, Christian Ostermeier, and Roger Benoit for helpful discussions.

References

1. Scharf, S.J., Horn, G.T., Erlich, H.A. (1986) Direct cloning and sequence analysis of enzymatically amplified genomic sequences. *Science*. **233**, 1076–78.
2. Costa, G.L., Grafsky, A., Weiner, M.P. (1994) Cloning and analysis of PCR-generated DNA fragments. *PCR Methods Appl*. **6**, 338–45.
3. Aslanidis, C., de Jong, P.J. (1990) Ligation-independent cloning of PCR products (LIC-PCR). *Nucleic Acids Res*. **18,** 6069–74.
4. Hsiao, K. (1993) Exonuclease III induced ligase-free directional subcloning of PCR products. *Nucleic Acids Res*. **21**, 5528–9.
5. Boyd, A.C. (1993) Turbo cloning: a fast, efficient method for cloning PCR products and other blunt-ended DNA fragments into plasmids. *Nucleic Acids Res*. **21**, 817–821.
6. Bubeck, P., Winkler, M., Bautsch, W. (1993) Rapid cloning by homologous recombination in vivo. *Nucleic Acids Res*. **21**, 3601–2.
7. Liu, Q., Li, M.Z., Leibham, D., Cortez, D., Elledge, S.J. (1998) The univector plasmid-fusion system, a method for rapid construction of recombinant DNA without restriction enzymes. *Curr Biol*. **8**, 1300–9.
8. Oliner, J.D., Kinzler, K.W., Vogelstein, B. (1993) *In vivo* cloning of PCR products in *E. coli*. *Nucleic Acids Res*. **21**, 5192–7.
9. Hartley, J.L., Temple, G.F., Brasch, M.A. (2000) DNA cloning using *in vitro* site-specific recombination. *Genome Res*. **10,** 1788–95.
10. Tillett, D., Neilan, B.A. (1999) Enzyme-free cloning: a rapid method to clone PCR products independent of vector restriction enzyme sites. *Nucleic Acids Res*. **27**, e26.
11. Chiu, J., March, P.E., Lee, R., Tillett, D. (2004) Site-directed, ligase-independent mutagenesis (SLIM): a single-tube methodology approaching 100% efficiency in 4h. *Nucleic Acids Res*. **32**, e174.
12. Kirsch, R.D., Joly, E. (1998) An improved PCR-mutagenesis strategy for two-site mutagenesis or sequence swapping between related genes. *Nucleic Acids Res*. **26**, 1848–50.
13. Sawano, A., Miyawaki, A. (2000) Directed evolution of green fluorescent protein by a new versatile PCR strategy for site-directed and semi-random mutagenesis. *Nucleic Acids Res*. **28**, e78.

14. Hanke, M., Wink, M. (1995) Direct DNA sequencing of PCR-amplified vector inserts following enzymatic degradation of primer and dNTPs. *Biotechniques.* **18**, 636.

15. Olsen, D.B., Eckstein, F. (1989) Incomplete primer extension during *in vitro* DNA amplification catalyzed by *Taq* polymerase; exploitation for DNA sequencing. *Nucleic Acids Res.* **23**, 9613–20.

Chapter 7

A Family of LIC Vectors for High-Throughput Cloning and Purification of Proteins[1]

William H. Eschenfeldt, Lucy Stols, Cynthia Sanville Millard, Andrzej Joachimiak, and Mark I. Donnelly

Summary

Fifteen related ligation-independent cloning vectors were constructed for high-throughput cloning and purification of proteins. The vectors encode a TEV protease site for removal of tags that facilitate protein purification (his-tag) or improve solubility (MBP, GST). Specialized vectors allow coexpression and copurification of interacting proteins, or *in vivo* removal of MBP by TVMV protease to improve screening and purification. All target genes and vectors are processed by the same protocols, which we describe here.

Key words: Structural genomics; High throughput; Protein purification; Ligation-independent cloning; Coexpression; *In vivo* proteolysis; Maltose-binding protein; TEV protease; TVMV protease

7.1. Introduction

A family of compatible ligation-independent cloning (LIC) vectors *(1)* was created to enable effective high-throughput cloning and purification of recombinant proteins for structural studies. All

[1] The submitted manuscript has been created by the University of Chicago as operator of Argonne National Laboratory under Contract No. W-31-109-ENG-38 with the U.S. Department of Energy. The U.S. government retains for itself, and others acting on its behalf, a paid-up, nonexclusive, irrevocable worldwide license in said article to reproduce, prepare derivative works, distribute copies to the public, and perform publicly and display publicly, by or on behalf of the government.

Sharon A. Doyle (ed.), *Methods in Molecular Biology: High Throughput Protein Expression and Purification, vol. 498*

Book doi: 10.1007/978-1-59745-196-3

the vectors contain a sequence encoding the tobacco etch virus (TEV) protease cleavage site *(2)* next to an *Ssp*I site used for LIC (**Fig. 7.1**). The base vector, pMCSG7 *(3)*, appends an N-terminal hexahistidine tag to proteins that is followed by the protease recognition sequence. Derivatives of the base vector (**Table 7.1**) add maltose-binding protein *(4, 5)*, glutathione-*S*-transferase *(6, 7)*, or a loop of GroES *(8–10)* to the leader, replace the his-tag with the S-tag *(7, 11)*, incorporate a second protease cleavage site for *in vivo* tag removal *(5, 12)*, or move the entire LIC region into different, compatible vectors to allow coexpression of proteins *(5, 7)*. In all cases, expression is driven by T7 polymerase under control of the lac promoter in specific host strains *(13)*. To introduce genes into the vector, LIC-compatible extensions are added through the use of specific primers during amplification by PCR *(1)*. Following appropriate processing, the PCR product can be introduced into any member of the family by a standard LIC protocol. This chapter describes the manual, nonhigh-throughput LIC of genes into the pMCSG vectors. **Chapter 8** describes plate-based methods for high-throughput applications.

The LIC process is identical for all members of the family (**Fig. 7.2**). The vector is first linearized by cleavage with *Ssp*I then treated with T4 polymerase in the presence of dGTP only. Exonuclease activity of the polymerase hydrolyzes nucleotides from the 3′ ends of the vector until it reaches a G residue, creating a 15-base, single-stranded 5′ overhang. Conversely, treatment of appropriate PCR products with polymerase in the presence of only dCTP generates a complementary overhang. The two treated fragments are combined, allowed to anneal, and introduced into a suitable host. The host's native enzymes ligate then propagate the plasmid. In order to place nucleotides encoding a TEV cleavage site as close as possible to the introduced gene,

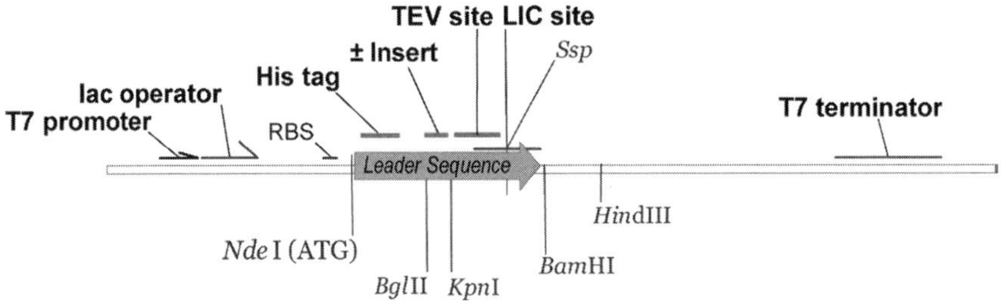

Fig. 7.1. Generalized organization of MCSG vectors. MCSG vectors encode an N-terminal leader sequence (*arrow*) that terminates in an LIC region centered on an *Ssp*I site. Restriction sites in and around the coding/cloning region allow insertion of protein or peptide modules into the leader (at *Bgl*II and/or *Kpn*I), replacement of the his-tag (*Nde*I to *Bgl*II or *Kpn*I), or transfer of the entire region to different backgrounds (*Nde*I to *Bam*HI, *Hin*DIII or *Xho*I). Derivatives (**Table 7.1**) add MBP, GST, or a loop of GroES (Sloop), replace the His-tag with the S-tag, and move these expression regions to pACYCDuet-1 and pCDFDuet-1, allowing cotransformation with two or three vectors and coexpression of multiple proteins. Another modification, pMCSG19, positions untagged MBP followed by the TVMV protease recognition sequence *(12)* ahead of the His-tag, allowing *in vivo* removal of MBP by coexpressed TVMV protease.

Table 7.1
MSCG vectors for high-throughput ligation-independent cloning

Vector	Parental vector	Antibiotic	Leader sequence	MW (leader)[a]	kb	Purpose
pMCSG7	pET-21a[b]	Amp	N-His-TEV-LIC	2,755	5,286	production
pMCSG8	pMCSG7	Amp	N-His-Sloop-TEV-LIC	4,399	5,341	toxicity
pMCSG9	pMCSG7	Amp	N-His-MBP-TEV-LIC	43,713	6,147	solubility
pMCSG10	pMCSG7	Amp	N-His-GST-TEV-LIC	29,046	5,961	solubility
pMCSG11	pACYC-Duet-1[c]	Cam	N-His-TEV-LIC	2,755	4,079	coexpression
pMCSG12	pACYC-Duet-1	Cam	N-His-Sloop-TEV-LIC	4,399	4,144	coexpression
pMCSG13	pACYC-Duet-1	Cam	N-His-MBP-TEV-LIC	43,713	4,940	coexpression
pMCSG14	pACYC-Duet-1	Cam	N-His-GST-TEV-LIC	29,046	4,754	coexpression
pMCSG17	pMCSG7	Amp	N-Stag-TEV-LIC	3,760	5,316	coexpression
pMCSG19	pMCSG7	Amp	N-MBP-TVMV-His-TEV-LIC	45,050/2,711[d]	6,441	production
pMCSG20	pMCSG17	Amp	N-Stag-GST-TEV-LIC	30,051	5,991	coexpression
pMCSG21	pCDFDuet-1[c]	Spec	N-His-TEV-LIC	2,755	3,852	coexpression
pMCSG22	pCDF-Duet-1	Spec	N-His-Sloop-TEV-LIC	4,399	3,906	coexpression
pMCSG23	pCDF-Duet-1	Spec	N-His-MBP-TEV-LIC	43,713	4,971	coexpression
pMCSG24	pCDF-Duet-1	Spec	N-His-GST-TEV-LIC	29,046	4,527	coexpression

All 15 vectors are processed for LIC the same way and accept the same properly prepared PCR products. *Abbreviations: Amp* ampicillin; *Cam* chloramphenicol; *Spec* spectinomycin; *His* hexahistidine tag; *TEV* tobacco etch virus protease recognition sequence; *LIC* ligation-independent cloning site centered on an *Ssp*I site; *Sloop* GroEL-binding loop of GroES; *MBP* maltose-binding protein; *GST* glutathione-*S*-transferase; *Stag* S-tag fragment of ribonuclease; *TVMV* tobacco vein mottling virus protease recognition sequence

[a]Molecular weight removed by cleavage with TEV protease

[b]Vector pET-21a is a product of Novagen, Inc. (Madison, WI)

[c]Vectors pACYCDuet-1 and pCDFDuet-1 are products of Novagen, Inc. (Madison, WI)

[d]First value is molecular weight removed by cleavage by TEV protease without prior treatment with TVMV protease. Second value is that removed after prior treatment with TVMV protease. TVMV protease removes 42,339 Da comprising untagged MBP and flanking amino acids

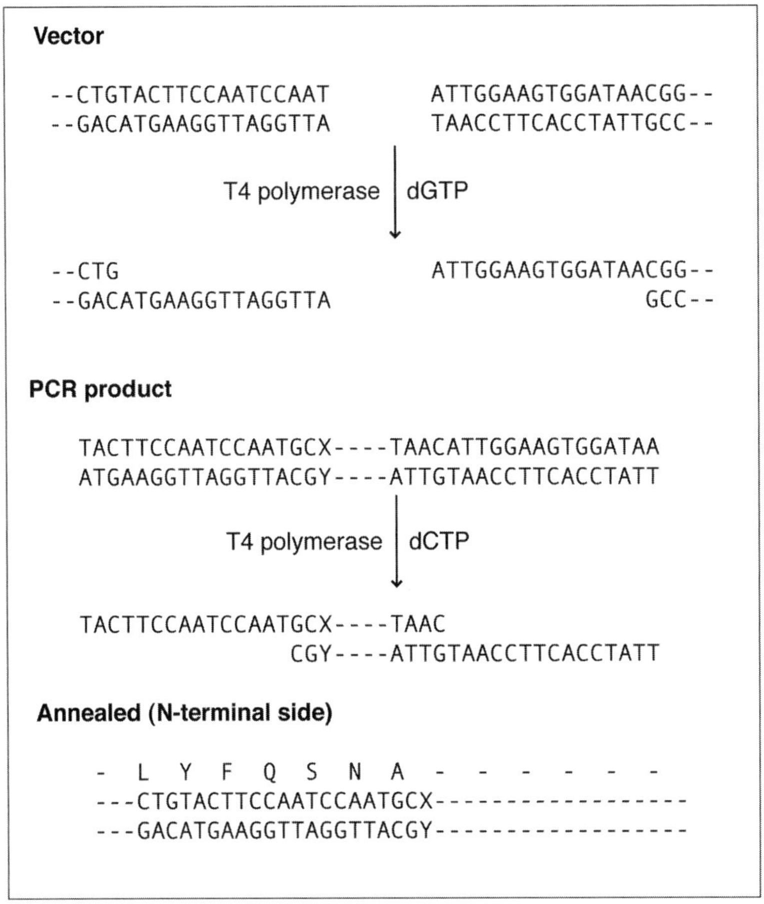

Fig. 7.2. LIC procedure using pMCSG vectors. All MCSG vectors contain an *Ssp*I site (AATATT) positioned immediately after the sequence encoding the TEV protease recognition site. Cleavage with *Ssp*I (a blunt cutter) followed by treatment with T4 DNA polymerase in the presence of only dGTP generates 15-base pair overhangs on both ends of the vector. PCR products must be generated using primers that begin with the complements of these overhangs followed by nucleotides required for LIC processing and proper expression. The sense primer must begin, TACTTCCAATCCAATGCC– – –, where *dashes* indicate nucleotides identical to the target gene, and requires the nucleotides GCC to stop the action of the polymerase and encode alanine in the correct reading frame. The antisense primer, TTATCCACTTCCAATGTTA– – –, requires a G, complement of the C that stops the endonuclease activity of T4 polymerase, and TTA, the complement of a stop codon. Annealing to treated vector positions the TEV protease site adjacent to the gene in the correct reading frame. The resulting protein has the residues SNA appended to its N terminus after TEV cleavage. Primers conforming to these restrictions can be designed manually or using commercial programs or online tools such as the Express Primer Tool ***(14)***, http://tools.bio.anl.gov/bioJAVA/jsp/ExpressPrimerTool/.

restraints were placed on the design of the LIC region *(3)*. The use of *Ssp*I at the center of the LIC region allows the resulting proteins to carry only three additional amino acids on their N termini after proteolysis – serine from the TEV site, asparagine from the *Ssp*I site, and alanine, the preferred choice for the limited number of amino acids possible due to the G used to stop the action of T4 polymerase in creating the overhang.

In addition to having the exact 16 base pairs at each end needed to create the complementary 15-base pair overhangs and stop the T4 polymerase, PCR primers must carry additional nucleotides (**Fig. 7.2**). For the sense primer, an additional two nucleotides are needed to maintain the proper reading frame and complete the codon begun with the required G. Because alanine is the most benevolent of the amino acids whose codons begin with G, the bases CC (or CA, CT, or CG) are added to give TACTTCCAATCCAATGCC. For the antisense primer, the required 16 nucleotides should be followed by the complement of a stop codon to prevent readthrough into vector sequences, giving TTATCCACTTCCAATGTTA. Following these sequences, primers terminate with nucleotides complementary to the gene being amplified, according to the requirements of the PCR conditions to be used. Primers may be designed manually or with software, such as the Express Primer tool (http://tools.bio.anl.gov/bioJAVA/jsp/ExpressPrimerTool/) *(14)*.

Use of highly purified vectors is not essential to successful LIC, but improves efficiencies. However, complete cutting of the vectors with *Ssp*I is critical and problematic. Simple restriction digests with commercial, normal-strength *Ssp*I resulted in high backgrounds – clones containing unmodified vector free of an inserted gene *(15)*. To circumvent this problem, vectors are cleaved with excess *Ssp*I using high concentration, LIC-qualified commercial enzyme, then purified on agarose gels to remove traces of uncut vector. When processed in this fashion, we routinely achieve efficiencies between 75 and 90%.

7.2. Materials

7.2.1. PCR Amplification of Genes and Processing for LIC

7.2.1.1. PCR

1. Sense primer: 5′-TACTTCCAATCCAATGCC– – – and Antisense primer: 5′-TTATCCACTTCCAATGTTA– – – obtained from oligonucleotide production service of choice. The dashes denote a series of nucleotides identical to those of the target gene (*see* **Section 7.1** and *(14)*)
2. Platinum *Pfx* DNA polymerase (Invitrogen, kit cat. no. 11708-013)
3. Platinum *Pfx* 10× buffer (included in kit)
4. $MgSO_4$ (50 mM) (included in kit)

7.2.1.2. Purification

1. Qiagen spin column (QIAquick PCR Purification Kit #28104)
2. Qiagen buffers (included in kit)

7.2.1.3. T4 Treatment

1. dCTP (100 mM) (Promega cat. no. U1221)
2. Dithiothreitol (DTT, 100 mM), molecular biology grade (Sigma cat. no. D-9779)
3. T4 DNA polymerase, LIC-qualified (Novagen cat. no. 70099)
4. 10× T4 polymerase buffer (included with polymerase)

7.2.2. Preparation of Vector for LIC

7.2.2.1. Vector Prep

1. LB Broth, Miller or other rich medium
2. LB Agar, Miller
3. Sterile polystyrene Falcon tube (14 ml)
4. QIAGEN Plasmid Midi Kit (Qiagen, Inc. cat. no. 12143 or 12145)
5. Ampicillin
6. Chloramphenicol
7. Spectinomycin
8. Oakridge centrifuge tubes (Nalgene 3118-0050)

7.2.2.2. SspI Treatment

1. *Ssp*I, Genome-Qualified, High Concentration, ca. 50 U/µl (Promega R4604) and *Ssp*I buffer (included with enzyme)
2. 50× TAE Buffer concentrate for gel: 2 M Tris/Acetate, pH 8.0, 50 mM EDTA
3. Ultrapure agarose for gel
4. Ethidium bromide, 50 µg/ml
5. Tracking dye: 0.25% bromophenol blue plus 0.25% xylene cyanol FF in 30% glycerol
6. Gel electrophoresis apparatus – Biorad Mini Sub Cell or other comparable equip
7. DNA molecular weight markers, 1-kb ladder
8. QIAEX II gel extraction kit (Qiagen)

7.2.2.3. T4 Treatment

1. dGTP (100 mM)
2. Dithiothreitol (DTT, 100 mM), molecular biology grade
3. T4 DNA polymerase, LIC-qualified (Novagen cat. no. 70099)
4. 10× T4 polymerase buffer (included with enzyme)

7.2.3. LIC Annealing and Transformation

1. Sterile polystyrene Falcon tube (14 ml)
2. DH5α cells (Invitrogen Library Competent, cat. no. 18263-012), or equivalent
3. SOC (included with DH5α cells)

7.3. Methods

7.3.1. PCR and Preparation of PCR Product for LIC

7.3.1.1. PCR

1. In a thin-walled PCR tube combine 5 µl 10x reaction buffer and primers at 1 µM with approximately 1 ng template DNA in a total volume of 48.5 µl.
2. Add 0.5 µl polymerase (1.25 U) and 0.5–1.5 µl 50 mM $MgSO_4$ (as needed).
3. Perform PCR at appropriate temperatures for amplifying specific targets. Typical reactions comprise an initial 3-min denaturation at 94°C, amplification through 35 cycles of 30 s at 94°C, 45 s at 55–60°C, and 1 min at 68°C, with a final extension for10 min at 68°C (*see* **Note 1**).

7.3.1.2. Purification of PCR Product

1. Apply PCR to Qiagen Spin column from QIAquick PCR Purification Kit.
2. Follow the protocol detailed in the kit.
3. Determine concentration spectrophotometrically (e.g., NanoDrop Technologies' ND-1000 spectrophotometer) or by another suitable method.

7.3.1.3. T4 Polymerase Treatment of PCR Product

1. To a 0.4-ml Eppendorf tube on ice add

	Volume
PCR product 20 ng	x µl ($x + y = 32$ µl)
T4 polymerase 10×reaction buffer	4 µl
dCTP (100 mM)	1 µl
DTT (100 mM)	2 µl
Sterile water	y µl ($x + y = 32$ µl)
T4 DNA polymerase	1 µl
	Total 40 µl

2. Incubate the reaction mix at room T for 30 min
3. Inactivate the T4 DNA polymerase by heating at 75°C for 20 min

7.3.2. Preparation of Vector for LIC

7.3.2.1. Isolation and Purification of Vector

1. Inoculate 2 ml of LB broth containing the appropriate antibiotic in a 14-ml Falcon tube from a glycerol stock of *E. coli* DH5α containing the desired vector. Incubate for several hours at 37°C, agitating at 250 rpm. Antibiotic concentrations: ampicillin, 100 µg/ml; chloramphenicol, 30 µg/ml; spectinomycin, 50 µg/ml. (*see* **Note 2**).

2. Inoculate 50 ml of LB plus antibiotic in a sterile, notched Erlenmeyer flask with the entire 2-ml culture and incubate overnight at 37°C at 250 rpm.
3. Transfer the culture to two 50-ml Oakridge centrifuge tubes and centrifuge at 6,000 × *g* for 15 min at 4°C. Pour off supernatant fluid and invert the tubes to drain.
4. Lyse cells and purify vector using QIAGEN Plasmid Midi Kit according to the instructions detailed in the QIAGEN Plasmid Purification Handbook that accompanies the product.
5. Dissolve the pellet resulting from the purification in 200–500 µl water or 10 mM Tris (pH 8.0). The DNA concentration can be estimated by running an aliquot on an agarose gel or spectrophotometrically with a Nanodrop instrument (*see* **Note 3**).

7.3.2.2. SspI Digestion

1. To a 1.5-ml Eppendorf tube on ice add

	Volume
Vector DNA 15 µg instrument	*x* µl (*x* + *y* = 52 µl)
*Ssp*I 10× reaction buffer	6 µl
Sterile water	*y* µl (*x* + *y* = 52 µl)
*Ssp*I	2 µl
	Total 60 µl

2. Incubate for 2 h at 37°C
3. Prepare a 50-ml, 0.8% TAE agarose gel containing ethidium bromide at 0.5 µg/ml in a 6.5 × 10-cm flat bed electrophoresis tray fitted with an 8-well comb
4. Add 10 µl of 6× tracking dye to the digestion mixture
5. Load 10–15 µl per well
6. Run the gel at 75 V for approximately 2.5 h or until the molecular weight ladder is well separated
7. Visualize plasmid with a UV light box and excise band containing cut vector and transfer to tared 1.5-ml sterile Eppendorf tubes (*see* **Note 4**)
8. Extract DNA from gel slices using a QIAEX II Gel Extraction kit following the instructions detailed in the product's manual (*see* **Note 5**)

7.3.2.3. T4 Polymerase Treatment

1. To a 0.4-ml Eppendorf tube on ice add

	Volume
Vector DNA 200 ng	x µl ($x + y = 32$ µl)
T4 polymerase 10× reaction buffer	4 µl
dGTP (100 mM)	1 µl
DTT (100 mM)	2 µl
Sterile water	y µl ($x + y = 32$ µl)
Novagen T4 DNA polymerase	1 µl
	Total 40 µl

2. Incubate the reaction mix at room T for 30′
3. Inactivate the T4 DNA polymerase by heating at 75°C for 20′ (*see* **Note 6**)

7.3.2.4. Large-Scale Preparation of Vector for Microtiter Plate Experiments

For stocks for several 96-well plates, all steps are scaled up 5- to 10-fold (*see* **Note 7**).

7.3.3. LIC Annealing and Transformation of Cells

1. In a 14-ml Falcon tube on ice mix

15 ng. vector DNA	(ca. 3 µl)
30–45 ng PCR product DNA	(ca. 2–3 µl)

7.3.3.1. LIC Annealing

Place the two aliquots together in a small droplet in the bottom of the tube in a total volume less than 8 µl.

2. Incubate for 30 min on ice

7.3.3.2. Transformation

1. To the tube, add 50 µl of Invitrogen Library Competent DH5α cells.
2. Incubate on ice for 30 min.
3. Heat shock 45 s at 42°C.
4. Chill on ice 2 min.
5. Add 0.45 ml SOC.
6. Incubate at 37°C, 250 rpm, 1 h.
7. Plate on LB agar containing the appropriate antibiotic(s). Antibiotic concentrations: ampicillin, 100 µg/ml; chloramphenicol, 30 µg/ml; spectinomycin, 50 µg/ml. Plate 100 µl of

the culture, then centrifuge the remaining 400 μl gently in an Eppendorf tube to concentrate the cells. Decant the supernatant, resuspend the pellet in the small volume of medium remaining (ca. 50 μl), and plate on a second LB agar plate.

7.4. Notes

1. Other proof reading DNA polymerases and PCR conditions may be used for amplification of genes. For high-throughput cloning into the MCSG vectors, PCR is routinely performed in 96-well plates using 1 U of KOD polymerase, 10 ng genomic target DNA, and primers at 0.2 μM in a 50-μl reaction.
2. If cultures are started in the morning, they should be slightly turbid by afternoon.
3. The DNA pellet usually is quite clear and difficult to see. Incubating the water/buffer over the position where the pellet is expected for several minutes to allow time for it to dissolve can improve yields moderately. Absorbance at 260 nm tends to overestimate the DNA concentration.
4. Compared to normal gel purifications, the gel is grossly overloaded and the DNA band often appears deformed. Run the gel out until the tracking dye approaches the end of the gel, so that the 5- to 6-kb vector band is well separated.
5. The agarose gel fragments should be cut into small pieces before extraction and can be stored overnight at 4°C prior to extraction if necessary. The extracted DNA can be frozen at –20°C if necessary.
6. Can store frozen, but precipitation will occur. Be sure to bring the solution to room temperature after storage and wait for cloudiness to clear.
7. To scale up preparation for cloning in microtiter plates (**Chapter 8**), start two 5-ml cultures of LB in the morning (inoculate heavily from glycerol stocks) and subculture into 500-ml medium for overnight incubation. Harvest cells in 250-ml centrifuge bottles and purify the vector using the QIAGEN Plasmid Maxi Kit, resuspending the final pellet in 0.5-ml water or buffer. This preparation typically yields about 500 μg of vector. For SspI digestion, incubate 100–200 μg vector with 200-U SspI for 2 h at 37°C in a volume of 600 μl. Add 100-μl tracking dye and purify the cut vector on a 150-ml agarose gel in a 11 × 14-cm tray fitted with a 14-tooth comb, loading 50 μl per well. Combine gel fragments and elute the extract the DNA using the QIAEX II Gel Extraction kit according to the instructions provided by the vendor.

Acknowledgments

This work was supported by grants from the NIH (GM62414, GM074942), A. Joachimiak, PI, and the U.S. Department of Energy, Office of Biological and Environmental Research under Contract W-31-109-ENG-38.

References

1. Aslanidis, C. & de Jong, P. J. (1990) Ligation-independent cloning of PCR products (LIC-PCR), *Nucleic Acids Res. 18*, 6069–74.
2. Parks, T. D., Leuther, K. K., Howard, E. D., Johnston, S. A. & Dougherty, W. G. (1994) Release of proteins and peptides from fusion proteins using a recombinant plant virus proteinase, *Anal. Biochem. 216*, 413–7.
3. Stols, L., Gu, M., Dieckman, L., Raffen, R., Collart, F. R. & Donnelly, M. I. (2002) A new vector for high-throughput, ligation-independent cloning encoding a tobacco etch virus protease cleavage site, *Protein Expr. Purif. 25*, 8–15.
4. Kapust, R. B. & Waugh, D. S. (1999) *Escherichia coli* maltose-binding protein is uncommonly effective at promoting the solubility of polypeptides to which it is fused, *Protein Sci. 8*, 1668–74.
5. Donnelly, M. I., Zhou, M., Millard, C. S., Clancy, S., Stols, L., Eschenfeldt, W. H., Collart, F. R. & Joachimiak, A. (2006) An expression vector tailored for large-scale, high-throughput purification of recombinant proteins, *Protein Expr. Purif. 47*, 446–54.
6. Nygren, P. A., Stahl, S. & Uhlen, M. (1994) Engineering proteins to facilitate bioprocessing, *Trends Biotechnol. 12*, 184–8.
7. Stols, L., Zhou, M., Eschenfeldt, W. H., Millard, C. S., Abdullah, J., Collart, F. R., Kim, Y. & Donnelly, M. I. (2007) New vectors for coexpression of proteins: structure of *Bacillus subtilis* ScoAB obtained by high-throughput protocols, *Protein Expr. Purif. 53*, 396–403.
8. Xu, Z., Horwich, A. L. & Sigler, P. B. (1997) The crystal structure of the asymmetric GroEL-GroES-(ADP)7 chaperonin complex, *Nature 388*, 741–50.
9. Dieckman, L. J., Zhang, W., Rodi, D. J., Donnelly, M. I. & Collart, F. R. (2006) Bacterial expression strategies for human angiogenesis proteins, *J. Struct. Funct. Genomics 7*, 23–30.
10. Donnelly, M. I., Stevens, P. W., Stols, L., Su, S. X., Tollaksen, S., Giometti, C. & Joachimiak, A. (2001) Expression of a highly toxic protein, Bax, in *Escherichia coli* by attachment of a leader peptide derived from the GroES cochaperone, *Protein Expr. Purif. 22*, 422–9.
11. Raines, R. T., McCormick, M., Van Oosbree, T. R. & Mierendorf, R. C. (2000) The S.Tag fusion system for protein purification, *Methods Enzymol. 326*, 362–76.
12. Nallamsetty, S., Kapust, R. B., Tozser, J., Cherry, S., Tropea, J. E., Copeland, T. D. & Waugh, D. S. (2004) Efficient site-specific processing of fusion proteins by tobacco vein mottling virus protease in vivo and in vitro, *Protein Expr. Purif. 38*, 108–15.
13. Studier, F. W., Rosenberg, A. H., Dunn, J. J. & Dubendorff, J. W. (1990) Use of T7 RNA polymerase to direct expression of cloned genes, *Methods Enzymol. 185*, 60–89.
14. Yoon, H. Y., Hwang, D. C., Choi, K. Y. & Song, B. D. (2000) Proteolytic processing of oligopeptides containing the target sequences by the recombinant tobacco vein mottling virus NIa proteinase, *Mol. Cells 10*, 213–9.
15. Dieckman, L., Gu, M., Stols, L., Donnelly, M. I. & Collart, F. R. (2002) High throughput methods for gene cloning and expression, *Protein Expr. Purif. 25*, 1–7.

Chapter 8

"System 48" High-Throughput Cloning and Protein Expression Analysis[1]

James M. Abdullah, Andrzej Joachimiak, and Frank R. Collart

Summary

We describe a plate-based cloning and expression strategy for efficient high-throughput generation of validated expression clones in *Escherichia coli*. The process incorporates 48- or 96-well plates at all stages including the cloning and colony selection phases which are often performed manually. A 48-grid agar growth plate has been integrated into the colony selection component to improve throughput at the cloning stage. The combinations of 48- and 96-well plate formats are compatible with automated liquid handlers and multichannel pipettes. This revised cloning and expression pipeline increases throughput significantly, and also results in a reduction in both time and material requirements. The system has been validated by the production and screening of several thousand clones at the Midwest Center for Structural Genomics.

Key words: Agar cloning; 48-Grid agar plate; Structural genomics; Automated cloning; High throughput; Protein expression; Automation; Robotics

8.1. Introduction

The abundance of data from genomic sequencing projects has led to an increased demand for the proteins encoded by these genomes *(1)*. This demand has stimulated the development of strategies for high-throughput (HTP) cloning and *in vitro*

[1] The submitted manuscript has been created by UChicago Argonne, LLC, Operator of Argonne National Laboratory (Argonne). Argonne, a U.S. Department of Energy Office of Science laboratory, is operated under Contract No. DE-AC02-06CH11357. The U.S. Government retains for itself, and others acting on its behalf, a paid-up nonexclusive, irrevocable worldwide license in said article to reproduce, prepare derivative works, distribute copies to the public, and perform publicly and display publicly, by or on behalf of the Government.

Sharon A. Doyle (ed.), *Methods in Molecular Biology: High Throughput Protein Expression and Purification, vol. 498*

Book doi: 10.1007/978-1-59745-196-3

expression approaches for protein production *(2, 3)*. An assortment of *in vivo (3, 4)* and cell-free *(5, 6)* expression strategies are available which are compatible with the HTP approach. These approaches incorporate different methods utilizing recombinational *(4)*, restriction enzyme *(7)*, or ligation-independent cloning (LIC) methods *(8)* for the preparation of cloned material. An essential requirement for these approaches is the ability to use commercially available liquid handlers and other hardware that are compatible with standard microplate formats. The ability to automate the cloning and expression components for protein expression is essential for minimization of handling/tracking errors associated with a myriad of individual samples, meeting throughput requirements and to insure cost effectiveness.

The production pipeline from target selection to validated expression clone can be organized into component processes such as target selection, amplification (PCR), cloning, and expression and solubility analysis. Each component typically consists of one or more automation methods that use liquid handlers or other robots to improve throughput and reduce costs. The selection of individual colonies (clones) represents a small fraction of the process for production of validated expression clones. For most applications, however, clonality is an essential requirement since a single clone is desirable for most experimental endpoints. Recent progress in cloning systems have led to high-efficiency approaches for generation of vector constructs that require screening of only a minimal number of clones to obtain the desired construct ***(9, 10)***. The effect of increased throughput at many steps in the cloning and expression process has increased demand for target proteins, resulting in HTP projects with requirements to screen clones for thousands of individual targets. However, most current cloning and expression processes usually transition from plate-based formats (automated) to individual culture plates (manual) at the cloning stage, resulting in a dramatic reduction in throughput. Several programs have described approaches to circumvent the cloning bottleneck by using plate formats to screen a block of twelve clones using specially constructed "cloning grills" ***(11)*** or plates ***(12)***. We have developed a method that uses commercially available 48-grid plates to circumvent in a large part, this "transition" process and the many steps related to it, via our novel cloning method.

8.2. Materials

Robot hardware and reagent mixes are defined for a single 96-well plate procedure. In practice, most laboratories typically utilize several plates requiring linear scaling of materials and reagent mixes.

8.2.1. Preparation of T4 Polymerase Treatment of Amplified DNA Fragments and Vector

1. Make LIC Reaction mix sufficient for one 96-well plate by combining the following reagents:
   - 10X T4 polymerase buffer (included with polymerase; 465 μl)
   - dCTP (465 μl of 25 mM), molecular biology grade
   - Dithiothreitol (DTT) solution (228 μl of 100 mM)
   - Water (60 μl)
   - T4 DNA Polymerase (250 units) (LIC quality, ~2.5 units/μl, EMD Biosciences/Novagen)
   - Keep this mixture on ice and add the T4 DNA polymerase just before use. Pipette up and down several times to uniformly distribute the enzyme in the reaction mix. (*see* **Note 1**)
2. Preparation of LIC-compatible vector is described in a **Chapter** 7.

8.2.2. Transformation and Selection of Individual Colonies

1. Prepare Ampicillin stock solution (50 mg/ml) by addition of 1 g of ampicillin to 20 ml of sterile water. Working concentration for Lauria Bertaini (LB) broth and agar is 100 μg/ml ampicillin.
2. Lauria Bertaini (LB) Agar (500–550 ml), Miller with ampicillin (LB/Amp agar).
3. Lauria Bertaini (LB) Broth (200 ml), Miller.
4. Two sterile Q Tray 48-grid plates (Cat. # X6029, Genetix USA Inc) are labeled to allow mapping of the agar wells to the corresponding well on the transformation plate. Choose a clean level surface and then proceed to pour sufficient LB/Amp agar (250–275 ml of LB/Amp agar per plate) into the two previously labeled 48-grid agar plates. The media should be poured slowly into one of the peripheral rectangular openings (any outside opening other than one of the actual 48-grid slots; refer to the shaded areas within the graphic in **Fig. 8.1**, stage 2) to minimize bubbling in agar. Allow to solidify at room temperature for at least 20–30 min prior to use (*see* **Note 2**).
5. Chemically competent cells are prepared using the Z-Competent™ E. coli Transformation Buffer Set (ZYMO Research) according to the procedure recommended by the vendor (*see* **Note 3** regarding efficiency and strain selection). Competent cells are stored as 5-ml aliquots (sufficient for transformation of a single plate) at –70°C.
6. Sterilized, 1.25-in. precut glass rod rollers ("Sterilized, 1.25-in. precut glass rod rollers" should be prepared in advance by cutting 130-mm Borosilicate Glass rod Petri Dish Spreaders into pieces 1.25 in. in length.)

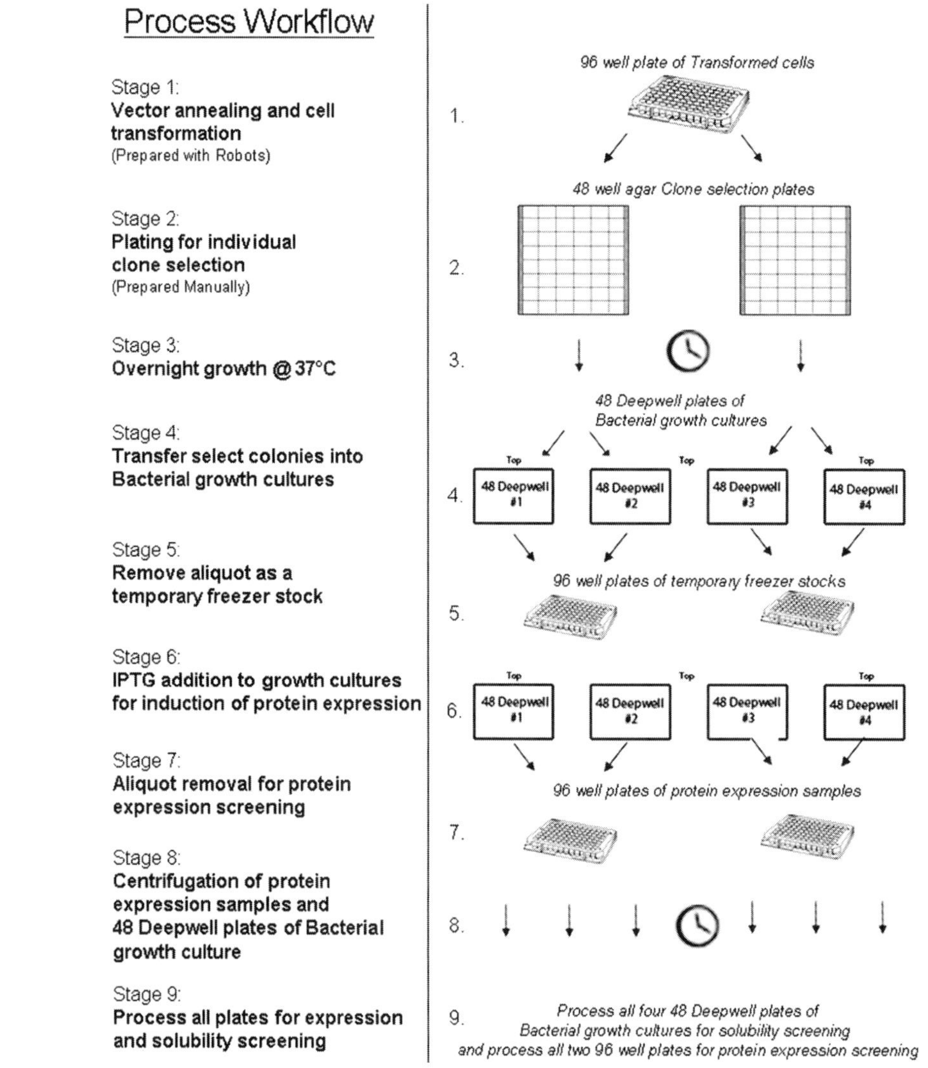
Process Workflow
Stage 1:
Vector annealing and cell transformation
(Prepared with Robots)
Stage 2:
Plating for individual clone selection
(Prepared Manually)
Stage 3:
Overnight growth @ 37°C
Stage 4:
Transfer select colonies into Bacterial growth cultures
Stage 5:
Remove aliquot as a temporary freezer stock
Stage 6:
IPTG addition to growth cultures for induction of protein expression
Stage 7:
Aliquot removal for protein expression screening
Stage 8:
Centrifugation of protein expression samples and 48 Deepwell plates of Bacterial growth culture
Stage 9:
Process all plates for expression and solubility screening
96 well plate of Transformed cells
1.
48 well agar Clone selection plates
2.
3.
48 Deepwell plates of Bacterial growth cultures
Top
48 Deepwell #1
48 Deepwell #2
48 Deepwell #3
48 Deepwell #4
4.
96 well plates of temporary freezer stocks
5.
6.
96 well plates of protein expression samples
7.
8.
9.
Process all four 48 Deepwell plates of Bacterial growth cultures for solubility screening and process all two 96 well plates for protein expression screening

Fig. 8.1. Illustration depicting the process flow for production and screening of expression clones.

7. A 96-microwell plate suitable for heat shock using a controlled temperature heat block (*see* **Note 4**).

8.2.3. Bacterial Growth and Preparation of Soluble Lysate

1. Lauria Bertaini (LB) Broth (500 ml), Miller (Fisher cat. no. BP-1426).
2. Isopropyl-β-D-thiogalactopyranoside (IPTG; 100 mM).
3. Four 48-deepwell plates (*see* **Note 5**).
4. Lysis buffer (sufficient for 100 samples): add 80 μl of Bacterial Protease Inhibitors (Cat. # P8849, Sigma Chemical, St. Louis, MO) to 20 ml of 50 mM Sodium phosphate, pH 8.0, 300 mM NaCl. Add 80 μl of recombinant stabilized T4 lysozyme solution (Epicentre Technologies, Madison,

WI), 20 µl of Benzonase (EMD Biosciences, San Diego, CA 92121), and mix.

8.2.4. SDS–Polyacrylamide Gel Electrophoresis (SDS–PAGE)

1. Tris–Glycine Gradient gels (4–20%) (Cambrex Bio Science Rockland, MD).
2. SDS–PAGE running buffer (Sigma Chemical).
3. 2× SDS–PAGE sample buffer (Sigma Chemical).
4. Simply Blue protein stain (Invitrogen).

8.3. Methods

Although several universal cloning site systems are presently available, we selected the Ligation-Independent Cloning (LIC) method *(13, 14)* for implementation into our high-throughput protein production pipeline. This approach is one of several universal cloning site systems presently available which provides flexibility of expression with different fusion tags and protease cleavage sites. The rationale *(2)* for selection of this HTP cloning and expression approach as well as robotic methods for fragment generation has been previously described *(3, 10)*. A library of LIC-compatible vectors *(15, 16)* is available for distribution, and the protocols for preparation and use of these vectors are described in **Chapter** 7. Automated methods using a Beckman FX, Multimek, and Biomek 2000 have been developed for procedures outlined in this chapter and are available by request from the authors.

8.3.1. Preparation of T4 Polymerase-Treated DNA Fragments

1. Array 10.4 µl of the LIC reaction mix into a polypropylene 96-well plate. This plate will be used for transformation via a heat shock procedure and should be compatible with a controlled temperature heat block (*see* **Note 4**).
2. Add 30 µl of purified and diluted PCR fragment (*see* **Note 6** regarding suggested dilutions of PCR fragments for the annealing reaction) to the LIC reaction mix and pipette up and down several times to mix. Incubate at room temperature for 30 min (*see* **Note** 7 regarding oil overlay or use of a thermocycler with heated lid).
3. Incubate on a heat block at 75°C for 20 min to inactivate the T4 DNA polymerase.
4. Following the heating process the LIC plates are stored in the refrigerator at 4°C until needed.

8.3.2. Vector Annealing and Cell Transformation (Fig. 8.1, Stage 1)

1. Take competent cells (BL21) and vector plates out of the freezer to thaw. Cells should be continuously stored on ice until use.

2. Array 4 µl of LIC-treated vector into a 96-well plate.
3. Add 4 µl of LIC-treated PCR fragment to the vector and pipette up and down several times to mix. After incubation at room temperature for 5 min, place the plate atop a cold block or in an ice bucket.
4. Add 45 µl of BL21 cold competent cells (*see* **Note 3**) to the plate, and BRIEFLY centrifuge this plate at 4°C to be sure the competent cells mix with the annealing reaction. Incubate at 4°C (usually on ice) for 5–15 min.
5. Heat shock the cells by placing the plate on the heat block set to 48°C for 1 min.
6. The plate is then transferred back to 4°C (usually on ice) for 2–10 min before adding 120 µl of LB to the transformation plate.
7. The plate is then incubated at 37°C for 30 min.

8.3.3. Plating for Individual Clone Selection (Fig. 8.1, Stage 2)

1. For one 96-microwell plate of transformants, label the bottom half of two empty 48-grid agar plates to correspond to the 96 samples of the transformation plate.
2. Predry the LB/Amp agar grid plates by incubation at 37°C for 15–20 min with the plate lids ajar. Remove the plates, uncover, and use flame-sterilized tweezers to place one sterilized, 1.25 in. precut glass rod roller into each grid location, preferably into the same uniform location (top, center, or bottom within the grid location). Then, for each well, place approximately 50–100 µl of the transformation plate solution within each corresponding grid location on the LB/Amp agar grid plate. It is recommended to dispense the transformant/solution onto the center of the rod, and to use the same single pipette tip to manipulate the same single sterilized, precut glass rod roller to spread the transformant uniformly throughout the grid location. Limit the spreading to 1–2 rolls, to maximize cell growth potential and alternatively to minimize cell destruction.
3. After processing the entire 48-grid, LB/Amp agar grid plate, cover the plate and incubate on the benchtop for 5–10 min at RT to allow the residual liquid to absorb into the agar. Uncover the plate and inspect each of the agar wells ensuring that there is very little liquid transformant remaining. If significant liquid remains, cover the plate and allow it to sit for another 10 min at RT, in order to *totally* absorb the transformant solution into agar. Once there is little-to-no liquid transformant remaining, quickly invert the entire 48-grid plate (lid and bottom) thus resulting in a bottom side up orientation.
4. Follow this up by gently tapping the bottom, in order to assist in the release of the glass rod rollers onto the inside of the

cover/lid. Lift the bottom of the 48-grid plate and place atop a new upside down, sterile 48-grid plate cover/lid.

5. Incubate at 37°C for 12–18h. Be sure to incubate the LB/Amp agar grid plate in an inverted orientation, with the plate lids offset (cracked) for ventilation. Plates with volumes greater than 300ml of agar and/or if volumes larger than 100μl of the transformation solution are to be plated, and should be left to incubate with their lids offset. Plates with volumes less than 300ml of agar should avoid being left to incubate with their lids offset.
6. Return to the used glass rod rollers lying inside of the upside down 48-grid plate cover/lid, and rinse both rod rollers and 48-grid plate cover/lid with H_2O. Follow this with a thorough 95% Ethanol rinse that should then be followed by another thorough H_2O rinse. Dry with paper towels or allow to air dry.

8.3.4. Colony Selection for Growth and Induction (Fig. 8.1, Stages 3 and 4)

1. For the 96-microwell plate of transformants, label four 48-deepwell plates (as illustrated in **Fig. 8.1**, stage 4) to correspond to the two colonies selected from the 48-grid, LB/Amp, agar plates.
2. To each of the four 48-deepwell plates, use a liquid-dispensing device (e.g., QFill2 station) to add 2ml/well of LB/Amp broth (medium for bacteria) containing ampicillin. Note: Add antibiotic each time.
3. Using a 10-μl pipette tip, select colonies from the two 48-grid, LB/Amp, agar plates (look for single colonies) by scraping part of a colony from the plate and placing the tip in the corresponding well containing 2-ml LB/Amp broth. Be sure to laterally mix the pipette tips in each well for up to 5s before incubating. Upon incubation, be sure to leave the tips in each of the wells to assist with agitation. Make sure the tips are fully immersed in the well and are touching the well bottom. Tips that are not flush with the well bottom may eject during incubation or adhere to the side of the well and may not properly mix in the LB/Amp. This may lead to a loss of sample, poor/no sample growth, and/or cause sample crosscontamination.
4. Place the cultures at 37°C in a shaking incubator, set at an rpm of no higher than 250.

8.3.5. Preparation of Temporary Freezer Stock (Fig. 8.1, Stage 5)

1. Temporary freezer stocks are prepared when the A_{600} of the bacterial growth culture is approximately 0.4 (*see* **Note 8**).
2. Remove the deepwell plates from the incubator and prepare a glycerol stock from the deepwell plate by manually removing 200μl from each well of the four 48 deepwell plates, and place into each corresponding location within two additional 96 round-bottom plates.

3. Centrifuge these 96 round-bottom plates for 10 min at 3,000 × *g* and remove the growth media by inverting the plate and blotting the top surface on a paper towel.
4. Resuspend the pellet in 200 µl of LB + 30% glycerol, mix 1–2 times, and follow by freezing.

8.3.6. Analysis of Protein Expression (Fig. 8.1, Stages 6 and 7)

1. Use an automated liquid handler or multichannel pipette to add 30 µl of 100 mM IPTG (induces bacteria to express protein) to each of the wells and return the deepwell plates to a shaker incubator for an additional 2 h.
2. While waiting, label two 96-deepwell plates (**Fig. 8.1**, Stage 7).
3. After incubation is complete, use an automated liquid handler or multichannel pipette to remove 400 µl from each well of the four 48-deepwell plates and place into a corresponding location of the two 96-deepwell plates (*see* **Note 9**).
4. Centrifuge the two 96-deepwell plates for 10 min at 3,000 × *g* and remove the growth media by inverting the plate and then blotting the top surface on a paper towel.
5. Resuspend the pellets in 70 µl of 1× SDS–PAGE buffer using an automated liquid handler or multichannel pipette. The plates are then sealed with foil and vortexed to mix.
6. Boil for 4 min and again vortex well. Prior to analysis by SDS–PAGE, samples should be briefly centrifuged to collect the liquids at the bottom of the well.

8.3.7. Preparation of a Soluble Lysate (Fig. 8.1, Stages 7 and 8)

1. The growth culture (1.4 ml) remaining in the four 48-deepwell plates is used for solubility analysis. Pellet the cells by centrifugation for 10 min at 3,000 × *g* and discard the supernatant. Dry the pellets by inverting the plate and blotting the plate on a paper towel for 15–30 s. The plates should be stored at −20°C until solubility analysis.
2. Lyse the bacteria by suspension of the pellets in 180-µl lysis buffer and incubation at room temperature for 5 min. During the incubation, vortex the samples 1–2 times.
3. Transfer the suspension to a labeled 96-microwell plate and centrifuge at 3,000 × *g*, at (4°C) for 10 min.
4. After centrifugation, remove 50 µl of the supernatant and add to an awaiting labeled 96-microwell plate containing 60 µl/well of 2× SDS–PAGE sample buffer and boil for 3 min.

8.3.8. SDS–PAGE Analysis of Expression and Solubility

1. For denaturing gel analysis, load 5 µl of low molecular weight marker for each gel.
2. Load 8 µl of each sample prepared from the total growth culture for analysis of total expressed protein.

3. Load 12–15 µl of each sample prepared from the lysate sample for analysis of soluble protein.
4. Remove from gel station and stain for protein. Gels can be stained anywhere from 60 min to overnight. (*see* **Note 10** expression/solubility scoring.

8.4. Notes

1. T4 DNA polymerase reaction buffers supplied by most vendors can be substituted for the LIC reaction buffer described in the text. Our comparison of various common T4 DNA polymerase reaction buffers shows less than a 25% difference in the cloning efficiency of the final product.
2. Q tray plates prepared in this manner can be stored for at least 2 weeks at 4°C. Freshly poured plates should be dried prior to use by removal of the plate cover and incubation in 37°C for 30 min
3. The method presented in this chapter uses direct transformation of an expression capable BL21 strain. Typical transformation efficiencies for BL21 strains are in the range of 0.5–1 $\times 10^6$ colonies/µg of plasmid DNA. The observed number of colonies on the Q tray plates is a function of the overall efficiency of the annealing reaction and the level of competency of the bacterial cells. It is recommended that the investigator place various amounts or dilutions of several representative target samples prior to multiple plate screening.
4. The well geometry of the plates should match the instrument (thermal cycler of heat block) used for the heat shock procedure. We optimize transformation efficiency for each strain and plate by testing different times and temperatures.
5. A variety of 48-deepwell plates are available from standard scientific supply vendors. We prefer pyramidal bottom plates as these more effectively retain the pellet after centrifugation steps and facilitate pipetting operations that involve removal operation near the bottom of the plate.
6. Our studies of various fragment-to-vector ratios *(2)* indicate a wide tolerance for variation in the amount of target DNA fragment on the annealing reaction. This latitude eliminates the need for normalization of fragment concentrations prior to annealing, thus conserving time and simplifying the process for implementation of the method as an automated process.

7. The use of a thermal cycler with a heated lid is preferred since there are no additional components to add to the mix and the number of pipetting steps is reduced. An alternative is to layer 20–40 µl of mineral oil on the annealing mix prior to heat treatment.
8. Cultures grown in plates typically reach an A_{600} of 0.4–2.5 h after inoculation. An exact A_{600} measurement of 0.4 is not necessary for the preparation of temporary freezer stocks in plates. These freezer stocks should be viable for at least a month when stored at −70°C. Stock culture for the positive expression clones can also be prepared by streaking out a loopful of bacterial growth culture from each well onto an agar plate.
9. Transfer from the 48-well to a 96-well plate can be accomplished using the span-8 tool of a Biomek FX instrument or by selective loading of tips on a Beckman Multimek. Tips are arrayed in six staggered columns in a tip box using a multichannel pipette (odd columns in one rack and even columns in another rack). These tip boxes are then used for successive transfer into the corresponding new location within the 96-deepwell plate.
10. Targets are scores as "no expression" or "insoluble" based on the absence of a detectable stained protein band of the correct molecular weight observed after SDS–PAGE analysis. Targets can be scored as positive, based on an observation of a protein stained band of correct molecular weight. We use a relative ranking scale that compares staining of the target relative to the general intensity most of the *E coli* proteins. Target bands that are visible but with intensity level less than most of the *E coli* proteins are scored as level 1 or low expression/solubility. Levels 2 and 3 (moderate and high expression/solubility) have staining intensity comparable to that of highly expressed *E. coli* proteins or more prominent than any *E. coli* protein, respectively.

Acknowledgments

The authors would like to thank Kendall Nettles (for technical assistance with using 48-grid plates), and Joseph Gregar (for preparation of the 130-mm Borosilicate Glass rod Petri Dish Spreaders). This work was supported by NIH grants GM62414 and GM074942, Andrzej Joachimiak, PI, and by the U.S. Department of Energy, Office of Science, under Contract DE-AC02-06CH11357 to UChicago Argonne LLC.

References

1. Brasch MA, Hartley JL, Vidal M. (2004) ORFeome cloning and systems biology: standardized mass production of the parts from the parts-list. Genome Res 14(10B):2001–9.
2. Dieckman LJ, Hanly WC, Collart FR. (2006) Strategies for high-throughput gene cloning and expression. Genet Eng (NY) 27:179–90.
3. Marsischky G, LaBaer J. (2004) Many paths to many clones: a comparative look at high-throughput cloning methods. Genome Res 14(10B):2020–8.
4. Walhout AJ, Temple GF, Brasch MA, et al. (2000) GATEWAY recombinational cloning: application to the cloning of large numbers of open reading frames or ORFeomes. Methods Enzymol 328:575–92.
5. Hoffmann M, Nemetz C, Madin K, Buchberger B. (2004) Rapid translation system: a novel cell-free way from gene to protein. Biotechnol Annu Rev 10:1–30.
6. Katzen F, Chang G, Kudlicki W. (2005) The past, present and future of cell-free protein synthesis. Trends Biotechnol 23(3):150–6.
7. Blommel PG, Martin PA, Wrobel RL, Steffen E, Fox BG. (2006) High efficiency single step production of expression plasmids from cDNA clones using the Flexi Vector cloning system. Protein Expr Purif 47(2):562–70.
8. Dieckman L, Gu M, Stols L, Donnelly MI, Collart FR. (2002) High throughput methods for gene cloning and expression. Protein Expr Purif 25(1):1–7.
9. Klock HE, White A, Koesema E, Lesley SA. (2005) Methods and results for semi-automated cloning using integrated robotics. J Struct Funct Genomics 6(2–3):89–94.
10. Moy S, Dieckman L, Schiffer M, Maltsev N, Yu GX, Collart FR. (2004) Genome-scale expression of proteins from *Bacillus subtilis*. J Struct Funct Genomics 5(1–2):103–9.
11. Mehlin C, Boni EE, Andreyka J, Terry RW. (2004) Cloning grills: high throughput cloning for structural genomics. J Struct Funct Genomics 5(1–2):59–61.
12. Hamilton CM, Anderson M, Lape J, Creech E, Woessner J. (2002) Multichannel plating unit for high-throughput plating of cell cultures. Biotechniques 33(2):420–3.
13. Aslanidis C, de Jong PJ. (1990) Ligation-independent cloning of PCR products (LIC-PCR). Nucleic Acids Res 18(20):6069–74.
14. Haun RS, Serventi IM, Moss J. (1992) Rapid, reliable ligation-independent cloning of PCR products using modified plasmid vectors. Biotechniques 13(4):515–8.
15. Donnelly MI, Zhou M, Millard CS, et al. (2006) An expression vector tailored for large-scale, high-throughput purification of recombinant proteins. Protein Expr Purif 47(2):446–54.
16. Stols L, Gu M, Dieckman L, Raffen R, Collart FR, Donnelly MI. (2002) A new vector for high-throughput, ligation-independent cloning encoding a tobacco etch virus protease cleavage site. Protein Expr Purif 25(1):8–15.

Chapter 9

Automated 96-Well Purification of Hexahistidine-Tagged Recombinant Proteins on MagneHis Ni^{2+}-Particles

Chiann-Tso Lin, Priscilla A. Moore, and Vladimir Kery

Summary

Functional genomics and the application of high-throughput (HT) approaches to solve biological and medical questions are the main drivers behind the increasing need for HT parallel expression and purification of recombinant proteins. Automation is necessary to facilitate this complex multistep process. We describe, in detail, an HT-automated purification of hexahistidine-tagged recombinant proteins using MagneHis™ Ni-Particles and the Biomek FX robot. This procedure is universally applicable to hexahistidine-tagged recombinant proteins with the tag positioned at either the N- or C-terminus. With minor modifications, the automated protein purification protocol presented in this chapter could be adapted to purify recombinant proteins bearing other tags than hexahistidine and/or other expression systems than *E. coli*.

Key words: Automated, Hexahistidine affinity tag, High throughput, Protein purification, Recombinant proteins.

9.1. Introduction

Now that many genomes, including the human genome, have been sequenced, the next milestone in fundamental biology is to identify the function of all proteins encoded by individual genes in the genomes. Although considerable effort has been dedicated to deciphering the protein function from its amino acid sequence *(1)*, the most recognized approaches to address fundamental questions about protein structure and function are based on protein purification and characterization *(2)*. Because each genome encodes many thousand proteins, genome-wide protein

Sharon A. Doyle (ed.), *Methods in Molecular Biology: High Throughput Protein Expression and Purification, vol. 498*

Book doi: 10.1007/978-1-59745-196-3

expression technologies *(3)* and proteome-scale affinity purification approaches *(4)* have been developed. Although using affinity tags significantly facilitates protein purification, it still remains a tedious multistep process. Automating the most laborious liquid handling steps is the key to streamlining the protein purification process and to achieving the throughput sufficient to purify a whole proteome in a reasonable timeframe.

In this chapter we present our protocol for automated protein purification (*see* **Note 1**) of recombinant proteins using hexahistidine affinity tags and MagneHis™ Ni^{2+}-Particles (Promega, Madison, WI) on the Biomek FX liquid-handling robot from Beckman Coulter, Inc. (Fullerton, CA). The protocol was optimized using *Escherichia coli* for protein expression. Autoinduction system from Promega (Madison, WI) was used for the bacterial protein expression to minimize manipulation with the expression plate thus preventing risk of contamination. Other expression systems could be used with minor alterations to the purification procedure.

9.2. Materials

The open reading frames encoding proteins from *Shewanella oneidensis*, a microorganism involved in environmental metal reduction *(5)*, were cloned into a pMCSG7 vector using a modified ligation-independent cloning protocol *(6)* and transfected into *E. coli* BL21 production strain. The fusion proteins were produced with an N-terminal hexahistidine affinity tag followed by a TEV protease recognition site ENLYFQS *(7)*. Biomek FX robot (Beckman Coulter, Fulerton, CA), equipped with two liquid transfer heads, a multichannel head for direct liquid transfer between 96-wells (Rod 1), and a Span 8 head with eight channels for transfer of variable volumes and sequential control was used for automated protein purification. FluorChem 9900 image system (Alpha Innotech, San Lenardo, CA) was used for imaging of gel pictures.

9.2.1. Protein Expression

1. 2X-YT medium
2. Carbenicillin
3. Overnight Express™ Auto Induction System 1 (Novagen, Madison, WI)
4. Marsh™ 2.2 mL round-bottom 96 deep-well plate (ABgene, Rochester, NY)
5. Thermal adhesive sealing film (E and K Scientific, Santa Clara, CA)

9.2.2. Cell Lysis

1. FastBreak™, 10× concentrated lysis solution (Promega, Madison, WI)
2. Dilution Buffer – for preparation of 20× Lysis-Aid Cocktail: 1× Tris–HCl buffered saline (TBS), 5% glycerol, 10 mM imidazole, pH 7.5
3. Lysis-Aid Cocktail 20× – freshly prepared: 100 μL DNAse from bovine pancreas, 100 μL RNAse, 400 μL PMSF, and 60 μL β-mercaptoethanol stock solutions as well as 40 mg of lysozyme and six tablets of Complete Protease Inhibitor Cocktail Tablets, EDTA free (Roche Applied Science) in 10 mL of Dilution Buffer

9.2.3. Protein Purification

1. MagneHis™ Ni^{2+}-Particles (Promega, Madison, WI) V8565
2. Round-bottom 96 deep-well plate (2.2 mL) (ABGene, Rochester, NY) Marsh AB0661
3. Round-bottom 96 deep-well plate (1.2 mL) (ABgene, Rochester, NY) Marsh AB0564
4. Innovative Microplate Pyramid bottom reservoir. Deep trough (Innovative Microplate, Chicopee, MA) S30014
5. Greiner U-bottom microtiter plate (Promega, Madison, WI) A9161
6. Tips for Robotic multiple channel head, Rod 1, (Axygen Scientific, Union City, CA) FXF-180-RS
7. Beckman Coulter Biomek FX Robotic Workstation (Beckman Coulter, Fullerton, CA)
8. Deep-Well MagnaBot 96 Magnetic Separation Device with 1/8-in. spacer (Promega, Madison, WI) V8151
9. BD Falcon tubes (15 mL), polypropylene
10. Binding Buffer (IMD-10): 10 mM imidazole, 20 mM Hepes, 500 mM NaCl, 2 mM $MgCl_2$, glycerol, pH 8.0
11. Washing Buffer (IMD-80): 80 mM imidazole, 500 mM NaCl, 20 mM HEPES, pH 7.0
12. Elution Buffer (IMD-500): 500 mM imidazole, 150 mM NaCl, 20 mM HEPES, pH 7.5

9.2.4. SDS–Polyacrylamide Gel Electrophoresis (SDS–PAGE)

1. Mother E-Base™ device (Invitrogen, Carlsbad, CA)
2. E-PAGE™ 96 6% gel Kit (Invitrogen, Carlsbad, CA)
3. E-PAGE™ Loading buffer 1 (6×concentrated) (Invitrogen, Carlsbad, CA)
4. NuPAGE™Sample Reducing Agent (Invitrogen, Carlsbad, CA)
5. BenchMark™ His-tagged Protein Standard (Invitrogen, Carlsbad, CA) and/or SeeBlue® (Invitrogen, Carlsbad, CA)

9.2.5. Gel Staining

1. InVision™ His-tag in-gel stain (Invitrogen, Carlsbad, CA)
2. Fixing Solution: 40 mL ethanol, 10 mL acetic acid, and 50 mL 18 MΩ water
3. Phosphate buffer (20 mM) at pH 7.8
4. GelCode Blue stain reagent (Pierce, Rockford, IL)

9.2.6. Protein Assay

1. Pierce Coomassie Plus Protein Assay Reagent™ (Pierce, Rockford, IL)
2. Albumin Standard, 2 mg/mL (Pierce, Rockford, IL)

9.3. Methods

9.3.1. Protein Expression

9.3.1.1. Seed Cultures

1. Prepare 2X-YT medium according to manufacturer's instructions. Prepare 1 L for seed cultures and 1 L for Auto Induction Media (enough for 1,000 samples) (*see* **Note 2**). Autoclave to sterilize.
2. Grow cells from different clones directly into a square, 2.2-mL 96 deep-well plate. Pipette 1 mL 2X-YT media containing 50 μg/mL carbenicillin into each well using a 1,200-μL multichannel pipette and sterile reservoir. Inoculate with 1 μL loop from glycerol stock, or take a colony from an agar plate.
3. Place thermal adhesive sealing film on the plate and seal with a hand roller. Incubate overnight at 37°C and 1,000 rpm (*see* **Note 3**).
4. Measure and record optical density at 600 nm (OD_{600}).

9.3.1.2. Protein Expression

1. Prepare autoinduction media. Sterile transfer the autoinduction reagents from the Promega autoinduction kit to cooled presterilized 2X-YT medium. DO NOT AUTOCLAVE AUTO INDUCTION MEDIA.
2. Transfer 100 mL autoinduction medium to a sterile 100-mL media bottle. Add carbenicillin (50 μg/mL) and invert to mix. Pour media into sterile reservoir.
3. Transfer 0.9 mL media into each of the 96 wells of a 2.2-mL deep-well 96 well plate using a multitip pipette.
4. To minimize crosscontamination, cover the plate with the sterile thermal adhesive sealing film.
5. Transfer approx. 10 μL of each seed culture to autoinduction medium using a fine tip (Rainin, 20 μL tip) pipette to pierce through the film and into well.
6. Clamp the plate to a high-speed shaker. Incubate 24–30 h, at 30°C and 1,000 rpm or greater.

Color Plates

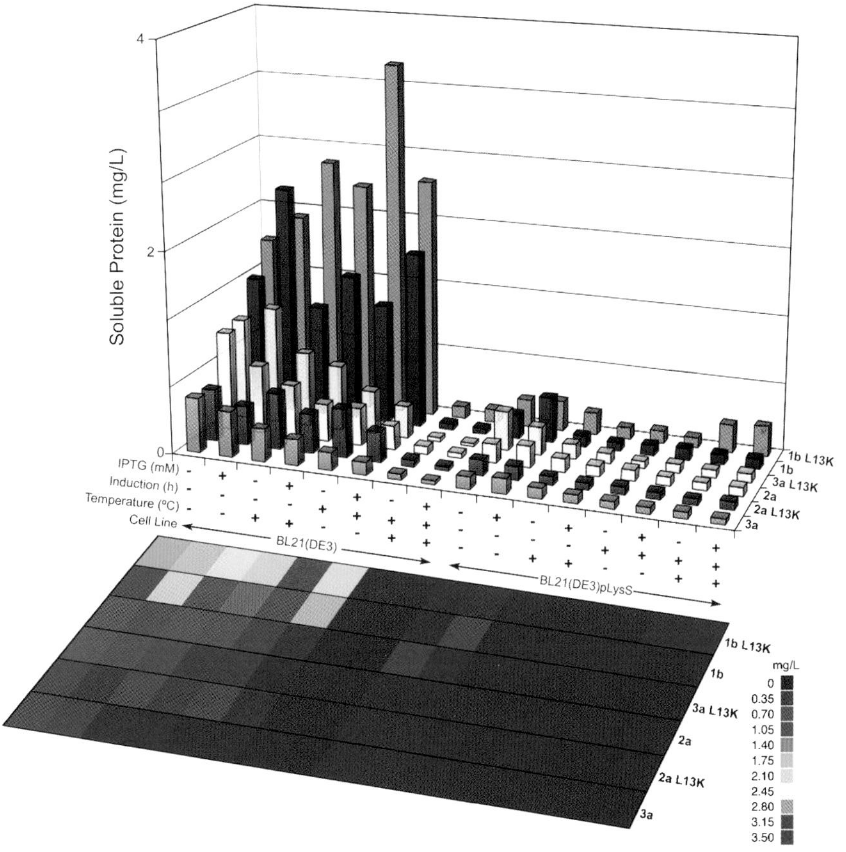

Plate 1, Fig. 2.2. Full factorial screening for effects on the soluble HCV NS3-prt expression in *E. coli.* Levels of expression are illustrated in a 3D Bar Chart and a Heat Map. Using a previously described high-throughput protein expression platform, each data point was obtained using the HT Protein Express 200 Chip run on the Caliper labchip 90 and measured in triplicate (see p. no. 25)

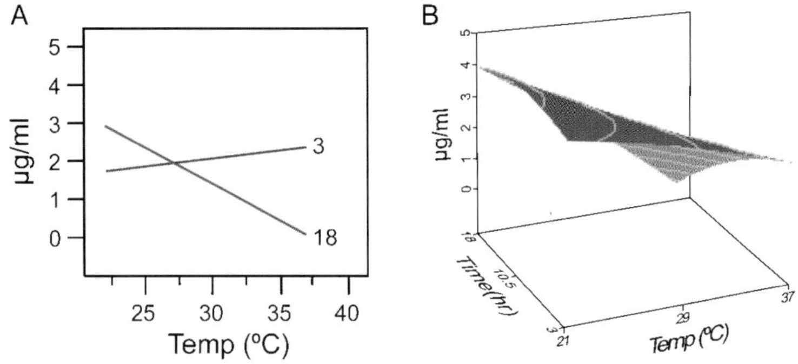

Plate 2, Fig. 2.3. (a) Interaction plot for temperature and time affects on the soluble expression of HCV NS3-prt (1b) L13K. The dependence of protein yield (y-axis) on temperature (x-axis) is plotted for two different times. The red line represents the 3-h data, while the purple line represents the 18-h data. (b) Response surface plot demonstrating expression as a function of time and temperature. The effect of time and temperature on protein yield at an optimal value of IPTG (0.78 mm) is depicted as a 3D surface using the statistical program JMP. The top surface is *colored purple* and the bottom is *colored gray*. Mapped to the surface in *blue* is the contour map of the same data (see p. no. 26)

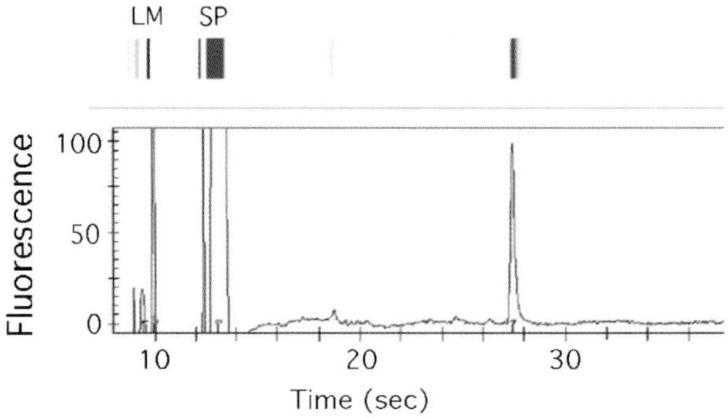

Plate 3, Fig. 10.3. Virtual gel image and electropherogram of protein generated by Caliper labchip 90. Protein was expressed in insect cells, grown in a deep well block, then purified and analyzed. The purified protein is readily identified on the gel and electropherogram, as is the lower marker (LM), an internal reference, and the system peak (SP) (see p. no. 152)

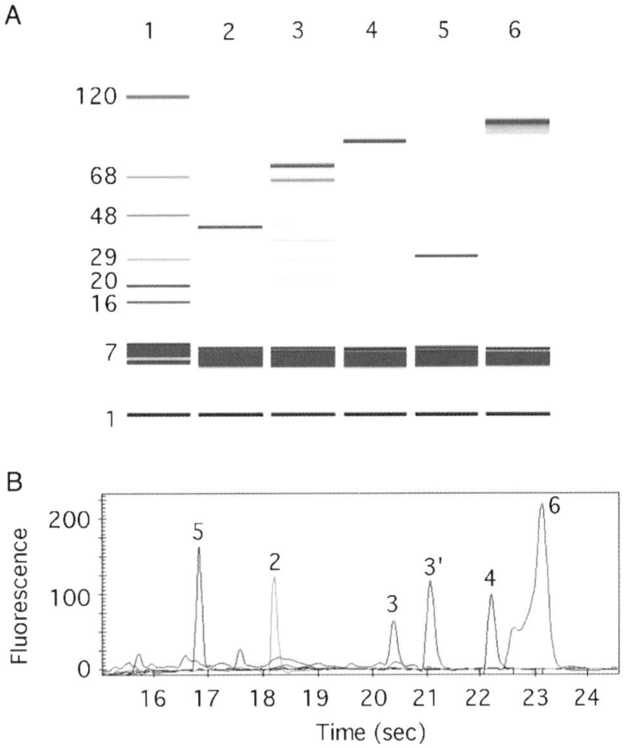

Plate 4, Fig. 10.4. Quantitative analysis of proteins purified from *E. coli* grown in deep well blocks. ***(a)*** The virtual gel image produced by the Caliper labchip 90. *Lane* 1, ladder; *lanes* 2–6, included a variety of different proteins. ***(b)*** Electropherogram of the same sample proteins (2–6), now overlaid, the system software generating both a molecular weight and a concentration value for assigned peaks. *N.B.* Doublet produced in *lane* 3, apparent in the virtual gel ***(a)***, shows up in the electropherogram (as peaks 3, 3´) (see p. no. 153)

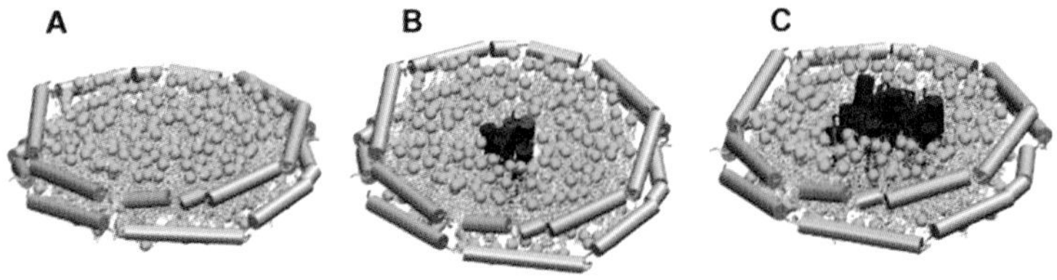

Plate 5, Fig. 18.1. *Models of NLPs with and without bacteriorhodopsin.* Models were constructed using molecular dynamic simulations (described elsewhere). (**A**) Model of a Nanolipoprotein particle (NLP) with a lipid bilayer in the middle and apolipoproteins encircling the hydrophobic portion of the lipids. (**B**) NLP modeled with a bacteriorhodopsin monomer inserted in the hydrophobic lipid core (**C**) NLP modeled with a bacteriorhodopsin trimer inserted in the hydrophobic lipid core (see p. no. 275)

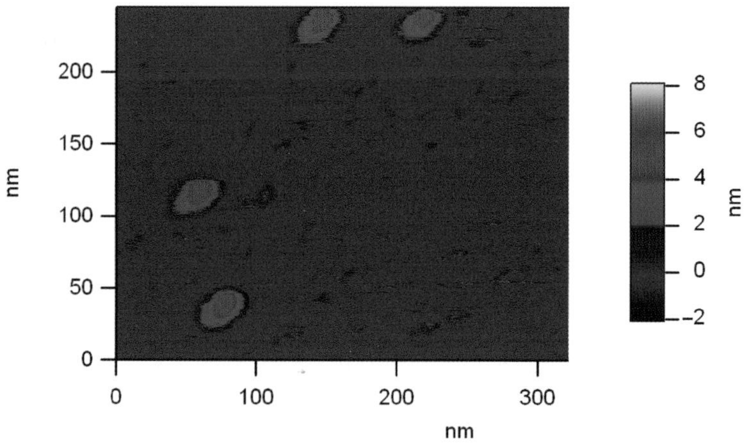

Plate 6, Fig. 18.7. *Atomic force microscopy of nanolipoprotein particles (NLPs).* NLPs consisting of cell-free produced apoE4 22K lipoprotein and DMPC. Particle dimensions are as follows; Height: 4.94 nm, std dev: 0.369 nm Width of top: 9.72 nm, std dev: 1.50 nm, Full width at half max: 20.4 nm std dev: 3.5 nm (see p. no. 287)

7. Measure and record OD_{600}. Proceed to automated protein purification. If storing a plate for later purification, then store tightly sealed cell plate at –20°C until needed.

9.3.2. Instrument Setup for Automated Protein Purification

9.3.2.1. Initial Instrument Setup

1. Use the Biomek FX robot, equipped with two liquid transfer heads, a multichannel head for direct liquid transfer between 96-wells (Rod 1), and a Span 8 head with eight channels for transfer of variable volumes and sequential control.
2. Use software programs built-in to the Biomek FX from the SAGIAN Software package (Beckman–Coulter), modified in accordance with experimental requirements. Translate the sequence of automated operations into an operation program and save in a special operating file.
3. The deck configuration for protein purification with Magne-His™ Ni-Particles is shown in **Fig. 9.1**. Each ALP is labeled with a designated functional position number from P0 to P13. Define the individual deck positions in the operating program.

9.3.2.2. Lab Ware Positions and Configuration

Refer to **Fig. 9.1** for the ALP arrangement on the robot deck.

1. Lysate tips – place one tip box on the Tip load box.
2. Washing tips – place one tip box on the ALP P5.
3. Elution 1 tips – place one tip box on the ALP P8.
4. Elution 2 tips – place one tip box on the ALP P12.
5. Cell plate – place a Marsh 2.2-mL deep-well plate with 900 µL lysate per well on the Orbital shaker (ALP P0) before starting.

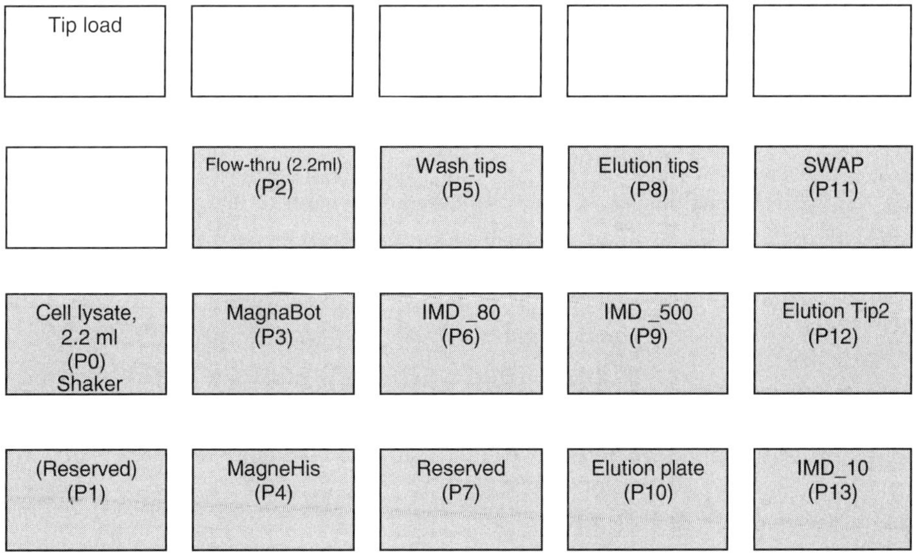

Fig. 9.1. The initial positions of labware for high-throughput purification on the deck of the Biomek FX.

6. Flow-thru – place an empty Marsh 2.2-mL deep-well plate on ALP P2 to receive the flow-thru fraction of lysate.
7. MagneHis™ plate – place a Marsh 1.2-mL deep-well plate containing 100 μl MagneHis™ Ni-particle suspension on ALP P4 to capture expressed His-tagged proteins.
8. Place a deep-well MagnaBot magnetic separation device with a 1/8-in. spacer (at the Working Plate Position) on ALP P3.
9. Working plate – place an empty Marsh 1.2-mL deep-well round-bottom plate, on ALP P3, directly on the top of the MagnaBot.
10. Binding Buffer (IMD-10) – add 180 mL of IMD-10 to an Innovative microplate, and place it on ALP P13.
11. Washing Buffer (IMD-80) – add 180 mL of IMD-80 to an Innovative Microplate Pyramid bottom reservoir, and place it on the ALP P6.
12. Elution Buffer (IMD-500) – add 30 mL of IMD-500 to an Innovative Microplate Pyramid bottom reservoir, and place it on ALP P9.
13. Final eluate (Product) – Place an empty Greiner U-bottom plate on ALP P10.
14. Positions P1 (Reserved), P7 (Reserved), and ALP P11 (Swap) should not have any plates on them before starting.

9.3.3. Automated Protein Purification Protocol

9.3.3.1. Cell Lysis

1. Place cell plate containing the cell suspension (0.9 mL) on the orbital shaker (P0).
2. Manually add 60 μL of 20× lysis aid cocktail to the lysate using a multichannel pipette.
3. Add 100 μL of 10× FastBreakTM cell lysis reagents to the cell suspension.
4. Shake at 1,100 rpm for 4 min, and then pause for 1 min. Repeat the shaking cycle five times for a total lysis time of 20 min (*see* **Note 4**).
5. Move the cell plate to the reserved ALP P1.

9.3.3.2. Conditioning the MagneHis™ Ni^{2+}-Particles

1. Move the MagneHis™ plate from ALP P4 to the orbital shaker, and move the empty Marsh 1.2-mL 96 deep-well plate (working plate) from ALP P3 to ALP P4.
2. Dispense 100 μL of binding buffer to the wells in the MagneHis™ plate, and shake for 30 s.
3. Move the MagneHis™ plate to the MagnaBot spacer (ALP3), and pause for 90 s.
4. Aspirate the clear solution from the captured beads to the flow-thru plate (Marsh 2.2-mL deep-well square plate on

ALP P2). Repeat the washing procedure three times. Move the MagneHis™ plate back to the orbital shaker (*see* **Note 5**).

Note: Including the time required for washing MagneHis Ni-Particles, the total lysis time is approximately 30 min.

9.3.3.3. Binding of His-Tagged Proteins to MagneHis™ Ni-Particles

1. Transfer 100 μL of lysate from the cell lysate plate to the MagneHis™ Ni^{2+}-particles on the orbital shaker (*see* **Note 6**). Repeat the transfer two more times.
2. Shake and transfer the suspension back to the cell plate (P1).
3. Repeat the transfer three times so that no magnet beads are left in the wells.
4. Move the working plate from ALP P4 back to the MagnaBot spacer (ALP P3); move the MagneHis™ plate to ALP P4.
5. Move the cell lysate plate to the orbital shaker.
6. Shake the cell lysate plate for 60 s at 1,100 rpm, and then pause for 3 min.
7. Repeat, alternatively shaking and pausing seven times, resulting in a total contact time of about 35 min including the time required for rod movements and liquid transfer (*see* **Note 6**).

9.3.3.4. Separation of Beads from Lysate

1. Mix the suspensions briefly, and aspirate 115 μL lysate and Ni^{2+}-particles suspension to the working plate on the MagnaBot (P3); repeat the aspiration five times to transfer the whole volume of suspensions (*see* **Note 7**).
2. Pause the working plate for 90 s to separate the beads from the supernatant.
3. Transfer the clear lysate from P3 to the flow-thru plate (P2) by repeatedly aspirating 100 μL volumes.
4. Repeat **steps 1–3** to transfer the residual lysates and Ni^{2+}-particles suspension from the cell plate to the working plate.
5. To rinse the residual beads in the cell plate, aspirate 2 × 90 μL of the lysate from the working plate back to the cell lysate plate. Mix by aspiration, dispensing the lysate in wells three times (*see* **Note 8**).
6. Immediately after mixing, transfer the suspension to the working plate.
7. Pause the plate again for 90 s.
8. Aspirate the remaining lysate in 100-μL increments from the working plate to the flow-thru plate.
9. Move the cell lysis plate to P7; move the working plate (MagneHis™ Ni^{2+}-particles with bound His-tagged proteins) to the orbital shaker.

9.3.3.5. Washing off Non-specific Proteins with IMD-80 Buffer

1. Change to new tips (this is important).
2. Transfer 500 μL IMD-80 from Wash I (P6) to the working plate in 100-μL increments.
3. Shake the mixture for 60 s.
4. Move the working plate from the orbital shaker to MagnaBot (P3).
5. Pause the working plate for 90 s.
6. Discard the washing solution into the cell plate, which is being used as waste container
7. Repeat washing procedure two more times.
8. Move the working plate from P3 to the orbital shaker P0.

9.3.3.6. Elution of Bound Proteins from the Beads with IMD-500 Buffer

1. Change to new tips (this step is important).
2. Transfer 60 μL of elution buffer, IMD-500 from Elution (P9, Innovative Pyramid bottom reservoir), to the working plate. Repeat the transfer one more time.
3. Mix the beads with 120 μL elution buffer at 1,100 rpm for 60 s.
4. Move the plate from the orbital shaker to the MagnaBot on P3.
5. Pause for 90 s.
6. Transfer the clear eluate to a clean Greiner U-bottom microtiter plate.
7. Unload the tips.
8. End the elution.
9. Move the product plate, i.e., the Greiner U-bottom microtiter plate, from P10 to ice. Take aliquots for protein assay and purity assessment.

9.3.4. Analysis of Purified Proteins

9.3.4.1. Protein Assay

1. Use a modified Bradford protein microassay. Load a 5-μL sample aliquot into a well of 96-well titer plate. Add 200 μL Coomassie Plus™ to the sample and read with a plate reader at 595 nm.
2. After shaking, read and analyze the data on a spectrophotometer (e.g., Spectra_{max}-Plus, using SOFT_{max}® -Pro).
3. Use bovine serum albumin as the standard for protein concentration calibration. Correct the absorption for imidazole (500 mM imidazole was present in the blank solution and all protein standards).

9.3.4.2. SDS–Polyacrylamide Gel Electrophoresis and Imaging of Purified Proteins

1. Mix 4 μg of protein in a sample solution containing 4 μL of 10× reducing reagent and 10 μL of 4× E-PAGE Loading buffer. Bring the volume for each sample to a total of 40 μL with deionized water (*see* **Note 9**).
2. Load all samples in a PCR 96-well plate and denature the samples at 95°C for 5 min under reducing conditions.

3. Remove the plastic comb from the purchased precast E-PAGE™ gel cassette.
4. Slide the gel cassette into the two electrode connections on the horizontal electrophoretic device, Mother E-Base™.
5. Preload 4 μL deionized water into all sample-slots of the gel, then add 25 μL sample. Add 5 μL SeeBlue® prestained protein markers into the standard wells.
6. Set the running time for 20 min.
7. Remove the gel cassette, open the cassette, rinse the gel twice with deionized water, and place the gel in fixing solution.
8. Fix for 4 h, with one change of fresh fixing solution. Rinse the gel twice with deionized water and stain with 25 mL (per gel) of InVision™ His-tag In-gel stain solution for at least 2 h (best effect is achieved by staining overnight).
9. Replace the stain solution with 20 mM phosphate buffer at pH 7.8.
10. Visualize and record the stained His-tag fusion proteins with the FluorChem 9900 image system (Alpha Innotech, San Lenardo, CA).
11. Rinse the gel with deionized water, and stain the gel with GelCode®Blue for 1 h.
12. Destain the gel with deionized water. Scan the gel picture, and save the picture for evaluation.
13. Use E-Editor™ software to analyze E-PAGE™ Gel images. An example of a set of purified proteins from *Shewanella oneidensis* with its visualization on polyacrylamide gel is shown in **Table 9.1** and **Fig. 9.2** (*see* **Note 10**).

9.4. Notes

1. Performance parameters of our automated purification protocol are shown in **Table 9.2**.
2. Deionized water with a resistance of 18 MΩ was used to prepare all solutions.
3. Make sure the cells are suspended in media while shaking during the protein expression. This provides maximal aeration for maximal growth. If traces of sticky residue (most likely DNA from lysed cells) start appearing at the bottom of the wells of the microplate during incubation, incubate for less than 30 h, or shake at higher speed.

Table 9.1
Gene names of proteins from Shewanella oneidensis purified on the Biomek FX and shown in Fig. 9.2

	1	2	3	4	5	6	7	8	9	10	11	12
A	SO3979	SO4095	SO4118	SO4221	SO4369	SO4502	SO4504	SO4563	SO4659	SO4713	SO012	SO018
B	SO0131	SO0138	SO0152	SO0167	SO0301	SO0304	SO0337	SO0338	SO0342	SO0358	SO0505	SO0527
C	SO0603	SO0620	SO0691	SO0861	SO088	SO1163	SO1183	SO1195	SO1237	SO1287	SO1306	SO1430
D	SO0155	SO1556	SO1583	SO1698	SO1898	SO1981	SO2017	SO2092	SO2201	SO2215	SO2267	SO2268
E	SO2435	SO2436	SO2669	SO2929	SO2947	SO2975	SO3138	SO3764	SO3888	SO3905	SO3920	SO3967
F	SO4164	SO4629	SO0740	SO0801	SO1878	SO2379	SO2098	SO0159	SO1611	SO2217	SO2444	SO3749
G	SO3948	SO4224	SO4374	SO4449	SO4521	SO1284	SO2944	SO2946	SO4462	SO4492	SO4507	SO0412
H	SO0492	SO0740	SO1068	SO1199	SO1738	SO2014	SO2055	SO2660	SO2827	SO2836	SO4257	SO0506

(A)

(B)

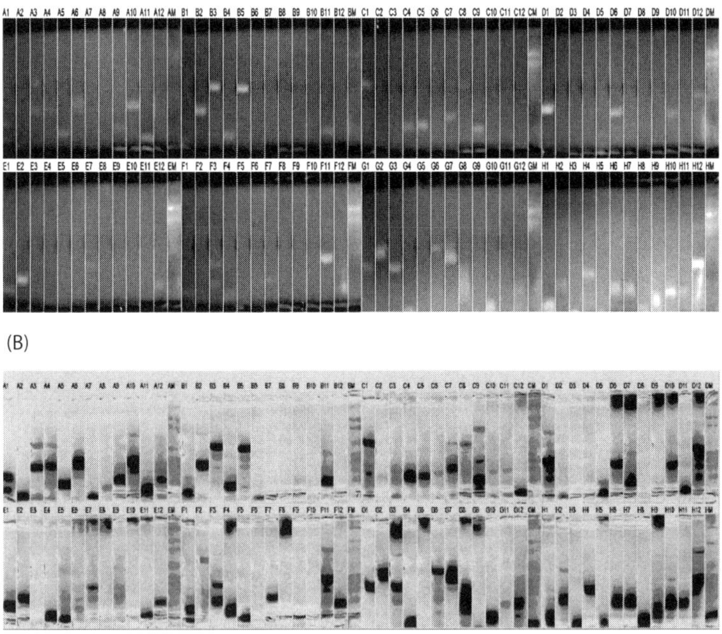

Fig. 9.2. High-throughput QC of His-tagged and total proteins using the E-PAGE system with invision™ and gelcode®Blue stains, respectively. **(a)** His-tag protein staining with invision™. **(b)** Total protein staining with gelcode®Blue stain.

Table 9.2
Performance parameters of our typical 96-well expression and automated purification

Total number of wells (lysates)	96
Average OD_{600} at harvest	19.1 ± 4.4
Average protein concentration[a] (mg/mL)	0.42 ± 0.27
Average yield of protein/well (μg)	84 ± 53
Purity of His-proteins[b] >90% (estimation on E-PAGE)	47
Specific yield of purified protein in μg/OD_{600} cell suspension	4.4 ± 2.8

[a]Protein concentration was determined directly from the eluted proteins in the 96-well plate
[b]Estimated based on the protein and In-gel His-tag staining pattern from E-PAGE

4. Bear in mind that the viscosity of the lysate and the buffer solution are different. Therefore, correct liquid transferring requires properly setting parameters on the robot.
5. Similar to the manual purification, MagneHis™ beads are pre-equilibrated with binding buffer at a relatively high pH (7.8) to achieve good binding efficiency. The nonspecifically

bound proteins are washed with a buffer containing a moderate concentration of imidazole (80 mM) and high concentration of salt (500 mM NaCl).

6. The MagneHis™ Ni-Particles have a binding capacity of 1 mg His-tagged proteins per 1 mL suspension (50% slurry). One hundred microliters of suspension used in experiments should maximally bind 100-μg His-tagged protein from cell lysate (0.9 mL) in each well, but we have shown that the yields could be higher if higher cell density suspension is used *(7)*.
7. To thoroughly mix the suspension or buffer and MagneHis™ beads, place the 96-well plate on the orbital shaker, or aspirate and dispense the solution within the well using the 96-tip robot head. To get better mixing, which is important, aspirate the liquid and bead mixture as near to the bottom of the well as possible and dispense the liquid to fill 50% of the well.
8. Proteins in lysate were protected against proteolysis by using a cocktail of inhibitors, including PMSF. We recommend adding DNase and RNase into the lysate to reduce its viscosity and facilitate the contact of proteins with beads.
9. E-PAGE is a newly available gel electrophoresis system from Invitrogen for visualizing protein distribution of samples. It does not require electrophoresis buffer. The running time is short (15–20 min). With a capacity of 49–98 samples per run, E-PAGE is a powerful tool for fast, high-throughput quality control of purified proteins. Gel loading can also be programmed on the robot. **Figure 9.2** shows gels stained for both hexahistidine affinity tags and protein.
10. Main advantages of this automated purification method are (a) direct lysis of cell suspension without centrifugation and removal of cell debris and (b) the process is quick, results is a high yield, and is reproducible *(7)*. To increase the yield, the cells could be grown in 25 mL suspension, then pelleted by centrifugation and resuspended in the 96-well microplate. Up to 30 OD_{600} per well could be achieved and processed, thus increasing the yield of purified protein per well.

Acknowledgments

This work was supported by US Department of Energy, Genomics: Genomes to Life Project, Grant #KP1102010. Pacific Northwest National Laboratory is operated for the DOE by Battelle Memorial Institute under Contract DEAC05-76RLO 1830. The publisher, by accepting the article for publication, acknowledges that the United States Government retains a non-exclusive, paid-up, irrevocable, worldwide license to publish or reproduce the published form of this manuscript, or allow others to do so, for United States Government purposes. The authors wish to thank to Dr. Christopher Cowan from Promega Corporation, for his help with programming the protein purification protocol on Biomek FX.

References

1. Minshull, J., Ness, J. E., Gustafsson, C., and Govindarajan, S. (2005) Predicting enzyme function from protein sequence. *Curr Opin Chem Biol.* 9: 202–209.
2. Braun, P., and LaBaer, J. (2003) High throughput protein production for functional proteomics. *Trends Biotechnol.* 21: 383–388.
3. Moy, S., Dieckman, L., Schiffer, M., Maltsev, N., Yu, G. X., and Collart, F. R. (2004) Genome-scale expression of proteins from *Bacillus subtilis. J Struct Funct Genomics.* 5: 103–109.
4. Braun, P., Hu, Y., Shen, B., Halleck, A., Koundinya, M., Harlow, E., and LaBaer, J. (2002) Proteome-scale purification of human proteins from bacteria. *Proc Natl Acad Sci USA.* 99: 2654–2659.
5. Heidelberg, J. F., Paulsen, I. T., Nelson, K. E., Gaidos, E. J., Nelson, W. C., Read, T. D., Eisen, J. A., Seshadri, R., Ward, N., Methe, B., Clayton, R. A., Meyer, T., Tsapin, A., Scott, J., Beanan, M., Brinkac, L., Daugherty, S., DeBoy, R. T., Dodson, R. J., Durkin, A. S., Haft, D. H., Kolonay, J. F., Madupu, R., Peterson, J. D., Umayam, L. A., White, O., Wolf, A. M., Vamathevan, J., Weidman, J., Impraim, M., Lee, K., Berry, K., Lee, C., Mueller, J., Khouri, H., Gill, J., Utterback, T. R., McDonald, L. A., Feldblyum, T. V., Smith, H. O., Venter, J. C., Nealson, K. H., and Fraser, C. M. (2002) Genome sequence of the dissimilatory metal ion-reducing bacterium *Shewanella oneidensis. Nat Biotechnol.* 20: 1118–1123.
6. Dieckman, L., Gu, M., Stols, L., Donnelly, M. I., and Collart, F. R. (2002) High throughput methods for gene cloning and expression. *Protein Expr Purif.* 25: 1–7.
7. Lin, C. T., Moore, P. A., Auberry, D. L., Landorf, E. V., Peppler, T., Victry, K. D., Collart, F. R., and Kery, V. (2006) Automated purification of recombinant proteins: combining high-throughput with high yield. *Protein Expr Purif.* 47: 16–24.

Chapter 10

E. coli and Insect Cell Expression, Automated Purification and Quantitative Analysis

Stephen P. Chambers, John R. Fulghum, Douglas A. Austen, Fan Lu, and Susanne E. Swalley

Summary

The production of recombinant proteins usually involves the exploration of a wide variety of expression and purification methodologies in the pursuit of a strategy tailored to a particular protein. The methods applied are reliant on exploiting individual differences between expression systems or the variations in specific protein properties. These bespoke strategies have not lent themselves to high-throughput methodologies. Ultimately the development of robust generic methods capable of simplifying and stabilizing the process, allowing automation, was necessary to increase throughput. This chapter describes a series of high-throughput methods used to express, purify, and quantify recombinant protein produced in *E. coli* or insect cells.

Key words: *E. coli* and insect cell, Protein expression screen, Automated purification, Quantitative analysis.

10.1. Introduction

Traditionally, a wide variety of expression and purification strategies have been pursued for the production of recombinant proteins. The methods applied rely on exploiting individual differences between expression systems or the variations in specific protein properties. These bespoke strategies have not lent themselves to high-throughput methodologies. Through miniaturization of *E. coli* and insect cell growth, automation of protein purification, and analysis we have developed a production process capable of high throughput (**Fig. 10.1**). The standardization necessary

Sharon A. Doyle (ed.), *Methods in Molecular Biology: High Throughput Protein Expression and Purification, vol. 498*

Book doi: 10.1007/978-1-59745-196-3

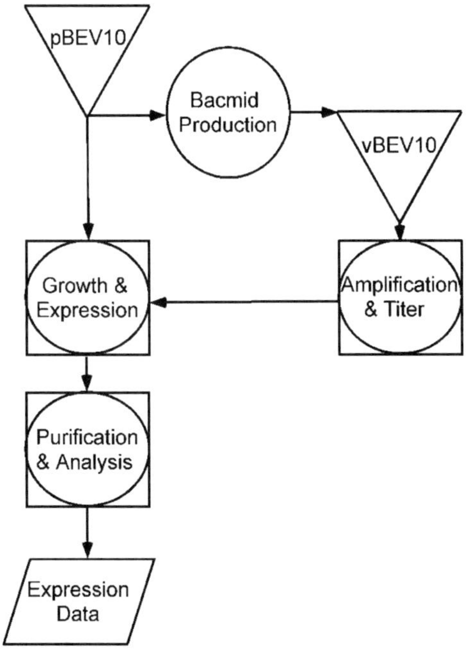

Fig. 10.1. Workflow describing *E. coli* and insect cell protein production. Genes cloned into the expression vector pbev10 are transformed into *E. coli* and expressed or used to produce a bacmid following recombination. Subsequent transfection into insect cells produces the recombinant baculovirus (vbev10). Operations within the process are performed by individual automated workstations capable of high throughput. American National Standards Institute (ANSI) standard symbols are used to describe the individual operations within the flow chart.

for automation was aided by the platform being designed around a single vector capable of directing both *E. coli* and insect cell expression *(1)*. Somewhat paradoxically, the restriction of expression vector options has allowed a much more thorough exploration of experimental conditions to be performed. The increased capacity of our platform allows the expression, purification, and analysis of multiple proteins or multiple variations of the same protein in different expression systems, under a diversity of growth and purification conditions.

10.2. Materials

10.2.1. Expression Systems

1. The *E. coli* and insect cell expression vector pBEV10 (**Fig. 10.2**) was used exclusively in this process.
2. pBEV10 utilizes a T7 promoter to direct recombinant expression in *E. coli*, which requires a host bearing the T7 RNA polymerase gene. For *E. coli* protein production we typically

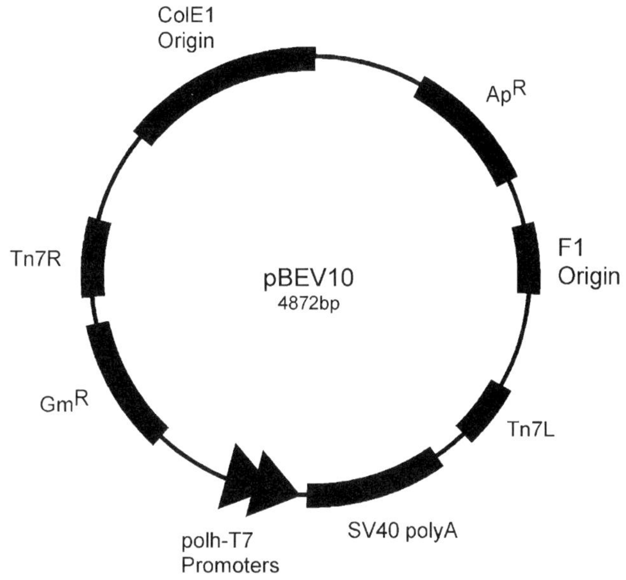

polh po ——→

TCATGGAGATAATTAAAATGATAACCATCTCGCAAATAAATAAGTATT

T7 po —

TTACTGTTTTCGTAACAGTTTTGTAATAAAAAAACCTATAAATAATAC

—— forward primer ——→

GACTCACTATAGGGGAATTGTGAGCGGATAACAATTCCCCTCTAGAAA

TAATTTTGTTTAACTTTAAGAAGGAGATATACC**ATG**GGCAGCAGCCAT
MetGlySerSerHis

His-Tag thrombin cleavage site

CATCATCATCATCACAGCAGCGGCCTGGTGCCGCGCGGCAGCCATATG
HisHisHisHisHisSerSerGlyLeuValProArgGlySerHisMet

←—reverse primer——

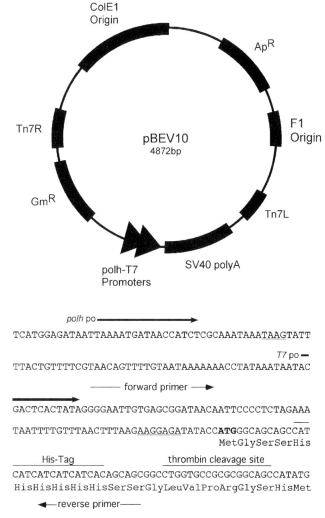

Fig. 10.2. Schematic map of the pbev10 expression vector. Based on the fastbac1 *(2)* backbone, pbev10 contains: *polh*: *T7* promoters, left and right arms of the bacterial transposon Tn7, SV40 polyadenylation signal, the cole1 origin of replication derived from the high copy number cloning vector puc8, the F1 origin for preparation of single-strand DNA for mutagenesis, gentamicin- and ampicillin-resistance genes for selection. The accompanying sequence identifies the *polh* and *T7* promoter regions (po), the ATG start codon (*in bold*), the His-tag and thrombin cleavage site, encoded by the open reading frame. The putative and actual ribosomal-binding sites of *polh* and T7, TAAG and AGGAG, respectively, are *underlined*. Also illustrated on the sequence are the forward and reverse primers used in qPCR.

use BL21 derivatives, like BL21(DE3) containing the λ phage lysogen DE3 encoding T7 RNA polymerase. Alternatively to overcome the inherent leakiness of expression in BL21(DE3) due to uninduced production of T7RNA polymerase, a coresident plasmid pLysS encoding T7 lysozyme, the natural inhibitor of T7 RNA polymerase, can be included in the host. Both BL21(DE3), BL21(DE3)pLysS (EMD Biosciences

Inc, Madison WI) are commercially available, as are other derivatives including BL21-CodonPlus (DE3)-RIPL (Stratagene, LaJolla, CA) developed to support overexpression of recombinant proteins by supplying rare codons. *E. coli* MAX Efficiency DH10 Bac cells were used for bacmid generation (Invitrogen, Carlsbad, CA).

3. *Trichoplusia ni* (High-5), *Spodoptera frugiperda* (Sf21 and Sf9) were used for the insect cell production of recombinant proteins. The latter was also used for the generation of recombinant viral stocks. Sf9 and Sf21 cells were obtained from American Type Culture Collection (ATCC). High-5 cells are available from Invitrogen, Carlsbad, CA.

10.2.2. Cell Growth and Harvest

1. 24-Deep well block, Uniplate (Whatman, Clifton, NJ).
2. HiGro incubator-shaker (Genomic Solutions, Ann Arbor, MI).
3. AirPore tape sheets (Qiagen, Valencia, CA).
4. CedexAS[20] analysis system (Innovartis, Bielefeld, Germany).
5. 96-Well plate centrifuge # 4-15C (Qiagen, Valencia, CA).

10.2.3. Baculoviral Production and Titer

1. *E. coli* MAX Efficiency DH10 Bac cells (Invitrogen, Carlsbad, CA).
2. Luria–Bertani (LB) agar plates containing 50 μg/ml kanamycin, 7 μg/ml gentamicin, 10 μg/ml tetracycline 40 μg IPTG and 100 μg/ml Bluo-gal (Sigma-Aldrich, St. Louis, MO).
3. Cellfectin (Invitrogen, Carlsbad, CA).
4. Sf-900 II Serum-Free Medium (SFM) (Invitrogen, Carlsbad, CA).
5. Cell culture dish, 100 mm × 20-mm style (Corning, Corning, NY).
6. 6-Well cell culture cluster (Corning, Corning, NY).
7. SeaPlaque GTG Agarose (Cambrex, Rockland, ME).
8. Grace's Insect Medium (Invitrogen, Carlsbad, CA), supplemented with 10% fetal bovine serum (Hyclone, Logan, UT), 1× Antibiotic-antimycotic solution (Invitrogen, Carlsbad, CA) and 0.1% Pluronic F-68 (Invitrogen, Carlsbad, CA).
9. BioRobot 3000 automated liquid-handling system (Qiagen, Valencia, CA).
10. R.E.A.L. Prep 96 BioRobot kit and DNeasy 96 tissue kit (Qiagen, Valencia, CA).
11. 96-Well MicroAmp optical reaction plate and caps (PE Applied Biosystems, Foster City, CA).
12. 7300 Real-Time PCR system (PE Applied Biosystems, Foster City, CA).

13. Vent DNA polymerase (New England BioLab, Beverley, MA).
14. TaqMan universal PCR master mix (PE Applied Biosystems, Foster City, CA).
15. Forward primer, 5′-GGGAATTGTGAGCGGATAAC-3′.
16. Reverse primer, 5′-CGCTGCTGTGATGATGATG-3′.
17. Dual-labeled probe, 5′-6FAM-ACCATGGGCAGCA-MGB-NFQ-3′ (PE Applied Biosystems, Foster City, CA).

10.2.4. Protein Expression

10.2.4.1. E. coli Protein Expression and Lysis

1. Standard K12 *E. coli* strains commonly used for recombinant protein production.
2. Brain Heart Infusion (BHI) media (Becton-Dickinson, Sparks, MD).
3. Isopropyl-β-D-thio-galactopyranoside (IPTG) stock: 100 mM.
4. Titer plate shaker (Lab-Line Instruments, Melrose Park, IL).
5. Lysis buffer: 10 mM Tris–HCl, pH 8.0, 40 mM KH_2PO_4, 500 mM NaCl, 5% glycerol, 0.25% Tween-20, 5 mM imidazole, 5 mM β-mercaptoethanol, 0.5 mg/ml lysozyme, 0.1% benzonase (Novagen, Madison, WI). Also included was Complete, EDTA-free protease inhibitor cocktail tablets (Roche Diagnostics, Indianapolis, IN).

10.2.4.2. Insect Cell Protein Expression and Lysis

1. Insect cell lines: *Trichoplusia ni* (High-5), *Spodoptera frugiperda* (Sf12) and (Sf9) were used for the expression of recombinant proteins.
2. EX-Cell 405 media with glutamine (JRH Bioscience, Lenxa, KA) was used for expression in High-5.
3. ESF921 media (Expression Systems, Woodland, CA) was used for expression in Sf9 and Sf21.
4. Titer plate shaker (Lab-Line Instruments, Melrose Park, IL).
5. Lysis buffer: same as the one used for *E. coli*, minus lysozyme.

10.2.5. Protein Purification and Analysis

1. His-Select™ HC Nickel Magnetic beads (Sigma-Aldnich, St. Louis, MO) (*see* **Note 1**).
2. Wash buffer: 10 mM Tris–HCl, pH 8.0, 40 mM KH_2PO_4, 500 mM NaCl, 5% glycerol, 0.25% Tween-20, 5 mM imidazole.
3. Elution buffer: 10 mM Tris–HCl, pH 8.0, 40 mM KH_2PO_4, 500 mM NaCl, 5% glycerol, 0.25% Tween-20, 1 M imidazole.
4. Genesis workstation 150 (Tecan, Research Triangle Park, NC).
5. Adhesive sealing sheets (ABgene, Epsom, UK).

6. Caliper AMS90SE/LabChip90 Instrument (Caliper Life Sciences, Hopkinton, MA).
7. LabChip kit: Protein Express 200 chip (Caliper Life Sciences, Hopkinton, MA).

10.3. Methods

The following methods describe (1) the generation of recombinant baculovirus and its titer, (2) the protein expression and purification from *E. coli* and insect cells, and (3) the quantitation of protein expression.

10.3.1. Baculoviral Production and Titer Using qPCR

Critical to high-throughput expression in insect cells is the ability to generate large numbers of baculoviruses. We believe equally important for the successful exploitation of insect cell expression is the utilization of clonal, high-titer virus for production. To that end, we have automated a number of the steps in baculoviral production to increase throughput. Central to this process is the expression vector pBEV10 which, following recombination in *E. coli* that forms an expression bacmid, is capable of producing baculovirus particles once transfected into insect cells *(3)*. Another piece of enabling technology was the use of qPCR to rapidly obtain accurate baculoviral titers.

10.3.1.1. Generation of Clonal Baculovirus

1. Transform pBEV10 vector-containing gene to be expressed into *E. coli* MAX Efficiency DH10 Bac cells (*see* **Note 2**). Plate transformants onto LB agar plates containing: 50 µg/ml kanamycin, 7 µg/ml gentamicin, 10 µg/ml tetracycline, 40 µg/ml IPTG, and 100 µg/ml Bluo-gal. Incubate plates for up to at least 24 h at 37°C (*see* **Note 3**).
2. Pick white colonies containing recombinant bacmids, from blue-white selection, and transfer to 3 ml LB medium containing: 50 µg/ml kanamycin, 7 µg/ml gentamicin and 10 µg/ml tetracycline. Grow overnight in a 24-deep well block shaken at 350 rpm and at 37°C.
3. Centrifuge at 2,000 × *g* for 5 min and remove supernatant. Purify recombinant bacmid DNA from cells using R.E.A.L Prep 96 BioRobot kit following the manufacturer's protocols.
4. Incubate 1 ml transfection solution containing 6 µl recombinant bacmid DNA and 6 µl Cellfectin in Sf-900 II SFM, for 45 min at room temperature.
5. Incubate Sf9 (9×10^5 cells) in 2 ml of Sf-900 II SFM for 1 h at 27°C in a cell culture dish and wash with Sf-900 II SFM, before use in transfection (*see* **Note 4**).

6. Add transfection solution to the preincubated Sf9 cells and incubate for 5 h at 27°C.
7. Remove media and add 2 ml of the fresh Sf-900 II SFM, supplemented with antibiotic-antimycotic, and culture cells for a further 72 h at 27°C.
8. Collect the medium now containing virus and isolate individual recombinant baculovirus clones by plaque purification on a monolayer of cells (*see* **Note 5**).
9. Culture 10 ml of Sf9 (6.5 × 10^6 cells) in supplemented Grace's insect medium for 1 h at 27°C to create a monolayer in a cell culture dish.
10. Infect monolayer with 0.1 ml of diluted virus (1:50 and 1:250) and incubate at 27°C for 1 h or longer, gently shaking at 30 rpm.
11. Remove media and overlay monolayer with 10 ml (1%) Sea-Plaque GTG agarose in supplemented Grace's insect medium and incubate for 5–6 days at 27°C.
12. Pick 4–8 plaques per construct, suspend each in 1 ml media for 4 h to elute virus.
13. Amplify 0.5 ml of eluted virus from each plaque in a 6-well cell culture cluster, seeded with 2 ml Sf9 (7 × 10^5 cells/mL). Incubate for a further 3 days at 27°C.
14. Perform PCR screening on amplified virus using primers that specifically identify the presence of the cloned gene. Recombinant virus once confirmed undergoes a further round of amplification.
15. Further expansion of 0.5 ml of virus, from primary amplification, takes place over 4–5 days in 25 ml of Sf9 (7 × 10^5 cells/mL) incubated at 27°C. This secondary amplification of virus provides the working stock for all future production.

10.3.1.2. Titer of Recombinant Baculovirus Using qPCR

1. Prepare high-titer baculoviral stocks with two rounds of amplification before titering the viruses by qPCR (*see* **Note 6**).
2. Purify baculoviral DNA for qPCR using the Qiagen DNeasy 96 tissue kit and the Qiagen BioRobot 3000 following the manufacturer's protocol.
3. The forward primer, 5′-GGGAATTGTGAGCGGATAAC-3′, binds downstream of the T7 promoter sequence. The reverse primer, 5′-CGCTGCTGTGATGATGATG-3′, anneals to the sequence encoding the start of the ORF including the 6× His-tag. The amplicon produced is 67 base pairs (bp) in length (**Fig. 10.2**).
4. The dual-labeled probe, 5′-6FAM-ACCATGGGCAGCA-MGBNFQ-3′, was custom prepared by ABI.

5. Carry out qPCR in a 96-well MicroAmp optical plate (*see* **Note 7**). Each well contains 10 μl viral DNA, 25 μl of TaqMan universal PCR master mix, 0.5 μM of forward and reverse primers, and 0.1 μM dual-labeled probe in a final volume of 50 μl.
6. Use the ABI 7300 Real-Time PCR system programmed to perform a 10-min denaturing step at 95°C, followed by annealing at 50°C for 2 min and 40 cycles of amplification at 95°C for 15 s and 1 min at 60°C.
7. Analyze the data using ABI 7300 Real-Time PCR system software.

10.3.2. Protein Expression and Purification

Earlier versions of these protocols have previously been published (**1, 4**) wherein details of the growth and expression conditions used for specific experiments have been described. Typically, a standard *E. coli* experiment would include the exploration of cells lines, growth temperatures, time and inducer concentration, but others factors can be added. There is a large body of literature claiming factors affecting bacterial expression and solubility; this list of variables is extensive. In some ways insect cells present a lesser challenge, with little precedent to follow and fewer factors to explore; therefore, the experimental design is much simpler.

Ultimately it would be for the individual experimenter to determine what and how many expression conditions they wish to examine. Experimental design will be more fully is covered in **Chapter 2**.

10.3.2.1. E. coli Expression

1. Transform *E. coli* with pBEV10 containing the gene to be expressed and plate out cells on LB agar containing the appropriate antibiotic selection. Grow *E. coli* overnight at 37°C. Typically the *E. coli* host used is one of the many BL21 derivatives, capable of supporting overexpression using the T7 promoter in pBEV10.
2. Pick a single fresh colony from the plate and inoculate 25 ml BHI media with the appropriate antibiotic selection. Grow overnight at 30°C and at 225 rpm (*see* **Note 8**).
3. Dilute overnight inoculums to an OD_{600} nm of approximately 0.4. Transfer 5 ml to a 24-deep well block. Grow cultures in a HiGro incubator/shaker for 30 min to 1 h at 30°C and at 350 rpm.
4. Induce with IPTG at OD_{600} nm 0.5–0.7. Concentrations of IPTG used in expression typically range from 0.1 to 1 mM.
5. Conditions for growth during expression, including temperature, length of induction, inducer concentration, etc., will be dictated by experimental design.
6. Harvest cells from 2 ml of culture by centrifugation between 4 and 6,000 × *g* for 5 min.

10.3.2.2. Insect Cell Expression

1. Grow insect cells used for expression to a density of 2×10^6 cells/ml in a fernbach flask (*see* **Note 9**). Transfer 2.5 ml of insect cells to a 24-deep well block
2. Infect insect cells with clonal, high-titer baculovirus. Using the methods described, we routinely produce titers greater than 2×10^8 virus/ml and use viral stock at this minimum concentration in production to ensure consistency.
3. Incubate insect cells in a 24-deep well block at 27°C for 72 h at 350 rpm using a HiGro incubator/shaker (*see* **Note 10**).
4. Harvest the insect cells for purification by centrifugation between 4 and 6,000 × *g* for 5 min.

10.3.2.3. Purification

1. Resuspend cell pellets in 24-deep well block using chilled lysis buffer. Use 1.1 ml and 1 ml lysis buffer for *E. coli* and insect cells, respectively (*see* **Note 11**).
2. Incubate on ice for 15–30 min, with shaking at room temperature for 15–30 min.
3. Finish cell lysis in an ice bath using a deep well cup horn sonicator programmed for 3 × 60 s cycles (1 s on/1 s off). Between cycles cells were chilled at least 1 min on ice.
4. Centrifuge the 24-deep well block at 6,000 × *g* for 5 min to separate cell debris from soluble material.
5. Transfer 1 ml of soluble cell lysate to a 96-deep well block for purification (*see* **Note 12**).
6. Add 30 μl of Ni magnetic beads (*see* **Note 1**) to soluble cell lysate and mix for 15 min at RT.
7. Place 96-deep well block onto a 96-well magnet for 5 min to separate the magnetic beads in solution. Remove the lysate and discard.
8. Add 1 ml of wash buffer to 96-deep well block, shake for 5 min to resuspend Ni magnetic beads, place again on the 96-well magnet and remove buffer. Repeat washing step.
9. Add 60 μl elution buffer and shake for 3 min. Spin plate in centrifuge to ensure that all beads come in contact with the elution buffer (*see* **Note 13**).
10. Place 96-deep well block on 96-well magnet to separate the magnetic beads and elution buffer. Transfer 60 μl eluted protein to a 96-well microtiter plate.
11. Store plate at –80°C, or use immediately for protein analysis.

10.3.3. Protein Analysis

The preceding methods described had the net effect of substantially increasing the number of protein samples to be analyzed, making standard SDS–PAGE analysis difficult. This problem was overcome with the incorporation of the LabChip 90 system into our expression and purification platform. The LabChip 90 system, a microfluildic device, utilizes the same electrophoresis separation principals as SDS–PAGE *(5)*. The LabChip 90 system software detects protein by analyzing the signal produced by the fluorescent dyes that bind SDS–protein complexes. The protein sample is depicted as either an electropherogram or a gel-like image (**Fig. 10.3**). The peak area and the retention time produced by the electropherogram provide a measure of the concentration and molecular weight of the protein. The assay itself relies on internal protein standards to determine protein sizes and concentration. The high sensitivity (5 ng/μl) of protein detection and low volume (1–4 μl) requirement for protein analysis combined with its ability to handle multiple samples (**Fig. 10.4**) in a 96-well format with run times of less than 90 min per plate make the LabChip 90 system ideal for our platform.

10.3.3.1. Quantitative Protein Analysis Using Labchip90 System

1. Prepare LabChips according to the manufacturer's instructions (*see* **Note 14**).
2. Prepare samples (*see* **Note 15**) in a 96-well microtiter plate by adding denaturant to the eluted protein sample in the ratio described by the manufacturer (*see* **Note 16**).
3. Seal microtiter plates with adhesive sealing sheet to prevent evaporation. Heat sample plate to 100°C for 5 min.
4. Mix on a plate shaker at 500 rpm for 30 s. Centrifuge plate at 1,200 × *g* for 15 s to remove condensation.
5. Remove adhesive sealing sheet. Add water to denaturant-protein sample and reseal plate.
6. Shake on a plate shaker at 500 rpm for 30 s.
7. Centrifuge plate at 1,200 × *g* for 15 s to make certain that all liquid is at the bottom of the well.

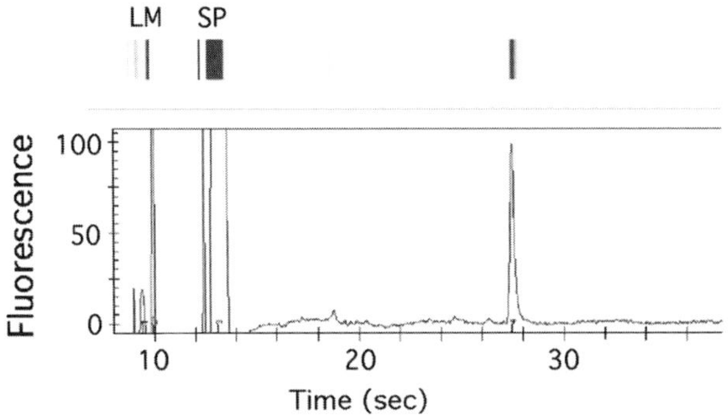

Fig. 10.3. Virtual gel image and electropherogram of protein generated by Caliper labchip 90. Protein was expressed in insect cells, grown in a deep well block, then purified and analyzed. The purified protein is readily identified on the gel and electropherogram, as is the lower marker (LM), an internal reference, and the system peak (SP) (*See Color Plate 3*).

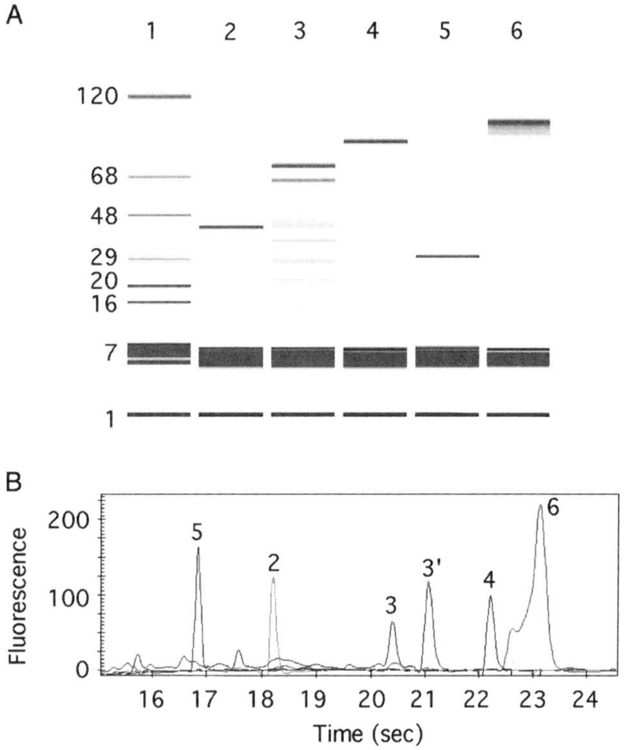

Fig. 10.4. Quantitative analysis of proteins purified from *E. coli* grown in deep well blocks. **(*a*)** The virtual gel image produced by the Caliper labchip 90. *Lane* 1, ladder; *lanes* 2–6, included a variety of different proteins. **(*b*)** Electropherogram of the same sample proteins (2–6), now overlaid, the system software generating both a molecular weight and a concentration value for assigned peaks. *N.B.* Doublet produced in *lane* 3, apparent in the virtual gel **(*a*)**, shows up in the electropherogram (as peaks 3, 3′) (*See Color Plate 4*).

8. The microtiter plate holding the prepared protein samples is placed in the LabChip 90 system.
9. The LabChip 90 chip is prepared as described in the manufacturer's instructions
10. Run assay.
11. Export expression data for statistical analysis (*see* **Note 17**).

10.4. Notes

1. Ni magnetic beads supplied in ethanol are repeatedly exchanged with wash buffer and used at a final concentration of 10% (v/v). Magnetic beads are also available for proteins tagged with Glutathione *S*-tranferase and maltose-binding

protein. The binding dynamics of these tags may necessitate alterations in the protocol as described, but should in general be compatible.

2. Transformation efficiencies were enhanced using DNA concentrated by isopropanol precipitation in transfections of *E. coli* MAX Efficiency DH10 Bac cells.
3. Incubate colonies for at least 24 h at 37°C to distinguish blue white selection on plates. Colonies tend to be very small and blue colonies may not be discernable prior to 24 h.
4. Determining cell concentration in cultures was automated using the CedexAS^{20} analysis system.
5. Plaque purification follows viral binding and adsorption on a monolayer of insect cells; the subsequent overlay of agarose restricts infection to adjacent cells. The resultant plaque, derived from a single infectious particle, provides a means of isolating an individual recombinant baculoviral clone. High-titer clonal virus stocks were used to produce all recombinant proteins minimizing any potential error due to nonclonal expression and ensuring consistency of protein production.
6. Multiple passages of baculoviral-infected insect cells can result in occurrence of defective interfering (DI) particles at the expense of the recombinant wild-type virus *(6)*. To minimize the accumulation of DI particles, baculoviral expansion is limited to two rounds of amplification to generate high-titer viral stocks.
7. 96-Well optical reaction plate and caps were used for qPCR, rather than regular plates, which can interfere with the fluorescent signal used to measure amplified product.
8. Overnight cultures grown at 30°C, compared to 37°C, produced a less saturated inoculum resulting in shorter lag and a quicker initiation of growth following inoculation.
9. Cells that have undergone multiple passages in serum-free medium showed reduced production yields compared to recently adapted cells *(7)*. Therefore, insect cell lines were periodically brought back from frozen stocks every 6 months or earlier if the cell lines deviate from normal growth or expression characteristics. These low passage number cell lines are revived in T-flasks, expanded as adherent cells until they have the correct morphology and are doubling in the expected time frame. Expected doubling times are 18–20 h for High-5 and 20–24 h for Sf9 and Sf21. Cells are then transferred and adapted to suspension culture. Once the correct morphology and doubling times have been re-established, these are expanded and ready to use.
10. Although a standard incubator-shaker fitted with a microplate holder (New Brunswick Scientific, Edison, NJ) worked

for *E. coli* growth in deep well blocks, insect cell growth and infection appeared to require the high torque provided by the HiGro incubator/shaker. Our experimentation had shown *E. coli* and insect cell growth in a deep well block, using a HiGro incubator/shaker, was equivalent to that in shake-flask using a conventional incubator-shaker *(1)*.

11. Bacterial pellets, in particular, can vary in size; therefore, a small excess of buffer can aid removal of the subsequent lysate from the pellet following sonication and centrifugation.
12. A custom script was written to perform the purification protocol using the Tecan Genesis. In practice, any liquid-handling robot capable of transferring liquid volumes in the range of 50 μl to 2.5 ml, and moving 2 ml × 96-deep well blocks around a deck and onto a magnetic separation device, could be programmed to carry out this task. This purification protocol has also been performed at the bench using a manual pipette and a block magnet.
13. The magnetic beads have a tendency to stick to the sides of the wells. A quick spin of 30 s at 5,000 rpm (4,000 × g) insures that all beads come in contact with the elution buffer.
14. The LabChip priming has been performed following the preparation procedure detailed in the manufacturer's instructions. The assay itself relies on internal protein standards to determine protein size and concentration, sizing accuracy is ±20%, relative quantization is ±30% CV, up to 120 kDa (Caliper Life Sciences, Hopkinton, MA). We also use a 50-ng/μl carbonic anhydrase (Sigma-Aldrich, St Louis, MO) standard, made up in wash buffer, represented in every row of every plate, to enable the evaluation of variability across each row and between different sample plates and LabChips.
15. If purified samples are frozen prior to analysis, thaw at room temperature and mix using an orbital plate shaker before preparation of sample for LabChip 90 system.
16. All the reagents required for LabChip priming and running samples are supplied with the LabChip kit (Caliper Life Sciences, Hopkinton, MA). The manufacturer recommends a ratio of 2:7:35 for protein, denaturing buffer and water mixture for samples run on the LabChip. We prepared samples containing threefold higher volumes (6:21:105) since the Tecan Genesis was not accurate to 2 μl. Any liquid-handling robot or manually operated multitip pipette capable of accurately transferring liquid volumes in the range of 2–200 μl could be used.
17. The LabChip 90 system software relies on peak-finding algorithms that utilize the molecular weight of the target proteins in each of the protein samples, so that information needs to be entered to enable export of the quantitative peak

data. Manual inspection of the electopherogram traces and virtual gels created by the LabChip 90 system insures that all peaks are called correctly. In instances where the default settings do not correctly call the unknown peaks, manual adjustment and modification of the default assumptions used by the algorithm are required. This can be done systematically and insures the integrity of the quantitative information exported from the plate.

References

1. Chambers, S. P., Austen, D. A., Fulghum, J. R., and Kim, W. M. (2004) High-throughput screening for soluble recombinant expressed kinases in *Escherichia coli* and insect cells. Protein Expr Purif. 36, 40–47.
2. Anderson, D., Harris, R., Polayes, D., Ciccarone, V., Donahue, R., Gerard, G., and Jessee, J. (1996) Rapid generation of recombinant baculoviruses and expression of foreign genes using Bac to Bac baculovirus expression system. Focus. 17, 53–58.
3. Luckow, V. A., Lee, S. C., Barry, G. F., and Olins, P.O. (1993) Efficient generation of infectious recombinant baculoviruses by site-specific transposon-mediated insertion of foreign genes into a baculovirus genome propagated in *Escherichia coli.* J Virol. 67, 4566–4579.
4. Swalley, S. E., Fulghum, J. R., and Chambers, S. P. (2006) Screening factors effecting a response in soluble protein expression: Formalized approach using design of experiments. Anal Biochem. 351, 122–127.
5. Bousse, L., Mouradian, S., Minalla, A., Yee, H., Williams, K., and Dubrow, R. (2001) Protein sizing on a microchip. Anal Chem. 73, 1207–1212.
6. Pijlman, G. P., van den Born, E., Martens, D. E., and Vlak, J. M. (2001) *Autographa californica* baculoviruses with large genomic deletions are rapidly generated in infected insect cells. Virology. 283, 132–138.
7. Donaldson, M.S., and Shuler, M.L. (1998) Effects of long-term passing of BTI-Tn5B1–4 insect cells on growth and recombinant protein production. Biotechnol Prog. 14, 543–547.

Chapter 11

Hexahistidine-Tagged Maltose-Binding Protein as a Fusion Partner for the Production of Soluble Recombinant Proteins in *Escherichia coli*

Brian P. Austin, Sreedevi Nallamsetty, and David S. Waugh

Summary

Insolubility of recombinant proteins in *Escherichia coli* is a major impediment to their production for structural and functional studies. One way to circumvent this problem is to fuse an aggregation-prone protein to a highly soluble partner. *E. coli* maltose-binding protein (MBP) has emerged as one of the most effective solubilizing agents. In this chapter, we describe how to construct combinatorially-tagged His_6MBP fusion proteins by recombinational cloning and how to evaluate their yield and solubility. We also describe a procedure to determine how efficiently a His_6MBP fusion protein is cleaved by tobacco etch virus (TEV) protease in *E. coli* and a method to assess the solubility of the target protein after it has been separated from His_6MBP.

Keywords: Maltose-binding protein; MBP; Inclusion body; Fusion protein; Solubility enhancer; TEV protease; Tobacco etch virus protease; Hexahistidine tag; His-tag; His_6MBP; Gateway cloning; Recombinational cloning

11.1. Introduction

In many cases, the poor solubility of recombinant proteins in heterologous hosts presents a major obstacle to their large-scale production for functional and structural studies *(1)*. Accordingly, researchers have long sought a generally effective means of improving protein solubility. One popular approach has been to fuse an aggregation-prone protein to a highly soluble partner. In many cases, this enables the aggregation-prone protein to be recovered in a soluble and properly folded form. It should be noted, however, that not all highly soluble proteins are equally

Sharon A. Doyle (ed.), *Methods in Molecular Biology: High Throughput Protein Expression and Purification, vol. 498*

Book doi: 10.1007/978-1-59745-196-3

effective solubilizing agents. Several comparative studies of putative solubility-enhancing fusion partners have yielded conflicting results *(e.g., 2–7)*. Although the reasons for these discrepancies remain unclear, there is ample evidence to indicate that *E. coli* maltose-binding protein (MBP) is among the more effective solubility enhancers. Additionally, to our knowledge, MBP is the only solubility-enhancing protein that is also a natural affinity tag. In principle, its affinity for amylose resin can be exploited to facilitate the purification of an MBP fusion protein *(8)*. In practice, however, we and others have observed that amylose affinity chromatography has several noteworthy disadvantages: the resin is fragile and comparatively expensive; some MBP fusion proteins do not bind efficiently to amylose resin; and, even when they do, this technique rarely yields samples of sufficient purity for structural studies *(9,10)*.

To circumvent the disadvantages of amylose affinity chromatography, a hexahistidine tag (His_6) can be added to the N-terminus of MBP and used for immobilized metal affinity chromatography instead *(11,12)*. We have shown that the addition of a His_6 tag to the N-terminus of MBP does not interfere with its ability to promote the solubility of its fusion partners *(11)*. Moreover, the dual His_6MBP tag can be exploited to purify proteins to homogeneity via an entirely generic and potentially automatable process *(11,12)*.

In this chapter, we focus on the utility of MBP (specifically His_6MBP) as a solubility-enhancing fusion partner. We describe in detail a method for constructing His_6MBP fusion protein expression vectors by Gateway recombinational cloning and provide illustrious examples of the most common outcomes. MBP fusion vectors (but not His_6MBP fusion vectors) designed for cloning by conventional methods (i.e., using restriction endonucleases and DNA ligase) can be obtained from New England Biolabs (Beverley, MA, USA) but will not be discussed here.

11.2. Materials

11.2.1. Recombinational Vector Construction

1. The Gateway destination vector pDEST-HisMBP (*see* **Fig. 11.1**), which can be obtained from AddGene (http://www.addgene.org) or the authors.
2. Reagents and thermostable DNA polymerase for PCR amplification (*see* **Note 1**).
3. Synthetic oligodeoxyribonucleotide primers for PCR amplification (*see* **Fig. 11.2**).
4. TE buffer: 10 mM Tris-HCl. pH 8.0, 1 mM EDTA.

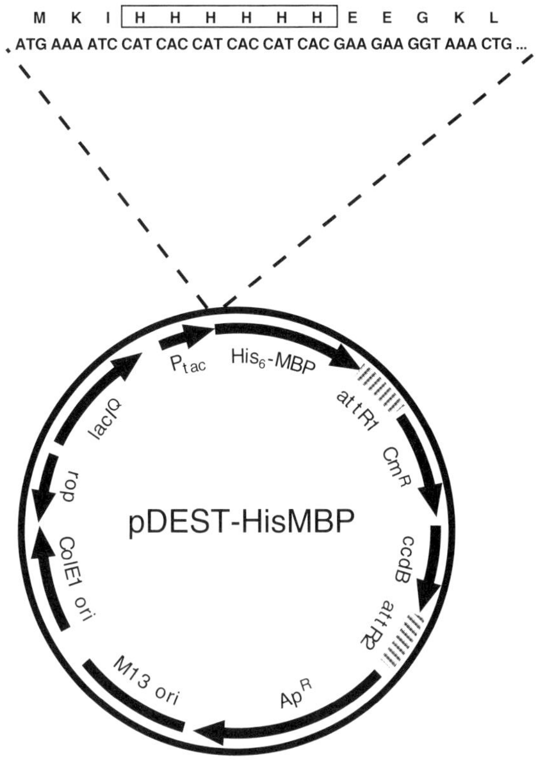

Fig. 11.1. Schematic representation of the Gateway destination vector pDEST-HisMBP. This vector can be recombined with an entry clone that contains an ORF of interest, via the LR reaction, to generate a His_6MBP fusion protein expression vector. The nucleotide and amino acid (single letter code) sequences at the beginning of the open reading frame are shown. Abbreviations: P_{tac} *tac* promoter; *attR1* and *attR2* recombination sites for Gateway cloning, Cm^R chloramphenicol acetyl transferase (chloramphenicol-resistance) gene; *ccdB* gene encoding DNA gyrase poison CcdB, Ap^R β-lactamase (ampicillin-resistance) gene, *M13 ori* origin of replication from bacteriophage M13, *ColE1 ori* origin of replication from ColE1 plasmid, *rop* repressor of primer gene, $lacI^Q$ gene encoding lactose repressor.

5. E-gels and an E-gel base (Qiagen, Valencia, CA, USA) for submarine gel electrophoresis of DNA (*see* **Note 2**).
6. QIAquick gel extraction kit (Qiagen, Valencia, CA, USA) for the extraction of DNA from agarose gels.
7. Chemically competent DB3.1 or "CcdB survival" cells (Invitrogen, Carlsbad, CA, USA) for propagating pDEST-HisMBP and pDONR221.
8. Competent *gyrA*$^+$ cells (e.g., DH5α, MC1061, HB101) (*see* **Note 3**).
9. The Gateway donor vector pDONR221 (Invitrogen, Carlsbad, CA, USA).
10. Gateway BP Clonase II (Invitrogen, Carlsbad, CA, USA).
11. Gateway LR clonase II (Invitrogen, Carlsbad, CA, USA).
12. LB medium and LB agar plates containing ampicillin (100 μg/ml). LB medium: Add 10 g bactotryptone, 5 g bacto yeast extract,

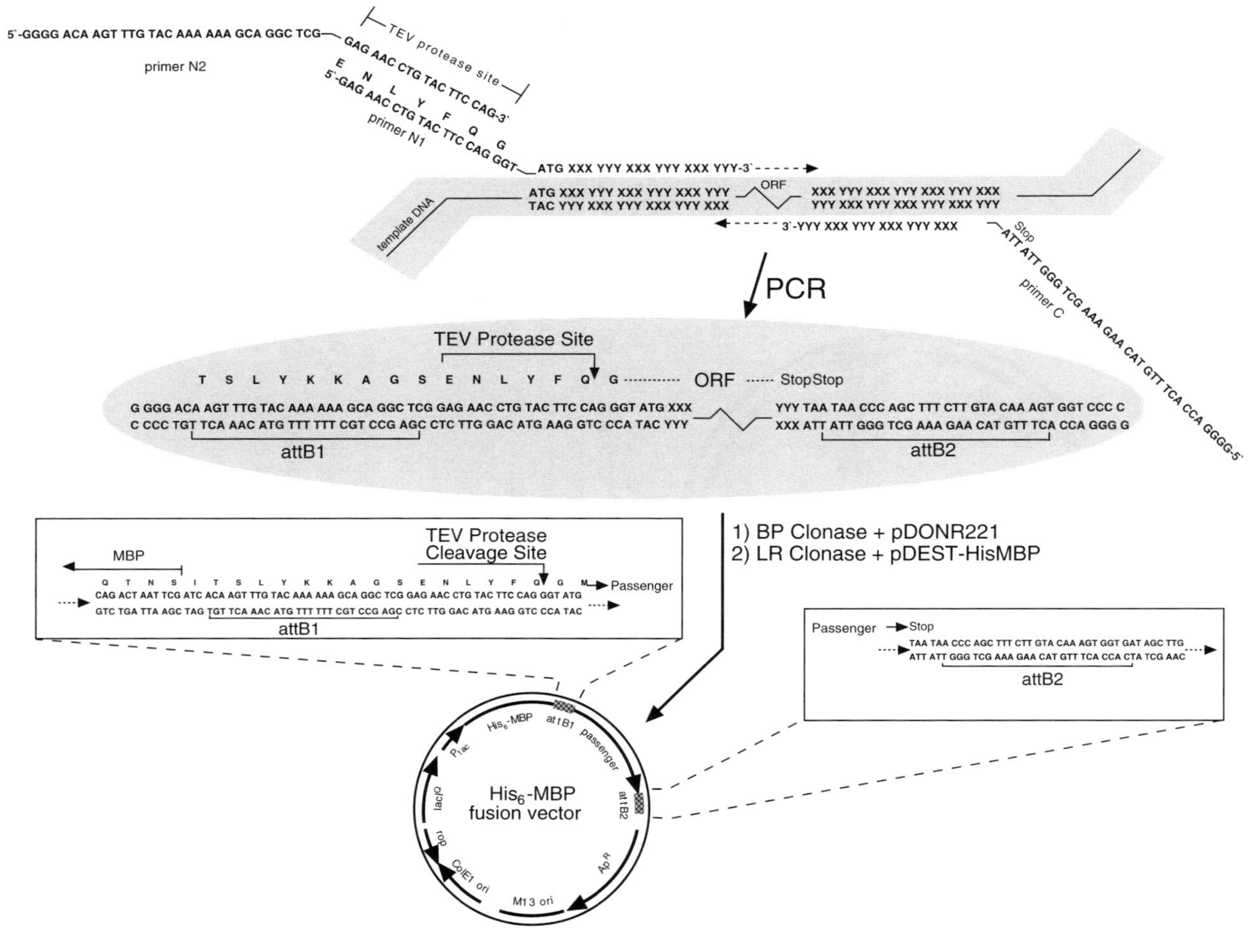

Fig. 11.2. Construction of a His$_6$MBP fusion vector using PCR and Gateway cloning technology. The ORF of interest is amplified from the template DNA by PCR, using primers N1, N2 and C. Primers N1 and C are designed to base-pair to the 5′ and 3′ ends of the coding region respectively and contain unpaired 5′ extensions as shown. Primer N2 base-pairs with the sequence that is complementary to the unpaired extension of primer N1. The final PCR product is recombined with the pDONR221 vector to generate an entry clone, via the BP reaction. This entry clone is subsequently recombined with pDEST-HisMBP using LR Clonase to yield the final His$_6$MBP fusion vector.

and 5 g NaCl to 1 L of H_2O and sterilize by autoclaving. For LB agar, also add 12 g of bactoagar before autoclaving. To prepare plates, allow medium to cool until flask or bottle can be held in hands without burning, then add 1 ml ampicillin stock solution (100 mg/ml in H_2O, filter sterilized), mix by gentle swirling, and pour or pipet ca. 30 ml into each sterile petri dish (100 mm dia.).

13. QIAprep Spin Miniprep Kit (Qiagen) for small-scale plasmid DNA isolation (*see* **Note 4**).
14. A microcentrifuge capable of operating at 14,000 rpm.
15. An incubator set at 37°C.

11.2.2. Protein Expression

1. Competent BL21-Pro cells (B&D Clontech, Palo Alto, CA, USA) containing the TEV protease expression vector pRK603 *(13)*. pRK603 can be obtained from AddGene (http://www.addgene.org) (*see* **Notes 5** and **6**).
2. A derivative of pDEST-HisMBP that produces a His_6-MBP fusion protein with a TEV protease recognition site in the linker between MBP and the passenger protein (*see* **Section 11.3.1**).
3. LB agar plates and broth containing both ampicillin (100 μg/ml) and kanamycin (35 μg/ml). *See* **Section 11.2.1 Item 10** for LB broth, LB agar, and ampicillin stock solution recipes. Prepare stock solution of 35 mg/ml kanamycin in H_2O and filter sterilize. Store at 4°C for up to 1 month. Dilute antibiotics 1,000-fold into LB medium or molten LB agar at no more than 50°C.
4. Isopropyl-thio-β-D-galactopyranoside (IPTG), analytical grade. Prepare a stock solution of 200 mM in H_2O and filter sterilize. Store at 4°C.
5. Anhydrotetracycline (ACROS Organics/Fisher Scientific, Springfield, NJ, USA). Prepare a 1,000× stock solution by dissolving in 50% ethanol at 100 μg/ml. Store in a foil-covered tube at −20°C.
6. Shaker/incubator.
7. Sterile baffle-bottom flasks (Bellco Glass, Inc., Vineland, NJ, USA)
8. Cell lysis buffer: 20 mM Tris-HCl, pH 8, 1 mM EDTA.
9. Sonicator (with microtip).
10. 2× SDS-PAGE sample buffer (Invitrogen) and 2-mercaptoethanol (Sigma-Aldrich, St. Louis, MO, USA).
11. SDS-PAGE gel, electrophoresis apparatus, and running buffer (*see* **Note 7**).
12. Gel stain (e.g. Gelcode® Blue from Pierce, Rockford, IL, USA, or PhastGel™ Blue R from Amersham Biosciences, Piscataway, NJ, USA).

13. Spectrophotometer.
14. 1.5 ml microcentrifuge tubes.

11.3. Methods

11.3.1. Construction of His$_6$MBP Fusion Vectors by Recombinational Cloning

The Gateway recombinational cloning system is based on the site specific recombination reactions that mediate the integration and excision of bacteriophage lambda into and from the *E. coli* chromosome, respectively. For detailed information about this system, consult the technical literature supplied by Invitrogen, Inc. and available online at: http://www.invitrogen.com/Content/Online%20Seminars/gateway/gatewayhome.html.

11.3.1.1. pDEST-HisMBP

To utilize the Gateway system for the production of His$_6$MBP fusion proteins, one must first construct or obtain a suitable "destination vector". Currently there is no commercial source for such a vector, but one (pDEST-HisMBP) can be obtained from the non-profit distributor of biological reagents AddGene, Inc. (http://www.addgene.org). A schematic diagram of pDEST-HisMBP is shown in **Fig. 11.1**. This plasmid was constructed by inserting an in-frame hexahistidine coding sequence between codons 3 and 4 of MBP in pKM596 *(12)*.

The Gateway cloning cassette in pDEST-HisMBP carries a gene encoding the DNA gyrase poison CcdB, which provides a negative selection against non-recombined destination and donor vectors so that only the desired recombinant is obtained when the end products of the recombinational cloning reaction are transformed into *E. coli* and grown in the presence of ampicillin or kanamycin, respectively. pDEST-HisMBP and other vectors that carry the *ccdB* gene must be propagated in a host strain with a *gyrA* mutation (e.g. *E. coli* DB3.1) or "CcdB survival" cells that are immune to the action of CcdB.

11.3.1.2. Gateway Cloning Protocol

To construct a His$_6$MBP fusion protein expression vector, one begins by amplifying the open reading frame of interest by PCR using a forward primer containing the TEV protease cleavage site as a 5′ unpaired extension (N1) and a reverse primer containing an attB2 recombination site as a 5′ unpaired extension (C) (*see* **Fig. 11.2**). The remainder of each primer (3′ ends) should consist of approximately 20–25 nucleotides that are complimentary to the ends of the open reading frame. To avoid the need for an excessively long forward primer, the first PCR amplicon is used as the template for a second PCR with a forward primer consisting of the TEV protease recognition site and an attB1 recombination site (N2) and the same reverse primer (C). The PCR reaction may contain all three primers (*see* **Note 8**). To favor the accumulation of the desired product, the *att*B-containing primers are used at typical concentrations for PCR but the concentration of the gene-specific N-terminal primer (N1) is 10–20 folds lower. The final PCR amplicon is inserted first into the donor vector pDONR221 by recombinational cloning with BP clonase and then into the destination vector pDEST-HisMBP in a second recombinational cloning reaction with LR clonase.

1. The PCR reaction mix is prepared as follows (*see* **Note 9**): 1 µl template DNA (~10 ng µl), 5 µl thermostable DNA polymerase 10× reaction buffer, 1.5 µl dNTP solution (10 mM each), 1.0 µl primer N1 (~30 ng), 3 µl primer N2 (~300 ng), 3.0 µl primer C (~300 ng), 1 µl thermostable DNA polymerase, 64.5 µl H_2O (to 100 µl total volume).
2. The reaction is placed in the PCR thermal cycler with the following program: initial melt for 5 min at 95°C; 35 cycles of 95°C for 30 s, 55°C for 30 s, and 72°C for 60 s (*see* **Note 10**); hold at 4°C.
3. Purification of the PCR amplicon by agarose gel electrophoresis (*see* **Note 2**) is recommended to remove *att*B primer-dimers.
4. To create the His_6MBP fusion vector, the PCR product is recombined first into pDONR221 to yield an entry clone intermediate (BP reaction), and then into pDEST-HisMBP (LR reaction; *see* **Note 11**). This is described in steps 5–13 below.
5. Add to a microcentrifuge tube on ice: 50–100 ng of the PCR product in TE or H_2O, 200 ng of pDONR221 DNA, and enough TE to bring the total volume to 8 µl. Mix well.
6. Thaw BP Clonase II enzyme mix on ice (2 min) and then vortex briefly (2 s) twice (*see* **Note 12**).
7. Add 2 µl of BP Clonase II enzyme mix to the components in (a) and vortex briefly twice.
8. Incubate the reaction at 25°C for at least 4 h (*see* **Note 13**).
9. Add to the reaction: 2 µl of the destination vector (pDEST-HisMBP) at a concentration of 150 ng/µl, and 3 µl of LR Clonase II enzyme mix (*see* **Note 12**). The final reaction volume is 15 µl. Mix by vortexing briefly.
10. Incubate the reaction at room temperature for 3–4 h.
11. Add 2.5 µl of the proteinase K solution and incubate for 10 min at 37°C.
12. Transform 2 µl of the reaction into 50 µl of chemically competent DH5α cells (*see* **Note 3**).
13. Pellet the cells by centrifugation, gently resuspend pellet in 100–200 µl of LB broth and spread on an LB agar plate containing ampicillin (100 µg/ml), the selective marker for

pDEST-HisMBP (*see* **Fig. 11.1**). Incubate the plate at 37°C overnight (*see* **Note 14**).

14. Plasmid DNA is isolated from saturated cultures started from individual ampicillin-resistant colonies and screened by PCR, using the gene-specific primers N1 and C, to confirm that the clones contain the expected gene. Alternatively, plasmids can be purified and screened by conventional restriction digests using appropriate enzymes. We routinely sequence clones that screen positive by either PCR or restriction digest to ensure that there are no PCR-induced mutations.

11.3.2. Protein Expression

To assess the yield and solubility of the fusion protein, the amount of total fusion protein produced in the crude cell extract is directly compared to the soluble fraction by visual inspection of a Coomassie blue stained gel. A parallel experiment is run to determine if the fusion protein is a good substrate for TEV protease and whether or not the cleaved target protein remains soluble after it is released from His_6MBP.

11.3.2.1. Selecting a Host Strain of E. coli

Sometimes delaying the induction of TEV protease until the fusion protein substrate has had time to accumulate in the cells results in greater solubility of the passenger protein after cleavage *(13, 14)*. However, to achieve regulated expression of TEV protease, the *in vivo* processing experiment must be performed in a strain of *E. coli* that produces the Tet repressor, such as BL21-Pro or DH5α–Pro (B&D Clontech, Palo Alto, CA, USA). The Tet repressor blocks the synthesis of TEV protease mRNA and allows the production of the enzyme to be regulated independently of the IPTG-inducible His_6MBP fusion vector. Independent production of TEV protease from the expression vector pRK603 *(13)* is initiated by adding anhydrotetracycline to the cell culture, usually two hours after induction of the fusion protein with IPTG. We prefer using BL21-Pro because of its robust growth characteristics and the fact that it lacks two proteases (Lon and OmpT) that are present in many *E. coli* K12 strains such as DH5α-Pro.

11.3.2.2. Protein Expression

1. Transform competent BL21-Pro or DH5α-Pro cells that already contain pRK603 (*see* **Notes 5** and **6**) with the His_6MBP fusion protein expression vector and spread them on LB agar plates containing ampicillin (100 μg/ml) and kanamycin (35 μg/ml). Incubate the plate overnight at 37°C.
2. Inoculate 2–5 ml of LB medium containing ampicillin (100 μg/ml) and kanamycin (35 μg/ml) in a culture tube or shake-flask with a single colony from the plate. Grow to saturation overnight at 37°C with shaking at 250 rpm.
3. The next morning, inoculate 50 ml of the same medium in a 250 ml baffled-bottom flask with 0.5 ml of the saturated overnight culture.

4. Grow the cells at 37°C with shaking to mid-log phase ($OD_{600\,nm}$ ~ 0.5).
5. Add IPTG (1 mM final concentration) and adjust the temperature to 30°C (*see* **Note 15**).
6. After 2 h, divide the culture into two separate flasks (ca. 20 ml in each). Label one flask "+" and the other "–".
7. Add anhydrotetracycline to the "+" flask (100 ng/ml final concentration).
8. After 2 more hours, measure the $OD_{600\,nm}$ of the cultures (dilute cells 1:10 in LB to obtain an accurate reading). An $OD_{600\,nm}$ of about 3–3.5 is normal, although lower densities are possible. If the density of either culture is much lower than this, it may be necessary to adjust the volume of the samples that are analyzed by SDS-PAGE.
9. Transfer 10 ml of each culture to a 15 ml conical centrifuge tube and pellet the cells by centrifugation (4,000 × *g*) at 4°C.
10. Resuspend the cell pellets in 1 ml of lysis buffer and then transfer the suspensions to a 1.5 ml microcentrifuge tube.
11. Store the cell suspensions at –80°C overnight. Alternatively, the cells can be disrupted immediately by sonication (without freezing and thawing) and the procedure continued without interruption, as described below.

11.3.2.3. Sonication and Sample Preparation

1. Thaw the cell suspensions at room temperature, and then place them on ice.
2. Lyse the cells by sonication (*see* **Note 16**).
3. Prepare samples of the total intracellular protein from the "+" and "–" cultures (T + and T–, respectively) for SDS-PAGE by mixing 50 µl of each sonicated cell suspension with 50 µl of 2× SDS-PAGE sample buffer containing 20% (v/v) 2-mercaptoethanol.
4. Pellet the insoluble cell debris (and proteins) by centrifuging the sonicated cell suspension from each culture at maximum speed in a microcentrifuge for 10 min at 4°C.
5. Prepare samples of the soluble intracellular protein from the "+" and "–" cultures (S + and S–, respectively) for SDS-PAGE by mixing 50 µl of each supernatant from step 4 with 50 µl of 2× SDS-PAGE sample buffer containing 20% (v/v) 2-mercaptoethanol.

11.3.2.4. SDS-PAGE

We typically use pre-cast Tris-Glycine or NuPAGE gradient gels for SDS-PAGE to assess the yield and solubility of MBP fusion proteins (*see* **Note 7**). Of course, the investigator is free to choose any appropriate SDS-PAGE formulation, depending on the protein size and laboratory preference.

1. Heat the T–, T +, S– and S + protein samples at 90°C for about 5 min and then spin them at maximum speed in a microcentrifuge for 5 min.
2. Assemble the gel in the electrophoresis apparatus, fill it with SDS-PAGE running buffer, load the samples (10 µl each), and carry out the electrophoretic separation according to standard lab practices. T and S samples from each culture ("+" and "–") are loaded in adjacent lanes to allow easy assessment of solubility. Molecular weight standards may also be loaded on the gel, if desired.
3. Stain the proteins in the gel with GelCode® Blue reagent, PhastGel™ Blue R, or a suitable alternative.

11.3.2.5. Interpreting the Results

The MBP fusion protein should be readily identifiable in the T- sample after the gel is stained since it will normally be the most abundant protein in the cells. Molecular weight standards can also be used to corroborate the identity of the fusion protein band. If the S- sample contains a similar amount of the fusion protein, this indicates that it is highly soluble in *E. coli*. If little or no fusion protein is observed in the S- sample, then this is an indicator of poor solubility. Of course, a range of intermediate states is also possible.

If the fusion protein is an efficient substrate for TEV protease, then little of it will be present in the T + and S + samples. Instead, one should observe a prominent band at ca. 42 kDa that corresponds to the His_6MBP moiety and another prominent band migrating with the expected mobility of the passenger protein. If the fusion protein is a poor substrate for the protease, then the "+" samples will look similar to the "–" samples.

If the passenger protein is soluble after it is released from His_6MBP, then a similar amount will be present in the T + and S + lanes. At this point, some or all of the passenger protein may precipitate. If a substantial fraction of the passenger protein is insoluble, then troubleshooting may be necessary. Alternatively, an acceptable yield might still be obtained by scaling up cell production.

Examples of several possible outcomes are illustrated in **Fig. 11.3**. In panel A, roughly equal amounts of the His_6MBP-DHFR fusion protein are readily visible in lanes T- and S-, indicating that fusion protein is overexpressed and highly soluble. The results of adding anhydrotetracycline two hours after the induction of the fusion protein with IPTG (to induce the production of TEV protease) are seen in lanes T + and S +. The band migrating at ~42 kDa corresponds to His_6MBP and the smaller band migrating at ~21.5 kDa is the cleaved DHFR. Since nearly equal amounts of DHFR are present in lanes T + and S +, this indicates that virtually all of the DHFR retains its solubility after being cleaved from the His_6MBP. Similarly, the His_6MBP-TIMP fusion protein

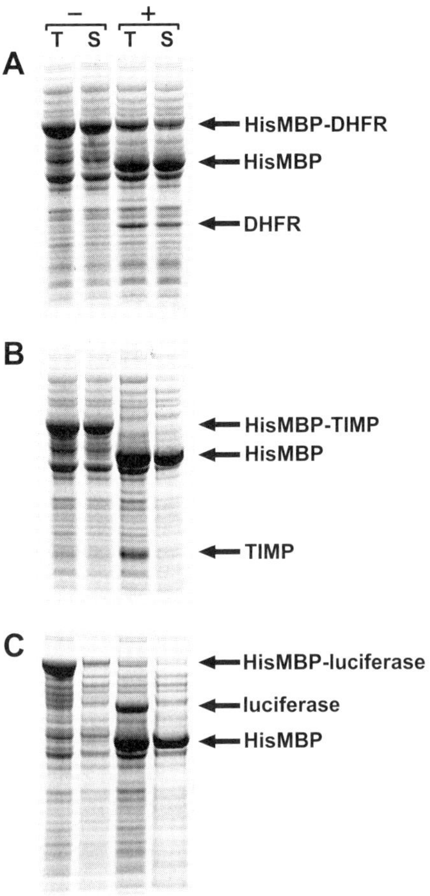

Fig. 11.3. Intracellular processing of His_6MBP fusion proteins by TEV protease. DHFR, TIMP and Luciferase were expressed from derivatives of pDEST-HisMBP in BL21-Pro cells that also contained the TEV protease expression vector pRK603 as described (*see* **Section 11.3.2**). "T" and "S" refer to the total and soluble fractions of the intracellular protein, respectively. All cultures were induced with IPTG to initiate the production of the His_6MBP fusion proteins. Samples marked "+" were induced with anhydrotetracycline to initiate the production of TEV protease 2 h after the addition of IPTG, whereas samples marked "–" were not induced with anhydrotetracyline. The results of this experiment are discussed in the Interpreting the Results section (*see* **Section 11.3.2.5**).

is also highly expressed and soluble (panel B), but in this instance the TIMP precipitates after digestion of the fusion protein *in vivo* with TEV protease. Hence, TIMP is an example of a passenger protein that is only temporarily soluble while it is fused to MBP. The His_6MBP-luciferase fusion protein is included here to illustrate the point that not all aggregation-prone proteins can be made soluble by fusing them to MBP (panel C).

Previous studies have shown that, despite its high solubility in *E. coli*, GST does not have the ability to promote the solubility of its fusion partners *(2, 15)*. Consequently, the solubility of a GST fusion protein is a good indicator of the solubility of a passenger protein in its unfused state. The results obtained when the three passenger proteins utilized here (DHFR, TIMP and luciferase)

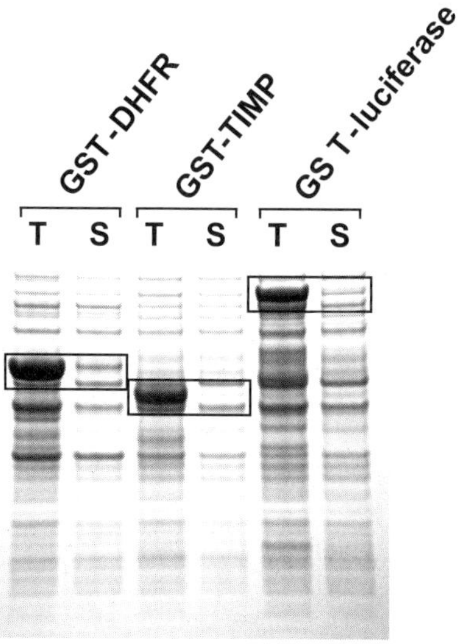

Fig. 11.4. Insolubility of GST fusion proteins. Entry clones encoding the three passenger proteins utilized in this study (DHFR, TIMP and Luciferase) were recombined into the destination vector pDEST3 (Invitrogen), which is designed to produce GST fusion proteins. Samples of the total (T) and soluble (S) intracellular protein from cells expressing the GST fusion proteins were prepared and analyzed by SDS-PAGE. The boxes indicate the positions of the fusion proteins on the Coomassie-stained gel.

were expressed as GST fusion proteins are shown in **Fig. 11.4**. Note that all three of the GST fusion proteins exhibit very poor solubility, indicating that the solubility of DHFR and TIMP was definitely enhanced by fusing them to His_6MBP.

Once it has been established that an aggregation-prone passenger protein can be rendered soluble by fusing it to His_6MBP, that the fusion protein can be cleaved by TEV protease, and that the passenger protein remains soluble after it is released from His_6MBP, then the passenger protein is ready to be purified on a large scale. Detailed instructions for how to accomplish this, using a generic IMAC-based protocol, have been described elsewhere *(12)*.

11.3.2.6. Checking the Biological Activity of the Passenger Protein

Occasionally, a passenger protein may accumulate in a soluble but biologically inactive form after intracellular processing of a His_6MBP fusion protein. Exactly why this occurs is uncertain, but we suspect that fusion to MBP somehow enables certain proteins to form soluble aggregates or evolve into kinetically trapped folding intermediates that are no longer susceptible to aggregation. Therefore, although solubility after intracellular processing is a useful indicator of a passenger protein's folding state in most cases, it is not absolutely trustworthy and can occasionally be misleading. For this reason, we strongly recommend that a biological assay (if available) or biophysical techniques be employed at an early stage to confirm that the passenger protein is in its native conformation.

11.4. Notes

1. We recommend a proofreading polymerase such as *Pfu* Turbo (Stratagene, La Jolla, CA, USA), Platinum *Pfx* (Invitrogen, Carlsbad, CA, USA), or Deep Vent (New England Biolabs, Beverly, MA, USA) to minimize the occurrence of mutations during PCR.
2. We typically purify fragments by electrophoresis using precast E-gels purchased from Invitrogen. However, suitable equipment and reagents for horizontal agarose gel electrophoresis can be purchased from a wide variety of scientific supply companies. DNA fragments are extracted from slices of the ethidium bromide-stained gel using a QIAquick gel extraction kit (Qiagen) in accordance with the instructions supplied with the product.
3. Any *gyrA*$^+$ strain of *E. coli* can be used. We prefer competent DH5α cells (Invitrogen) because they are easy to use and have high transformation efficiencies.
4. We prefer the QIAprep Spin miniprep kit (Qiagen), but similar kits can be obtained from a wide variety of vendors.
5. While any method for the preparation of competent cells can be used (e.g., $CaCl_2$) *(16)*, we prefer electroporation because of the high transformation efficiency that can be achieved. Detailed protocols for the preparation of electrocompetent cells and electrotransformation procedures can be obtained from the electroporator manufacturers (e.g., Bio-Rad, BTX, Eppendorf). Briefly, the cells are grown in 1 L of LB medium (with antibiotics, if appropriate) to mid-log phase (OD_{600} ~0.5) and then chilled on ice. The cells are pelleted at 4°C, resuspended in 1 L of ice-cold H_2O and then pelleted again. After several such washes with H_2O, the cells are resuspended in 3–4 ml of 10% glycerol, divided into 50 µl aliquots, and then immediately frozen in a dry ice/ethanol bath. The electrocompetent cells are stored at –80°C. Immediately prior to electrotransformation, the cells are thawed on ice and mixed with 10–100 ng of DNA (e.g., a plasmid vector or a Gateway reaction). The mixture is placed into an ice-cold electroporation cuvette and electroporated according to the manufacturers recommendations (e.g., a 1.5 kV pulse in a cuvette with a 1 mm gap). 0.450 ml of SOC medium *(16)* is immediately added to the cells and they are allowed to grow at 37°C with shaking (ca. 250 rpm) for 1 h. 5–200 µl of the cells are then spread on an LB agar plate containing the appropriate antibiotic(s).
6. If the open reading frame encoding the passenger protein contains codons that are rarely used in *E. coli* (http://www.doe-mbi.ucla.edu/cgi/cam/racc.html), this can adversely affect

the yield of an MBP fusion protein. In such cases, it is advisable to introduce an additional plasmid into the host cells that carries the cognate tRNA genes for rare codons. The pRIL plasmid (Stratagene, La Jolla, CA, USA) is a derivative of the p15A replicon that carries the *E. coli argU*, *ileY*, and *leuW* genes, which encode the cognate tRNAs for AGG/AGA, AUA and CUA codons, respectively. pRIL is selected for its resistance to chloramphenicol. In addition to the tRNA genes for AGG/AGA, AUA and CUA codons, the pRARE accessory plasmid in the Rosetta™ host strain (Novagen, Madison, WI, USA) also includes tRNAs for the rarely used CCC and GGA codons. Like pRIL, the pRARE plasmid is a chloramphenicol-resistant derivative of the p15A replicon. Both of these tRNA accessory plasmids are compatible with derivatives of pDEST-His-MBP. On the other hand, they are incompatible with the vector pRK603 that we use for intracellular processing experiments (*see* **Section 11.3.2**). Nevertheless, because pRK603 and the tRNA accessory plasmids have different antibiotic resistance markers, it is possible to force cells to maintain both plasmids by simultaneously selecting for kanamycin and chloramphenicol resistance. Alternatively, the kanamycin-resistant TEV protease expression vector pKM586, a pRK603 derivative with the replication machinery of a pSC101 replicon, which can be obtained from the authors, can be stably maintained in conjunction with p15A-type tRNA plasmids.

7. We find it convenient to use pre-cast gels for SDS-PAGE (e.g., 1.0 mm × 10 well, 10–20% Tris-Glycine gradient), running buffer, and electrophoresis supplies from Invitrogen (Carlsbad).
8. Alternatively, the PCR reaction can be performed in two separate steps, using primers N1 and C in the first step and primers N2 and C in the second step. The PCR amplicon from the first step is used as the template for the second PCR. All primers are used at the typical concentrations for PCR in the two-step protocol.
9. The PCR reaction can be modified in numerous ways to optimize results, depending on the nature of the template and primers (*see* **(16)** (Vol. 2, Chapter 8) for more information).
10. PCR cycle conditions can also be varied. For example, the extension time should be increased for especially long genes. A typical rule-of-thumb is to extend for 60 s/kb of DNA.
11. This "one-tube" Gateway protocol bypasses the isolation of an "entry clone" intermediate. However, the entry clone may be useful if the investigator intends to experiment with additional Gateway destination vectors, in which case the BP and LR reactions can be performed sequentially in separate steps; detailed instructions are included with the Gateway

PCR kit. Alternatively, entry clones can easily be regenerated from expression clones via the BP reaction, as described in the instruction manual.

12. Clonase enzyme mixes should be thawed quickly on ice and then returned to the −80°C freezer as soon as possible. It is advisable to prepare multiple aliquots of the enzyme mixes the first time that they are thawed in order to avoid repeated freeze-thaw cycles.
13. At this point, we remove a 5 μl aliquot from the reaction and add it to 0.5 μl of proteinase K solution. After 10 min at 37°C, we transform 2 μl into 50 μl of competent DH5α cells (*see* **Note 3**) and spread 100–200 μl on an LB agar plate containing kanamycin (35 μg/ml), the selective marker for pDONR221. From the number of colonies obtained, it is possible to gauge the success of the BP reaction. Additionally, entry clones can be recovered from these colonies in the event that no transformants are obtained after the subsequent LR reaction.
14. If very few or no ampicillin-resistant transformants are obtained after the LR reaction, the efficiency of the process can be improved by incubating the BP reaction overnight.
15. 30°C is the optimum temperature for TEV protease activity. At 37°C, the protease does not fold properly in *E. coli* and little processing will occur. Reducing the temperature also improves the solubility of some MBP fusion proteins.
16. We routinely break cells in a 1.5 ml microcentrifuge tube on ice with two or three 30 s pulses using a VCX600 sonicator (Sonics & Materials, Newtown, CT, USA) with a microtip at 38% power. The cells are cooled on ice between pulses.

Acknowledgement

This research was supported by the Intramural Research Program of the NIH, National Cancer Institute, Center for Cancer Research.

References

1. Chayen, N. E. (2004) Turning protein crystallization from an art into a science. *Curr. Opin. Struct. Biol.* 14, 577–583
2. Kapust, R. B., and Waugh, D. S. (1999) *Escherichia coli* maltose-binding protein is uncommonly effective at promoting the solubility of its fusion partners. *Protein Sci.* 8, 1668–1674
3. Shih, Y.-P., Kung, W.-M., Chen, J.-C., Yeh, C.-H., Wang, A. H.-J., and Wang, T.-F. (2002) High-throughput screening of

soluble recombinant proteins. *Protein Sci.* 11, 1714–1719

4. Hammarstrom, M., Hellgren, N., van Den Berg, S., Berglund, H., and Hard, T. (2002) Rapid screening for improved solubility of small human proteins produced as fusion proteins in *Escherichia coli*. *Protein Sci.* 11, 313–321
5. Dyson, M. R., Shadbolt, P. S., Vincent, K. J., Perera, R. L., and McCafferty, J. (2004) Production of soluble mammalian proteins in *Escherichia coli*: identification of protein features that correlate with successful expression. *BMC Biotech.* 4, 32 doi:10.1186/1472–6750/4/32
6. Busso, D., Delagoutte-Busso, B., and Moras, D. (2005) Construction of a set of gateway-based destination vectors for high-throughput cloning and expression screening in *Escherichia coli*. *Anal. Biochem.* 343, 313–321
7. Korf, U., Kohl, T., van der Zandt, H., Zahn, R., Schleeger, S., Ueberle, B., Wandschneiderr, S., Bechtel, S., Schnolzer, M., Ottleben, H., Wiemann, S., and Poustka, A. (2005) Large-scale protein expression for proteomics research. *Proteomics* 5, 3571–3580
8. Riggs, P. (2000) Expression and purification of recombinant proteins by fusion to maltose-binding protein. *Mol. Biotechnol.* 15, 51–63
9. Pryor, K. A., and Leiting, B. (1997) High-level expression of soluble protein in *Escherichia coli* using a His6-tag and maltose binding protein double-affinity fusion system. *Protein Expr. Purif.* 10, 309–319
10. Routzahn, K. M. and Waugh, D. S. (2002) Differential effects of supplementary affinity tags on the solubility of MBP fusion proteins. *J. Struct. Funct. Genomics* 2, 83–92
11. Nallamsetty, S., Austin, B. P., Penrose, K. J., and Waugh, D. S. (2005) Gateway vectors for the production of combinatorially-tagged His6-MBP fusion proteins in the cytoplasm and periplasm of *Escherichia coli*. *Protein Sci.* 14, 2964–2971
12. Tropea, J. E., Cherry, S., Nallamsetty, S., Bignon, C., and Waugh, D. S. (2007) A generic method for the production of recombinant proteins in *Escherichia coli* using a dual hexahistidine-maltose-binding protein affinity tag. *Methods Mol. Biol.* 363, 1–19
13. Kapust, R. B. and Waugh, D. S. (2000) Controlled intracellular processing of fusion proteins by TEV protease. *Protein Expr. Purif.* 19, 312–318
14. Fox, J. D. and Waugh, D. S. (2003) Maltose-binding protein as a solubility enhancer. *Methods Mol. Biol.* 205, 99–117
15. Fox, J. D., Routzahn, K. M., Bucher, M. H., and Waugh, D. S. (2003) Maltodextrin-binding proteins from diverse bacteria and archaea are potent solubility enhancers. *FEBS Lett.* 537, 53–57
16. Sambrook, J. and Russell, D. W. (2001) *Molecular cloning: a laboratory manual.* Cold Spring Harbor Laboratory Press, Cold Spring Harbor, NY(2002) High-throughput screening of soluble recombinant proteins. *Protein Sci.* 11, 1714–1719

Chapter 12

PHB-Intein-Mediated Protein Purification Strategy

Alison R. Gillies, Mahmoud Reza Banki, and David W. Wood

Summary

A method has been developed that eliminates the need for complex chromatographic apparatus in the purification of recombinant proteins expressed in *Escherichia coli*. This method is similar to conventional affinity-tag separations, but the affinity resin is replaced by polyhydroxybutyrate (PHB) particles produced *in vivo* in the *E. coli* expression host during protein expression. A PHB-binding protein known as a phasin is genetically fused to the product protein via an engineered pH and temperature dependent self-cleaving intein linker. Thus the phasin–intein fusion acts as a self-cleaving purification tag, with affinity for the co-expressed PHB granules. The PHB particles and tagged target protein are purified by lysing the cells and washing the granules with sequential rounds of centrifugation and resuspension. The native target protein is then released from the bound tag through an intein-mediated self-cleavage reaction, induced by a mild pH shift. A final round of centrifugation removes the granules and associated tag, allowing the purified target to be recovered in the supernatant. This method has been shown to yield 35–40 µg of purified product per milliliter of liquid cell culture and is likely to be applicable to a wide range of expression hosts.

Key words: Affinity tag purification; Inteins

12.1. Introduction

The purification of recombinant proteins is a routine but often difficult task in molecular biology. This task can be simplified by genetically fusing the target protein sequence to one encoding an easily purified affinity tag. By contacting the expressed fusion protein with a suitable affinity resin, simple recovery and purification of the attached target protein is accomplished *(1)*. Although this method is highly effective and is ubiquitous at laboratory scale, some drawbacks remain. One of these is the significant expense associated with conventional affinity chromatography, while another is the required proteolytic removal of the affinity

Sharon A. Doyle (ed.), *Methods in Molecular Biology: High Throughput Protein Expression and Purification, vol. 498*

Book doi: 10.1007/978-1-59745-196-3

tag to acquire a native target protein. Proteolytic tag removal in particular can be inefficient, or can lead to unwanted destruction of the target and is prohibitively expensive for scale-up. This work describes a method wherein a self-cleaving affinity tag is used to purify an expressed target protein through association with co-expressed intracellular polymeric granules, thus eliminating both the expense and complexity of conventional affinity separations with proteolytic tag removal *(2)*.

The granules are composed of polyhydroxybutyrate (PHB), which is a form of polyhydroxyalkanoate produced naturally by a number of bacterial species *(3)*. The enzymes required for PHB production have been cloned into *E. coli*, where they lead to the production of granular inclusion bodies *in vivo (4)*. These macroscopic granules are easily recovered by centrifugation or other mechanical means, suggesting their use as an affinity carrier for a tagged target *(5)*. The expressed target protein is genetically fused to a recently discovered PHB-binding protein, known as a phasin *(6)*. The phasin exhibits specific affinity for PHB granules, allowing it to act as an affinity tag for the target protein. Finally, the method also makes use of an engineered self-cleaving protein segment known as an intein *(7,8)*, which undergoes a self-cleaving reaction at its C-terminus in response to a small shift in pH or temperature. Insertion of the intein between the phasin and target protein effectively makes the phasin self-cleaving, thus eliminating the need for protease treatment to acquire the native target *(9)*.

12.2. Materials

12.2.1. Cloning the Product Protein Gene

1. Luria–Bertani (LB) medium: 1% tryptone, 0.5% yeast extract, 1% NaCl (w/v) at pH 7.0, sterilized by autoclaving at 15 psi for 20 min. Store at room temperature.
2. Ampicillin (25 mg/ml), sterilized by filtration. Store in 1 ml aliquots at −20°C.
3. 10× ExTaq polymerase buffer (TaKaRa Bio Inc, Shiga, Japan). Store at −20°C.
4. Forward PCR primer (10 pmol/μl) (*see* **Note 1**, **Note 2**). Store at −20°C.
5. Reverse PCR primer (10 pmol/μl) (*see* **Note 3**). Store at −20°C.
6. Deoxynucleotide triphosphates (dNTP Mix) (TaKaRa Bio Inc, Shiga, Japan): 2 mM final concentration each of dATP, dCTP, dGTP, dTTP. Store at −20°C.
7. Taq Polymerase (TaKaRa Bio Inc, Shiga, Japan). Store at −20°C.

8. Restriction enzymes and concentrated buffers (New England Biolabs) (as required for cloning strategy *see* **Note 1, Note 3**). Store at –20°C.
9. Agarose, agarose gel running buffer, ethidium bromide staining solution.
10. T4 DNA Ligase (New England Biolabs, Beverly, MA).
11. 4× T4 DNA Ligase Buffer (New England Biolabs, Beverly, MA).
12. Competent strain of *E. coli* to be used for cloning (i.e. XL-1 blue or DH5α). Available from various sources.
13. pET phasin–intein expression vector pET-PPPI:M (contact authors for availability) (**Fig. 12.1**).

12.2.2. Co-Transformation, Expression and PHB Production

1. *Luria–Bertani Medium supplemented with lactate*: 1% tryptone, 0.5% yeast extract, 1% NaCl, 2% sodium lactate, pH 7.0, sterilized by autoclaving at 15 psi for 20 min. Store at room temperature.

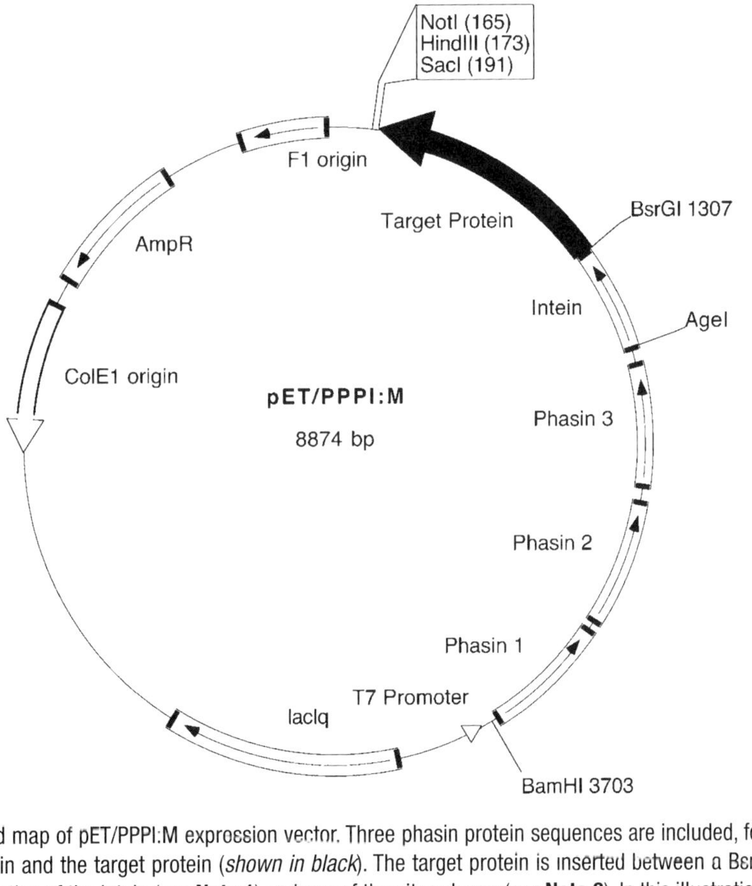

Fig. 12.1. Plasmid map of pET/PPPI:M expression vector. Three phasin protein sequences are included, followed by the self-cleaving intein and the target protein (*shown in black*). The target protein is inserted between a BsrG I site within the C-terminal portion of the intein (*see* **Note 1**) and one of the sites shown (*see* **Note 3**). In this illustration, the maltose binding protein has been used as an example target protein.

2. *IPTG (Isopropyl-beta-D-thiogalactopyranoside)*: 0.1 M in deionized water, sterilized by filtration. Store in 1 ml aliquots at −20°C.
3. Ampicillin (25 mg/ml), sterilized by filtration. Store in 1 ml aliquots at −20°C.
4. Kanamycin (10 mg/ml), sterilized by filtration. Store in 1 ml aliquots at −20°C.
5. Competent *E. coli* expression strain, typically BLR containing pJM9131 (*see* **Note 4**).

12.2.3. Purification

1. Modified lysis buffer: 20 mM Tris, 20 mM Bis-Tris, pH 8.5, 50 mM NaCl, 1 mM DTT, 2 mM EDTA, 2.5 μg/ml lysozyme. Store at 4°C.
2. Wash buffer: 20 mM Tris, 20 mM Bis-Tris, pH 8.5, 50 mM NaCl, 1 mM DTT, 2 mM EDTA. Store at 4°C.
3. Cleavage buffer: 20 mM Tris, 20 mM Bis-Tris, pH 6.5 or pH 6.0 (*see* **Note 5**), 50 mM NaCl, 1 mM DTT, 2 mM EDTA. Store at 4°C.

12.2.4. SDS-Polyacrylamide Gel Electrophoresis Analysis

1. 40% Acrylamide/bis solution (29:1 with 3.3% C) (Bio-Rad, Hercules CA) (Caution: Acrylamide monomer is a neurotoxin and can be absorbed through skin. Avoid direct contact.) Store at 4°C.
2. TEMED (Fisher Biotech, Fairlawn NJ). Store at 4°C.
3. *Ammonium persulfate*: Prepare a 10% (w/v) solution in water. Can be used for one week if stored at 4°C in small sealed container.
4. *4× separating gel buffer*: 36.3 g Tris base, 8.0 ml 10% SDS, 190 ml water, add HCl until pH drops to 8.8, add water to 200 ml final volume. Store at room temperature.
5. *4× stacking gel buffer*: 12.0 g Tris-base, 8 ml 10% SDS, 190 ml water, add concentrated HCl until pH drops to 6.8, add water to 200 ml final volume. Store at room temperature.
6. *Saturated isobutanol*: Mix equal volumes of isobutanol and 1× separating gel buffer (prepared by diluting one part 4× separating gel buffer with three parts water). Shake in a glass bottle and allow to settle. Saturated isobutanol forms the top layer. Store at room temperature.
7. *10× protein gel running buffer*: 30 g Tris-Base, 144 g glycine, 10 g SDS, add de-ionized water to 1 L. Store at room temperature.
8. *Coomassie Brilliant Blue stain solution*: 450 ml methanol, 2.5 g brilliant blue G-250, dissolve brilliant blue in methanol, then add 100 ml glacial acetic acid, 450 ml of water, mix, filter particulates using Whatman Number 1 filter paper with a Buchner funnel. Store at room temperature.

9. *SDS-PAGE destain solution*: 300 ml methanol, 100 ml glacial acetic acid, 600 ml water. Store at room temperature.
10. *1× SDS-gel loading buffer*: 10% glycerol (v/v), 0.1% bromophenol blue (w/v), 2% SDS (w/v), 50 mM Tris-HCl, pH 6.8, 100 mM dithiothreitol. Store at 4°C.

12.3 Methods

The phasin–intein affinity tag is used in PHB-producing bacterial strains to provide a very simple and inexpensive method for the purification of recombinant proteins. In practice, a plasmid encoding three PHB production enzymes is initially transformed into a suitable expression strain. The resulting strain is then co-transformed with a second plasmid encoding the target protein, tagged with two or three phasin sequences and an intein. This strain is then grown on a lactose-supplemented medium to allow expression of the PHB granules and tagged target protein. Lysis and recovery of the granules takes place at pH 8.5, which suppresses the intein cleaving reaction. Once the granules have been washed by several cycles of centrifugation and resuspension, the intein is induced to self-cleave by a pH shift to 6.5. The cleaving reaction releases the native target from the granule surface, allowing it to be easily separated from the granules and associated cleaved tag. Tests on several proteins indicate that this method is effective for recovering 30–50 mg of cleaved, native target per liter of shake-flask culture *(2)*. Further, although the present method is limited to expression in *E. coli*, PHB granules have been produced in bacteria *(10)*, yeast *(11)* and transgenic plants *(12,13)*, and thus any of these hosts might be adapted to this method.

12.3.1. Cloning the Product Protein Gene (see Note 6)

1. *Prepare a 100 µl PCR reaction*: 60 µl sterile water, 10 µl Forward PCR Primer, 10 µl Reverse PCR Primer, 10 µl dNTP Mix, 10 µl 10× ExTaq Polymerase buffer, 1 µl Taq polymerase and 1 µl of target protein template DNA.
2. *Load reactions into a PCR machine (heated lid) and run with the following protocol*: 2 min at 95°C, 25 cycles at 95°C for 1 min, 50°C for 1 min, 72°C for 1 min and then a final incubation at 72°C for 10 min.
3. The pET/PPPI:M plasmid (Amp^R) and the PCR product are digested in 40 µl reactions with the appropriate restriction enzymes and buffers. For a typical 40 µl digest use 20 µl sterile water, 4 µl NEB 10× restriction enzyme buffer, 4 µl 10× BSA (if required – *see* vendor information on each enzyme to determine) or 4 µl sterile water (if BSA is not required),

1 µl of each restriction enzyme and 10 µl of plasmid or PCR product DNA to be digested. Incubate at vendor recommended enzyme temperature for at least 3 h (**Fig. 12.2**).

4. Separate the digested DNA fragments on a 1% agarose gel and stain by soaking in ethidium bromide solution. Cut the desired insert and plasmid bands from the gel on a UV gel box using a clean razor blade. Recover insert and vector DNA fragments from the gel using the GeneClean gel extraction kit (QBioGene, Carlsbad, CA) as per kit instructions.
5. Quantify the recovered fragments by separating on a 1% agarose gel and stain by soaking in ethidium bromide solution (*see* **Note 7**).
6. Set up a ligation reaction with a 3:1 molar ratio of product gene to plasmid. For a typical 20 µl reaction use: 11 µl sterile water, 5 µl 4× T4 DNA Ligase buffer, 1 µl extracted plasmid DNA (from **Step 4**), 3 µl extracted insert DNA (from **Step 4**), 0.5 µl T4 DNA Ligase (*see* **Note 8**).
7. Incubate at 16°C for at least 3 h.
8. Transform 1 µl of ligation reaction into electrocompetent XL1-Blue (Stratagene, La Jolla, CA) and place on an LB-agar plate containing 100 µg/ml ampicillin. Incubate overnight at 37°C.

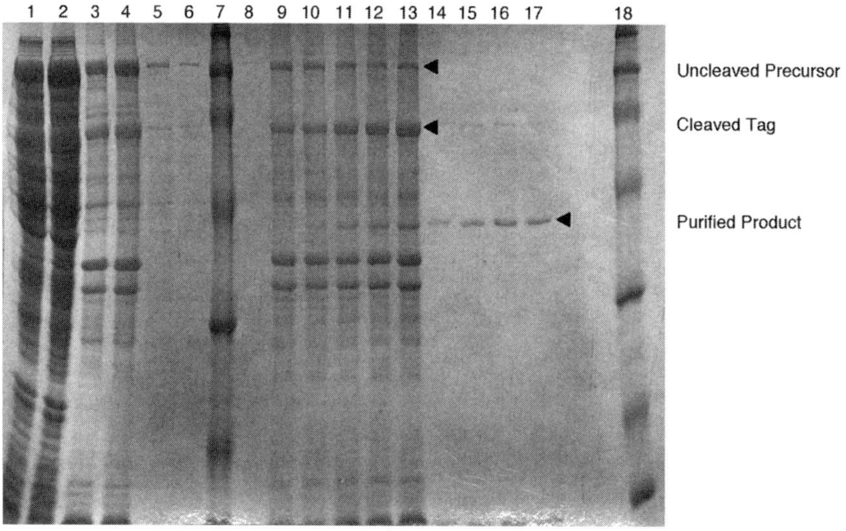

Fig. 12.2. Typical purification results. The maltose binding domain is purified as an example protein. Lanes: 1 = soluble portion of lysate from pre-induced cells; 2 = soluble portion of lysate after induction; 3 = insoluble portion of lysate before induction; 4 = insoluble portion of lysate after induction; 5, 6 and 8 = supernatants of three wash cycles of granules; 7 and 18 = molecular weight markers; 9–13 = total samples at various times during cleaving (both granule-bound and soluble material are shown); 14–17 = soluble material corresponding to lanes 10–13 showing cleaved material in the supernatant. It can be seen that significant material builds up in the cells during growth, but that additional induction of protein expression by IPTG increases overall yields of uncleaved material. In this case, significant prematurely cleaved material is evident, but this can be minimized by good aeration and expression at lower temperature.

9. Screen colonies by digestion of mini-prep DNA with BsrG I and Hind III/ Not I/ Sac I enzymes. Correct clones will show presence of appropriately sized fragment.

12.3.2. Co-Transformation, Expression and PHB Production

1. Transform *E. coli* BLR (DE3) cells containing the pJM9131 plasmid (*see* **Note 4**) with the pET/PPPI:M expression plasmid and plate on LB agar supplemented with 50 µg/ml kanamycin, and 100 µg/ml ampicillin.
2. Inoculate a 5 ml overnight culture of LB medium plus 50 µg/ml kanamycin and 100 µg/ml ampicillin with a single colony of the double-transformed expression strain.
3. Dilute overnight cultures 1:100 into 5 ml of LB medium supplemented with lactate plus 50 µg/ml kanamycin and 100 µg/ml ampicillin.
4. Incubate with shaking for 30 h at 37°C to allow PHB production (*see* **Note 9**).
5. Induce overexpression of tagged fusion protein by addition of 1 mM isopropyl-beta-D-thiogalactopyranoside (IPTG) for an additional 4–8 h at 37°C (*see* **Note 10**).
6. Harvest cells from 5 ml cell broth by centrifugation for 10 min at 5,000 × *g* and 4°C.

12.3.3. Purification

1. Resuspend the cell pellets in 300 µl of modified lysis buffer.
2. Lyse the cells by ultrasonic disruption at 4°C. (*see* **Note 11**).
3. Centrifuge the cell lysate at 14,000 × *g* and 4°C for 10–30 min.
4. Discard the supernatant and resuspend the cells in 300 µl of wash buffer (*see* **Note 12**).
5. Centrifuge at 14,000 × *g* and 4°C for 10–30 min and discard the supernatant. Repeat **Steps 4 and 5** as necessary, but at least three times using fresh wash buffer at pH 8.5.
6. After the last wash, resuspend the pellet in cleavage buffer at pH 6.0 and centrifuge at again 14,000 × *g* and 4°C for 10–30 min.
7. Resuspend the pellet in cleavage buffer and incubate at 20°C for 25 h to allow cleavage to proceed (*see* **Note 13**).
8. Centrifuge at 14,000 g and 4°C for 10–30 min.
9. Remove the supernatant to a fresh tube. This contains the purified soluble product (*see* **Note 14**).

12.3.4. SDS-PAGE Analysis

1. Add 5 µl of classic load buffer to a 15 µl sample of purified protein and heat for 3 min at 95°C (in-process samples can then be frozen for later use).
2. The volumes assume the use of a Gibco BRL model V16 vertical gel electrophoresis apparatus and can be modified as per

manufacturer instructions to any full-size, pre-cast or mini-gel electrophoresis system. If an alternate is used, consult manufacturer instructions for proper volumes and assembly instructions. In all cases, it is critical that the glass plates be thoroughly cleaned before use by first scrubbing with a rinsable detergent and then rinsing with water and 95% ethanol.

3. Assemble gel plates according to manufacturer instructions, with 1 mm spacers between plates.
4. Prepare 20 ml of a 12% polyacrylamide separating gel solution by mixing 9 ml of water with 5 ml of 4× separating buffer and 6 ml of 40% acrylamide/bis solution.
5. Add 20 μl of TEMED and 100 μl APS to the solution and mix to initiate polymerization. Quickly pour the gel into the assembled plates and add a layer of saturated butanol to cover the top of the gel during polymerization.
6. When the separating gel has set (typically 10–30 min), rinse off the butanol with DI water. Place a gel comb at the top of the gel and prepare 10 ml of stacking gel solution by combining 6.25 ml of water, 2.5 ml of 4× stacking buffer, and 1.25 ml of 40% acrylamide.
7. Add 20 μl of TEMED and 100 μl of APS and mix to initiate polymerization. Quickly pour the gel around the comb until it reaches the top of the gel apparatus.
8. When the gel is set, remove the comb and complete assembly of the gel and buffer tanks. Add 1 L of protein gel running buffer and ensure that there are no leaks between the buffer tanks.
9. Carefully load the samples prepared above and attach the gel apparatus to an appropriate power supply (consult manufacturer instructions).
10. Run the gel at 35 mA per gel for 2 h.
11. Remove the gel from the apparatus and stain using Coomassie Brilliant Blue stain for 1h with slow shaking (*see* **Note 15**).
12. Destain the gel overnight using Coomassie destaining solution. The SDS-PAGE gel can then be dried for storage. Observation of the gel is used to monitor cleavage progress as well as to determine the degree of purification and recovery of the product protein *(2)*.

12.4. Notes

1. Cloning of product proteins into the intein is generally done using a unique BsrG I site (TGTACA) at the end of the intein and one of several sites downstream from the product

protein (typically Hind III). It is critical that the intein peptide sequence must end with the amino acids VVVHN, to be followed immediately by the initial methionine residue of the product protein. To make this possible without adding any additional residues to the product protein, a translationally silent BsrG I restriction site was created close to the end of the intein *(14)*. Therefore, the upstream PCR primer used to amplify the target protein gene must include both the BsrG I site and the last few nucleotides of the intein (CAAC) that lie between the intein BsrG I site and the initial amino acid of the target protein. Thus a typical upstream primer will start with the sequence 5′-GTT **GTT GTA CAC** AAC-3′, to be immediately followed by the annealing portion of the primer for the 5′ end of the target protein (typically 18–24 DNA additional bases). Note that the BsrG I site (shown in bold) is not in frame with the intein.

2. In all of our reported work, the initial amino acid of the target protein (encoded by the upstream primer) is methionine. This amino acid is not chemically involved in the cleaving reaction and in principle can be substituted with other amino acids. However, it is likely that any replacement will affect the structure and charge density of the cleaving junction and may have unpredictable effects. Previously reported work with other C-terminally cleaving inteins suggests that changes in this amino acid can accelerate, slow or abolish cleaving *(15)*.
3. The 3′ end of the target gene can be cloned using Sac I, Hind III or Not I. In each case, the stop codon of the target protein must precede the restriction site to ensure it will lie at the end of the target protein sequence.
4. We have successfully produced PHB granules in *E. coli* strains XL1-Blue, ER2566, BL21 (DE3) and BLR (DE3). However, the T7 RNA polymerase gene present in BLR (DE3) ensured strong expression of the tagged product protein from the T7 promoter on the pET-PPPI:M vector (derived from pET-21). If using a different promoter, any of the above strains should work. In each case, the desired expression strain is made competent and transformed with the pJM9131 vector (Kan^R). The transformed strain is then made competent again for transformation of the protein expression plasmid pET/PPPI:M (Amp^R).
5. Cleaving can take place at a range of pH values. If a product protein is less stable at pH 6.0, then the cleaving buffer pH can be raised. The cleaving rate will slow down at pH values above 6.5, but cleaving as high as pH 7 is possible.
6. The target protein can be inserted into the pET/PPPI:M vector using any conventional cloning method. The method shown here is typical of what we use.

7. This step is optional, but allows visual confirmation that the fragments have been recovered and allows a more precise estimation of insert to plasmid ratio for the ligation step.
8. The relative volumes of plasmid and insert should be adjusted to provide roughly 3 moles of insert per mole of plasmid. Changes in their relative volumes can be compensated by changing the volume of water used to maintain the reaction total of 20 μl.
9. Aeration of the cultures seems to have a significant effect on premature cleaving of the intein tag. It is recommended that the aeration be as high as possible to maximize uncleaved precursor and final yield. This is typically not a problem at test tube scale, but can be significant as increased scales are used in shake-flasks.
10. The inductions can be performed at 20°C as well, which can decrease premature cleaving of the intein during expression. In this case, grow for approximately 8 h in the shaker bath.
11. It is important to keep the cells as cold as possible during this step. Therefore, sonication should be performed on ice. Sonicate in 10 s intervals, followed by at least 10 s of icing until the cells are completely disrupted. This will typically take 30–60 s of sonication depending on the cell density. To aid sonication, lysozyme can be added to the lysis buffer to a final concentration of 1 μg/ml. Also, do not allow the lysate suspension to foam during sonication.
12. Since intein activity is pH dependent, the pH of the wash buffer is very important. Ensure that pH is 8.5.
13. Cleavage should proceed at any temperature between 18–23°C. Cleavage progress can be monitored by taking 15 μl samples at various time points for SDS-PAGE analysis.
14. This process generally yields 35–40 μg of purified product per ml of culture *(2)*. The activities of purified maltose binding protein and beta-galactosidase have been confirmed by simple tests.
15. Both the Coomassie Brilliant Blue and destaining solutions can be saved and reused. The destaining solution can be regenerated by adding Kim Wipes to the solution to absorb excess stain.

Acknowledgement

This work was partially supported by a National Science Foundation Graduate Student Fellowship to MRB and Army Research Office grant W911NF-04–1–0056.

References

1. Terpe, K. (2003) Overview of tag protein fusions: from molecular and biochemical fundamentals to commercial systems. Appl Microbiol Biotechnol 60, 523–533
2. Banki, M. R., Gerngross, T. U., and Wood, D. W. (2005) Novel and economical purification of recombinant proteins: intein-mediated protein purification using in vivo polyhydroxybutyrate (PHB) matrix association. Protein Sci 14, 1387–1395
3. Anderson, A. J., and Dawes, E. A. (1990) Occurrence, metabolism, metabolic role, and industrial uses of bacterial polyhydroxyalkanoates. Microbiol Rev 54, 450–472
4. Choi, J. I., Lee, S. Y., and Han, K. (1998) Cloning of the Alcaligenes latus polyhydroxyalkanoate biosynthesis genes and use of these genes for enhanced production of Poly(3-hydroxybutyrate) in Escherichia coli. Appl Environ Microbiol 64, 4897–4903
5. Moldes, C., Garcia, P., Garcia, J. L., and Prieto, M. A. (2004) In vivo immobilization of fusion proteins on bioplastics by the novel tag BioF. Appl Environ Microbiol 70, 3205–3212
6. Wieczorek, R., Pries, A., Steinbuchel, A., and Mayer, F. (1995) Analysis of a 24-kilodalton protein associated with the polyhydroxyalkanoic acid granules in *Alcaligenes eutrophus*. J Bacteriol 177, 2425–2435
7. Chong, S., Mersha, F. B., Comb, D. G., Scott, M. E., Landry, D., Vence, L. M., Perler, F. B., Benner, J., Kucera, R. B., Hirvonen, C. A., Pelletier, J. J., Paulus, H., and Xu, M. Q. (1997) Single-column purification of free recombinant proteins using a self-cleavable affinity tag derived from a protein splicing element. Gene 192, 271–281
8. Perler, F. B., Davis, E. O., Dean, G. E., Gimble, F. S., Jack, W. E., Neff, N., Noren, C. J., Thorner, J., and Belfort, M. (1994) Protein splicing elements: inteins and exteins–a definition of terms and recommended nomenclature. Nucleic Acids Res 22, 1125–1127
9. Wood, D. W., Wu, W., Belfort, G., Derbyshire, V., and Belfort, M. (1999) A genetic system yields self-cleaving inteins for bioseparations. Nat Biotechnol 17, 889–892
10. Fidler, S., and Dennis, D. (1992) Polyhydroxyalkanoate production in recombinant *Escherichia coli*. FEMS Microbiol Rev 9, 231–235
11. Leaf, T. A., Peterson, M. S., Stoup, S. K., Somers, D., and Srienc, F. (1996) *Saccharomyces cerevisiae* expressing bacterial polyhydroxybutyrate synthase produces poly-3-hydroxybutyrate. Microbiology 142 (Pt 5), 1169–1180
12. John, M. E., and Keller, G. (1996) Metabolic pathway engineering in cotton: Biosynthesis of polyhydroxybutyrate in fiber cells. Proc Natl Acad Sci U S A 93, 12768–12773
13. Hahn, J. J., Eschenlauer, A. C., Sleytr, U. B., Somers, D. A., and Srienc, F. (1999) Peroxisomes as sites for synthesis of polyhydroxyalkanoates in transgenic plants. Biotechnol Prog 15, 1053–1057
14. Wood, D. W., Derbyshire, V., Wu, W., Chartrain, M., Belfort, M., and Belfort, G. (2000) Optimized single-step affinity purification with a self-cleaving intein applied to human acidic fibroblast growth factor. Biotechnol Prog 16, 1055–1063
15. Southworth, M. W., Amaya, K., Evans, T. C., Xu, M. Q., and Perler, F. B. (1999) Purification of proteins fused to either the amino or carboxy terminus of the *Mycobacterium xenopi* gyrase A intein. Biotechniques 27, 110–104, 16, 18–20

Chapter 13

High-Throughput Biotinylation of Proteins

Brian K. Kay, Sang Thai, and Veronica V. Volgina

Summary

One of the more useful tags for a protein in biochemical experiments is biotin, because of its femtomolar dissociation constant with streptavidin or avidin. Robust methodologies have been developed for other the *in vivo* addition of a single biotin to recombinant protein or the *in vitro* enzymatic or chemical addition of biotin to a protein. Such modified proteins can be used in a variety of experiments, such as affinity selection of phage-displayed peptides or antibodies, pull-down of interacting proteins from cell lysates, or displaying proteins on arrays. We present three complementary approaches for biotinylating proteins *in vivo* in *Escherichia coli* or *in vitro* using chemical or enzymatical reactions all of which can be scaled up to tag large numbers of proteins in parallel.

Key words: Affinity selection; Biotin; Biotin ligase; BirA; *E. coli*; Phage-display; Protein labeling; Protein–protein interactions; Streptavidin; Streptavidin coated magnetic beads

13.1. Introduction

Recombinant proteins are typically overexpressed in heterologous hosts (i.e., bacteria, insect cells, mammalian cells, plants) attached to "fusion tags," which are short peptides, protein domains, or entire proteins and can be fused to proteins of interest, with the goal of imparting the biochemical properties of the fusion tag to the protein of interest. This is done at the genetic level by fusing the gene of interest to the gene encoding the fusion tag of interest, resulting in the expression of a single protein fused to the tag. In general, the type of fusion tag used is dictated by its application. Short peptide tags (e.g., six-histidine, epitopes, StrepTag, calmodulin-binding peptide) regularly serve to permit

Sharon A. Doyle (ed.), *Methods In Molecular Biology: High Throughput Protein Expression and Purification, vol. 498*

Book doi: 10.1007/978-1-59745-196-3

facile purification of the recombinant protein, permit detection of the fusion protein, or to direct interaction of the recombinant protein with other proteins or inert surfaces. Larger fusion partners, such as protein domains (e.g., chitin-binding domain) or proteins (e.g., cutinase, green fluorescent protein (GFP), glutathione-S-transferase (GST), intein, maltose binding protein (MBP), are commonly used to promote folding, solubility, purification, labeling, chemical ligation, or immobilization of the recombinant protein. If desired, the fusion tag can be detached from the protein of interest by cleavage of a linker region with a site-specific protease, which does not cleave the protein of interest.

One popular tag for detecting recombinant and native proteins is the small molecule biotin, which is a component of the vitamin B_2 complex. It binds with high affinity to the chicken egg white protein, avidin, and the fungal protein, streptavidin. (Note that a deglycosylated, recombinant form of avidin, with near-neutral isoelectric point (i.e., pI = 6.3) and minimizes nonspecific interactions, is commercially distributed as "neutravidin.") Avidin and streptavidin are tetrameric proteins that bind four molecules of D-biotin extremely tightly (i.e., dissociation constant of $\sim 10^{-15}$ M) *(1, 2)*. Proteins can be modified (i.e., biotinylated) with biotin very easily *in vitro* with chemical reagents, which are linked to biotin, under relatively mild conditions that do not affect protein stability, three-dimensional structure and function.

However, two drawbacks of *in vitro* chemically biotinylated target protein is that the number of biotin added is not uniform and the modification of certain lysine residues may lead to inactivation of the binding site(s). In *E. coli*, the biotin carboxy carrier protein (BCCP) is biotinylated by BirA *(3)*, a biotin ligase which covalently attaches a biotin to the amino group of the lysine residue present within the recognition sequence of BCCP *(4)*. A minimal biotinylation sequence has been found from screens of combinatorial peptide libraries; this 13 amino acid peptide *(5)*, along with a 15 amino acid long variant *(5)*, termed the AviTag™, have been identified as effective *in vivo* and *in vitro* substrates for the BirA enzyme. When target proteins are fused to the AviTag and co-expressed *in vivo* along with BirA, they can be biotinylated in bacteria *(6–8)*, yeast *(9–11)*, insect *(12)*, or mammalian cells *(13, 14)*. Furthermore, when recombinant proteins are fused to the AviTag and incubated *in vitro* with purified BirA, they can be biotinylated efficiently on the central lysine residue in the AviTag *(15, 16)*.

To generate biotinylated proteins in the laboratory, one has several options. If the protein is not available but is found to express well in *E. coli*, then it may be expedient to construct recombinant DNA in which its coding region is fused with the AviTag at its N- or C-terminus. (While there is a single biotin attached per protein molecule, the efficiency of biotinylation typically ranges between 50 and 80%.) Alternatively, if the protein is already available in sufficient amounts, then one can chemically biotinylate the protein prior to affinity selection experiments. (Typically, 100% of the molecules will be labeled, with one or more biotins on one of the lysine residues.) Finally, one can construct recombinant DNA with the AviTag fused at the protein's N- or C-terminus, express it and purify it from *E. coli*, and then biotinylate the protein *in vitro* with purified BirA. (Typically, 80–100% of the target protein is biotinylated *in vitro*.) All three approaches are described herein.

13.2. Materials

13.2.1. Reagents

1. Ampicillin (Sigma-Aldrich Chemical Company, St. Louis, MO)
2. Anti-Fab-HRP antibody (Jackson ImmunoResearch Laboratories, West Grove, PA)
3. Autoinducing medium (described in *(17)*; can be purchased from Novagen, Madison, WI)
4. D-biotin (Sigma-Aldrich)
5. BirA enzyme (Avidity, Boulder, CO)
6. BugBuster detergent (Novagen)
7. *E. coli* strain BL21 (DE3) (Novagen)
8. EZ-Link® Sulfo-NHS-LC-biotin (Pierce Chemical Company, Rockford, IL; MW = 557 daltons)
9. Immobilized metal affinity chromatography (IMAC) resin (Qiagen, Valencia, CA)
10. LB + ampicillin: Luria Broth (10 g yeast extract, 10 g peptone and 5 g NaCl in 1 L of water; autoclaved) plus 100 μg/mL ampicillin
11. LB + ampicillin + chloramphenicol: LB + ampicillin plus 12.5 μg/mL chloramphenicol
12. MagnaBind™ Streptavidin Beads (Pierce Chemical Company)
13. PBirA Cm^r biotinylation plasmid (Avidity)
14. PMCSG16 and pMCSG17 vectors (described in *(8)*; available upon request)
15. Phosphate Buffered Saline (PBS: 137 mM NaCl, 3 mM KCl, 8 mM Na_2HPO_4 1.5 mM KH_2PO_4)
16. *p*-Nitrophenyl phosphate (Sigma-Aldrich)
17. Qiagen Gel Extraction Kit (Qiagen, Valencia, CA)

18. Slide-A-Lyzer™ dialysis cassettes (Pierce)
19. *SmaI* Restriction Enzyme (New England Biolabs, Waverly, MA)
20. Streptavidin (Sigma-Aldrich Chemical Company, St. Louis, MO)
21. Streptavidin-alkaline phosphatase (Sigma-Aldrich)
22. T4 DNA polymerase (Promega Corporation, Madison, WI)
23. Zeba spin-columns (Pierce)

13.2.2. Construction of Recombinant Plasmids Encoded Protein Fusions to the AviTag

To generate biotinylated proteins for affinity selection experiments, one can transfer the open reading frame (ORF) of a protein of interest into plasmids that contain both the AviTag biotinylation sequence and a six-histidine tag, at either the N- or C- terminus of the ORF. The AviTag encodes the peptide sequence, GLNDIFEAQ<u>K</u>IEWHE, where the underlined lysine residue is biotinylated by BirA. The two bacterial expression vectors, pMCSG16 and pMCSG17, also contain a ligation independent cloning (LIC) site for efficient cloning of the ORFs, which allows high-throughput cloning, expression, *in vivo* biotinylation, purification, and streptavidin/avidin immobilization of target proteins for affinity selection of phage-displayed libraries *(8)*. Expression of the target-AviTag fusion protein is under the control of the T7 RNA polymerase promoter, which is under the transcriptional control of the LacZ promoter in *E. coli* strain BL21 (DE3). To produce enough BirA in the bacterial cells, they also contain the pBirA Cm^r plasmid *(15)*, which carries resistance to chloramphenicol and a compatible origin of replication. A protocol for generating the recombinants in pMCSG16 and pMCSG17 is briefly described below. (Formore extensive protocols found elsewhere *(8)*, see Dr. Frank Collart's publication in this book).

13.2.2.1. Preparation of the Vector DNA

1. Digest 5 μg of pMCSG16 and pMCSG17 DNA with the restriction enzyme, Ssp I, which linearizes the plasmid DNA in the center of the LIC site. Check an aliquot for complete digestion by agarose gel electrophoresis.
2. Purify the linearized DNA by passing it through a YM-100 column (Millipore) for enzyme removal and buffer exchange.
3. Treat the linearized DNA with T4 DNA polymerase (1 unit/μg of DNA) for 2 hours (h) in the appropriate buffer with 2.5 mM dGTP. Based on the nucleotide sequences adjacent to the *Ssp* I site in either vector, the proof-reading exonuclease activity of the enzyme will trim back 15 nucleotides from the 3′ termini of the linearized DNA.
4. Purify the treated vector DNA in a 0.5% agarose gel and recover DNA using the Qiagen Gel Extraction Kit. Quantify recovery of the DNA spectroscopically and store at –20°C.

13.2.2.2. Preparation of the Insert

1. Amplify the coding region of the target protein by polymerase chain reaction (PCR). Design the oligonucleotide primers with the sequence 5′-TACTTCCAATCCAATGGC-3′ followed by the nucleotides encoding the target protein. The anti-sense primers should begin with the sequence 5′-TCCACTTCCAATGGA-3′ followed by the reverse complement of the 3′ end without a stop codon (TAA). The same PCR products, without the TAA codon in the LIC overhang, can be cloned in either pMCSG16 or pMCSG17 vector, where the AviTag is N- or C-terminal to the cloning site, respectively. (Generally, both vectors are often used to construct recombinant DNA, just in case the N-terminal or C-terminal fusion interferes with protein folding or access to a binding site in the target protein.)
2. Resolve the PCR product on an agarose gel and purify the desired fragment with the Qiagen Gel Extraction Kit. Quantify recovery of the DNA fragment spectroscopically and store at –20°C.
3. Incubate the fragment with T4 DNA polymerase in the appropriate buffer with 2.5 mM dCTP for 2 h, at 37°C.
4. The T4 DNA polymerase-treated PCR product is purified and stored in 10 mM Tris-HCl (pH 8.0).

13.2.2.3. Construction of Recombinant Plasmids

1. Mix the T4 polymerase-treated PCR products with treated vector, heat to 70°C for 5 min and allow to cool to room temperature for 30 min. Annealed vector plus insert is then transformed into the *E. coli* strain BL21 (DE3), which contains the biotin ligase expression plasmid, pBirA Cm^r.
2. Select transformants on Petri plates containing LB + ampicillin + chloramphenicol. Grow overnight at 37°C.
3. Confirm recombinants by PCR and agarose gel electrophoresis. Under ideal conditions, the rate of recombinants is >90%.
4. Verify the construction of the recombinant plasmids by DNA sequencing. Store bacterial clones in LB + ampicillin + chloramphenicol with 20% glycerol, at –80°C.

13.3. Methods

13.3.1. In Vivo Enzymatic Biotinylation of Proteins

To biotinylate target proteins in *E. coli*, it is important to grow the expression plasmids in bacteria that contain the pBirA Cm^r biotinylation plasmid *(15)*. Without this plasmid, the levels of endogenous BirA enzyme are inadequate to get more than 5% of the overexpressed protein biotinylated. It is also important to add

D-biotin to the culture medium at the time of overexpression of the recombinant protein, to ensure that sufficient amounts of this molecule are available for post-translational modification of the target protein. A typical protocol for *in vivo* labeling of AviTagged proteins in bacteria **(Fig. 13.1)** consists of the following:

1. Inoculate a 2 mL culture of LB + ampicillin + chloramphenicol with bacteria from an individual colony. Shake overnight at 37°C.
2. Transfer 1 mL of the stationary phase culture to a flask containing 100 mL of autoinducing medium, 100 μg/mL ampicillin, 12.5 μg/mL chloramphenicol and 20 μM D-biotin. Shake overnight at 37°C.
3. Pellet the cells and discard the culture medium.
4. Lyse the cell pellet with BugBuster detergent, according to the manufacturer's instructions.
5. Recover the recombinant protein by immobilized metal affinity chromatography (IMAC).
6. If desired, dialyze away the free imidazole and store the protein in PBS.
7. Determine the percent of the protein sample that is biotinylated **(see Section 3.5)**.

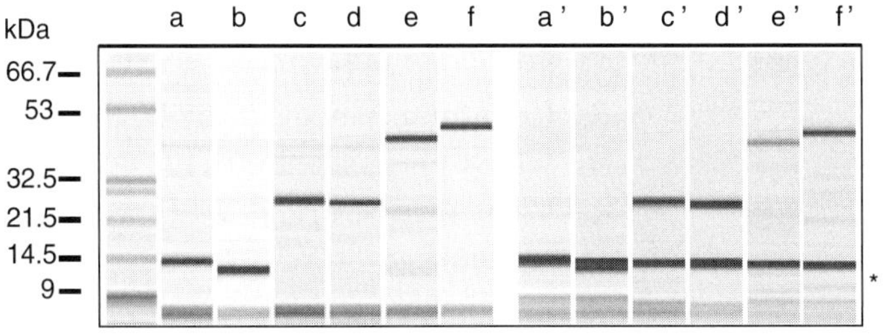

Fig. 13.1. Fractionation of three different AviTagged and His_6-tagged proteins. Three different recombinant proteins in pMCSG16 and pMCSG17 vectors, which contain a His_6-tag at the N-terminus and AviTag at either the N-terminus or C-terminus, respectively, were expressed in *E. coli* with the pBirA Cm^r plasmid present. Total bacterial cell lysates were mixed with streptavidin-coated magnetic beads, tumbled, washed and bound material resolved by capillary electrophoresis with an Agilent Bioanalyzer 2100. Lanes a–b, c–d, and e–f correspond to human SrcSH3 domain, human superoxide dismutase (SOD) and the *Bacillus subtilis* APC1259 protein with AviTags at their N- and C-termini, respectively, which were purified by IMAC. Lanes a′–f′ contain the same protein samples, excepted purified on streptavidin-coated magnetic beads. The asterisk refers to streptavidin monomer released from the streptavidin coated magnetic beads when boiled. This figure is modified from another publication ***(8)***.

13.3.2. In Vitro Chemical Biotinylation of Proteins

1. Dissolve the protein sample in PBS or some amine-free buffer. If the sample is in a Tris or amine-containing buffer such as imidazole, exchange it for an amine-free buffer by dialysis or with the use of a spin column.
2. Calculate the amount of biotin solution to add to the protein sample. The extent of biotinylation can be controlled by the molar ratio of biotin to protein. As a general rule, for protein samples at 2-10 mg/mL, ≥ 12-fold excess of biotin should be added; whereas for samples at ≤ 2 mg/mL, ≥20-fold excess of biotin should be added. It is important to know the volume (mL), concentration (mg/mL), and size (Daltons or mg) of the protein sample.

   First, determine the millimoles of biotin reagent to add to the protein.

   $$\text{mmol Biotin} = \text{mL of protein} \times \frac{\text{mg protein}}{\text{mL protein}} \times \frac{\text{mmol protein}}{\text{mg protein}} \times \frac{12\text{–}20\,\text{mmol Biotin}}{\text{mmol protein}}$$

   Then, calculate the volume of 10 mM biotin solution to add to the reaction.

   $$\mu\text{L Biotin} = \text{mmol Biotin} \times \frac{1{,}000{,}000\,\mu\text{L}}{\text{L}} \times \frac{\text{L}}{10\ \text{mmol}}$$
3. Before use, remove biotin reagent from freezer and bring to room temperature before opening to prevent excess moisture from entering the vial. Immediately before use, prepare a 10 mM solution of the biotin reagent in water.
4. According to the above calculations, add the appropriate volume of biotin solution to the protein sample.
5. Incubate on ice for 2 h or at room temperature for 30 minutes.
6. Stop the biotinylation reaction with the addition of 10 mM glycine (pH 7.2).
7. Eliminate free biotin **(see Protocol 3.4)**.
8. Determine the protein concentration by optical absorbance or through SDS-PAGE **(see Protocol 3.5)**.

13.3.3. In Vitro Enzymatic Biotinylation of Proteins

1. For subsequent BirA biotinylation exchange purified protein to 10 mM Tris, pH 8.0. Glycerol, NaCl. As ammonium sulfate inhibits BirA activity, this chemical should be eliminated or minimized through dialysis.
2. Incubate 100 µg of protein in 50 mM bicine, pH 8.3, 10 mM ATP, 10 mM $Mg(OAc)_2$, 50 mM biotin with 15 units of BirA for 1 h at 30°C. At protein concentrations higher than 40 mM, additional biotin should be supplemented to keep the molar ratio of protein to biotin at ~1:1.

3. Eliminate free biotin (**see Protocol 3.4**).
4. Determine the protein concentration by optical absorbance or through SDS-PAGE.
5. Determine the percent biotinylation of the protein sample (**see Section 3.5**; **Fig. 13.2**).

13.3.4. Elimination of Free Biotin

13.3.4.1. By Dialysis

Pierce has developed a Slide-A-Lyzer™ dialysis cassette that can conveniently remove low molecular weight contaminants and salts. Determine the molecular weight and available volume of your protein to choose an appropriate dialysis cassette size and membrane capacity for the molecular weight cut-off (MWCO). Avoid choosing a membrane with a MWCO that is too close to the size of your protein; this can cause some loss of sample. To protect the physical integrity of the dialysis membrane, use precaution when handling the dialysis cassette and touch only the plastic frame.

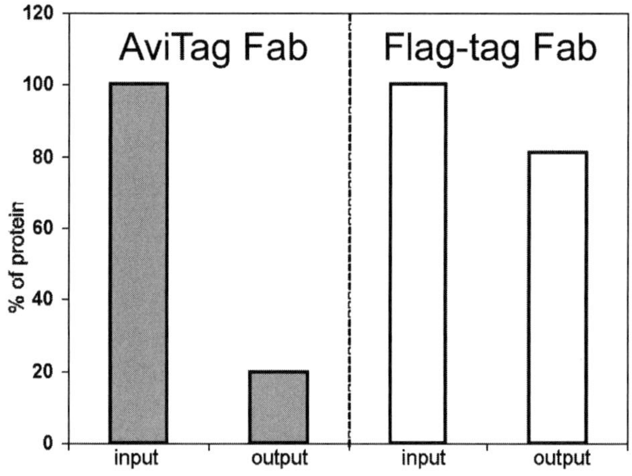

Fig. 13.2. BirA can efficiently attach biotin *in vitro* to a protein carrying an AviTag. In one form of the Herceptin Fab, which binds to the ectodomain of the Erb-B2 membrane protein ***(18, 19)***, the AviTag is at the C-terminus of the light chain and in another form, an epitope-tag, in which the Flag Tag ***(20)*** replaces the AviTag. Both forms of Fab were overexpressed in *E. coli*, purified by Protein G chromatography, incubated with BirA *in vitro*, mixed with magnetic streptavidin-coated beads (input), tumbled for 1 h at 4°C, centrifuged and unbound fractions (output) were collected. (Under these conditions, the AviTagged, but not the Flag-tagged, form of Fab should become biotinylated.) The protein input and output samples were then incubated in microtiter plate wells coated with Erb-B2 (gracious gift of Dr. Daniel Leahy, Johns Hopkins University, Baltimore, MD) for 2 h and immune complexes in the wells were detected with anti-Fab-HRP antibody. The Y-axis refers to the percent of protein bound to antigen in the output relative to the input (100%) fractions, as tested by ELISA. In the histograms, ~80% of the Fab–AviTag fusion could be biotinylated *in vitro* with little or no negative effect on binding ***(21, 22)*** and we interpret the small amount of binding of the non-biotinylated form of the protein (i.e., Fab-Flag tag) to the streptavidin coated magnetic beads to be due to non-specific binding.

1. As a quality assurance that there is no leakage in the cassette, inject sterile distilled water into the cassette to check for leaks and remove after visual inspection prior to the addition of protein sample.
2. If the dialysis cassette size requires hydration or a low volume of sample is used, remove the cassette from its pouch, slip it into the accompanying buoy, and immerse in dialysis buffer for 30 s. Remove the cassette from the buffer and gently tap the edge on a paper towel to remove excess liquid.
3. Attach an 18-gauge, 1 in. beveled needle (or 21-gauge, 1-in. beveled needle) to the Luer-Lock of the syringe by screwing it into place.
4. Remove the protective sheath from the needle and draw a small volume of air into the syringe to void the syringe's dead volume. Immerse the needle into your sample and then slowly draw back on the syringe piston.
5. Remove the dialysis cassette from the buoy and penetrate the gasket with the needle through one of the syringe ports located at the top corners of the cassette. The needle should penetrate the gasket to a minimal extent to avoid puncturing the membrane. Inject the sample slowly to avoid foaming.
6. Before removing the needle from the cassette, draw up on the syringe piston to remove air from the cassette cavity. This ensures the sample solution contacts the greatest amount of surface area. At the same time avoid contacting the needle to the membrane. Mark the corner of the cassette with a permanent marker to identify the port injected.
7. Reattach the buoy and float in a beaker with 200–500 times the volume of your sample of dialysis buffer. Dialyze at room temperature for 2 h, change the dialysis buffer and dialyze for another 2 h. Finally, change the dialysis buffer and dialyze overnight at 4°C. Keeping the solution in constant stirring can speed up the dialysis process.
8. To remove the sample after dialysis, use a new syringe and needle and fill the syringe with air at least equal to the volume of your sample size. Penetrate the gasket at an unused syringe port. Discharge the air into the cavity to separate the membrane to prevent from piercing the membrane. Insert only the tip of the needle through the syringe port.
9. Rotate the cassette so that the syringe and the sample are on the bottom and slowly draw back on the syringe piston to capture the dialyzed sample. Remove the needle and discard the dialysis cassette. Expel the sample into a new tube.

13.3.4.2. With a Microcon Device

Determine the molecular weight of your protein to choose an appropriate filter membrane capacity for the nominal molecular weight limit (NMWL). Avoid choosing a membrane with a NMWL that is too close to the size of your protein; this can cause some loss of sample. It should be noted that generally some of the protein sample is lost due to sticking to the membrane.

1. Assemble the microcon centrifugal filter device by inserting the sample reservoir into the microfuge tube.
2. Pipette sample solution into sample reservoir. Do not add more than 500 μL into the sample reservoir. Avoid touching the membrane at the bottom of the sample reservoir with pipette tip. Close the tube with attached cap.
3. Place assembled filter device with sample into a compatible centrifuge aligning the cap strap toward the center of the rotor.
4. Centrifuge using appropriate spin times according to the NMWL at 4°C. Repeat two times adding more solvent to 500 μL and discarding the flow through before centrifugation.
5. Remove assembled filter device from centrifuge and separate the sample reservoir from the tube. Carefully invert the sample reservoir into a new tube.
6. Centrifuge for 3 min at 1,000 × *g* to collect concentrated sample. Notice that the attached cap will not close when the sample reservoir is inverted. To prevent the cap from breaking off, align the cap to the center of the rotor so that it is flush against the rotor.

13.3.5. Verification of Protein Biotinylation

Biotinylation of proteins can be monitored by three different means:

1. An enzyme linked immunosorbent assay (ELISA) format is a quick method for demonstrating that the target protein is biotinylated. First, biotinylated proteins are immobilized onto polystyrene microtiter plates, and after washing away the free protein, non-specific binding of proteins is blocked with excess BSA, and the wells incubated with streptavidin alkaline-phosphatase (1 μg/mL in PBST, 1 h incubation). The amount of streptavidin-enzyme conjugate retained in the wells is detected using p-nitrophenyl phosphate, with the yellow color (measured at 405 nm wavelength) proportional to the amount of biotinylated protein adsorbed to the microtiter plate well surface.
2. The number of biotins per protein molecule can be determined using a kit (Pierce Chemical Co., # 28005). In the kit, the 4′-hydroxyazobenzene-2-carboxylic acid (HABA) dye binds to avidin to produce a yellow-orange color, which absorbs at 500 nm wavelength. Biotin conjugated to a protein will displace the dye and cause the absorbance to decrease; a standard curve can be established using the free

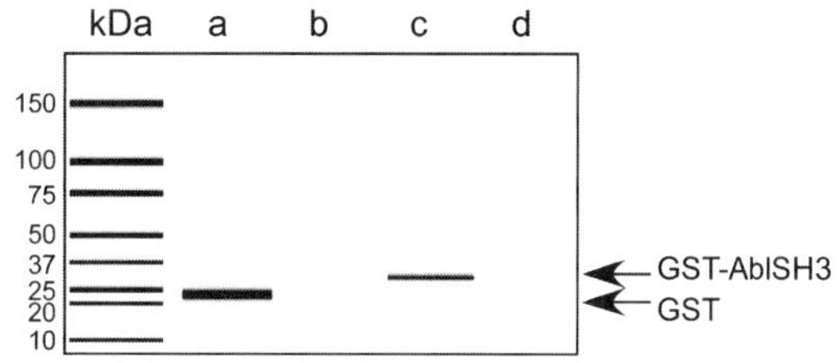

Fig. 13.3. Chemical biotinylation of two proteins *in vitro* and confirmation that they bind to streptavidin coated magnetic beads. Purified Glutathione-S-transferase (GST) and GST-Abl SH3 domain fusion proteins were chemically biotinylated with EZ-Link Sulfo-NHS-LC-Biotin (Sulfosuccinimidyl-6-(biotinamido) hexanoate). After quenching the reagent with excess glycine and spinning the sample in a microconcentrator to remove free biotin, the proteins were separately mixed with streptavidin coated magnetic beads. Input and bound material was then resolved by capillary electrophoresis using a biorad Experion. The left hand lane contains molecular weight standards listed in kilodaltons (kDa). Lanes a–d resolves (**a**) biotinylated GST, (**b**) biotinylated GST that did not bind to the streptavidin coated magnetic beads (i.e., after it was tumbled with streptavidin-coated beads), (**c**) GST-AblSH3 domain fusion protein, and (**d**) GST-AblSH3 domain fusion protein that did not bind to streptavidin-coated magnetic beads.

biotin, permitting the number of moles of biotin incorporated (after biotinylating a protein) to be estimated. A similar kit, but more sensitive (based on the displacement of a ligand tagged with a quencher dye from the biotin-binding sites of Biotective™ Green reagent), is available from Invitrogen (# F30751).

3. The presence of non-biotinylated proteins in a sample can be estimated by following the binding of the sample to streptavidin coated magnetic beads by sodium dodecyl sulfate-polyacrylamide gel electrophoresis (SDS-PAGE). A sample can be incubated with streptavidin coated magnetic beads for 30 min and the amount of non-bound protein can be compared to input (**Fig. 13.3**). In a successful biotinylation experiment, 100% of the biotinylated protein should be removed from sample, assuming that the streptavidin coated magnetic beads are in excess to the amount of biotin attached to the protein, compared to a much lower value for non-biotinylated protein, which is non-specifically stuck to the beads.

13.4. Notes

1. From our experience, we find we can achieve 50–80% biotinylation of our *in vivo* expressed AviTagged proteins.

Generally, this level of biotinylation is sufficient for our needs, as one can conveniently capture the biotinylated target protein from lysed *E. coli* directly on streptavidin- or avidin-coated microtiter plate wells or streptavidin coated magnetic beads, without prior purification *(8)*. If 100% biotinylation of a particular protein is required, we recommend either enzymatic biotinylation of an AviTagged protein *in vitro* or chemical biotinylation, although the number of biotins attached to the protein will differ. It should be noted that BirA can be purchased (Avidity) or purified from bacteria carrying an overexpression plasmid *(23)*.

2. For labeling purposes, it is possible to incorporate different forms of biotin or biotin containing labeling agents. For example, the biotin analog, desthiobiotin, which retains the uredio ring of biotin, but lacks the 5-member sulfur-containing thiophene ring, can be a substrate for efficient *in vitro* desthiobiotinylation by BirA *(16)*. The use of desthiobiotin offers a gentle method for affinity purifying desthiobiotinylated proteins, as they can be eluted off avidin or streptavidin matrices by competition with biotin *(24)*. There are also several analogs of biotinylation reagents available from Pierce Chemical Company available for use in chemical biotinylation, such as EZ-Link NHS-Chromogenic Biotin, which contains a chromophore that absorbs strongly at 354 nm and allows accurate measurement of the number of biotins attached to a protein sample, and EZ-Link NHS-SS-Biotin, which allows release of the biotin tag with exposure to reducing agents. Finally, it should be noted that there are a variety of other strategies for adding biotin to a protein (reviewed in *(25)*), including translation with biotinylated puromycin *(26, 27)*, intein-mediated site-specific biotinylation *(28)* and use of novel peptide tags *(29)* or protein domains *(30)*, which are substrates for post-translational modification with biotin-derivatized compounds.

Acknowledgments

The authors acknowledge intellectual contributions by Mr. Michael Scholle and Dr. Frank Collart, and financial support from the National Institutes of Health (1R01 GM079096, P01 GM075913, 1U54 CA119343).

References

1. Wilchek M., Bayer E.A., and Livnah O. (2006). Essentials of biorecognition: the (strept)avidin-biotin system as a model for protein–protein and protein–ligand interaction. Immunol. Lett. *103*, 27–32
2. Laitinen O.H., Nordlund H.R., Hytonen V.P., and Kulomaa M.S. (2007). Brave new (strept)avidins in biotechnology. Trends Biotechnol. *25*, 269–277
3. Chapman-Smith A., and Cronan J.E.J. (1999). Molecular biology of biotin attachment to proteins. J. Nutr. *129*, 477S–484S
4. Smith P.A., Tripp B.C., DiBlasio-Smith E.A., Lu Z., LaVallie E.R., and McCoy J.M. (1998). A plasmid expression system for quantitative in vivo biotinylation of thioredoxin fusion proteins in *Escherichia coli*. Nucleic Acids Res. *26*, 1414–1420
5. Beckett D., Kovaleva E., and Schatz P.J. (1999). A minimal peptide substrate in biotin holoenzyme synthetase-catalyzed biotinylation. Protein Sci. *8*, 921–929
6. Ashraf S.S., Benson R.E., Payne E.S., Halbleib C.M., and Gron H. (2004). A novel multi-affinity tag system to produce high levels of soluble and biotinylated proteins in *Escherichia coli*. Protein Expr. Purif. *33*, 238–245
7. Penalva L.O., and Keene J.D. (2004). Biotinylated tags for recovery and characterization of ribonucleoprotein complexes. BioTechniques *37*, 604, 606, 608–10
8. Scholle M.D., Collart F.R., and Kay B.K. (2004). In vivo biotinylated proteins as targets for phage-display selection experiments. Protein Expr. Purif. *37*, 243–252
9. Athavankar S., and Peterson B.R. (2003). Control of gene expression with small molecules: biotin-mediated acylation of targeted lysine residues in recombinant yeast. Chem. Biol. *10*, 1245–1253
10. Parthasarathy R., Bajaj J., and Boder E.T. (2005). An immobilized biotin ligase: surface display of *Escherichia coli* BirA on *Saccharomyces cerevisiae*. Biotechnol. Prog. *21*, 1627–1631
11. Scholler N., Garvik B., Quarles T., Jiang S., and Urban N. (2006). Method for generation of in vivo biotinylated recombinant antibodies by yeast mating. J. Immunol. Methods. *317*, 132–143
12. Duffy S., Tsao K.L., and Waugh D.S. (1998). Site-specific, enzymatic biotinylation of recombinant proteins in *Spodoptera frugiperda* cells using biotin acceptor peptides. Anal. Biochem. *262*, 122–128
13. de Boer E., Rodriguez P., Bonte E., Krijgsveld J., Katsantoni E., Heck A., Grosveld F., and Strouboulis J. (2003). Efficient biotinylation and single-step purification of tagged transcription factors in mammalian cells and transgenic mice. Proc. Natl. Acad. Sci. U S A *100*, 7480–7485
14. Viens A., Mechold U., Lehrmann H., Harel-Bellan A., and Ogryzko V. (2004). Use of protein biotinylation in vivo for chromatin immunoprecipitation. Anal. Biochem. *325*, 68–76
15. Cull M.G., and Schatz P.J. (2000). Biotinylation of proteins in vivo and in vitro using small peptide tags. Methods Enzymol. *326*, 430–440
16. Wu S.C., and Wong S.L. (2004). Development of an enzymatic method for site-specific incorporation of desthiobiotin to recombinant proteins in vitro. Anal. Biochem. *331(2)*, 340–348
17. Studier F.W. (2005). Protein production by auto-induction in high density shaking cultures. Protein Expr. Purif. *41*, 207–234
18. Cho H.S., Mason K., Ramyar K.X., Stanley A.M., Gabelli S.B., Denney D.W., Jr., and Leahy D.J. (2003). Structure of the extracellular region of HER2 alone and in complex with the Herceptin Fab. Nature *421*, 756–760
19. Vajdos F.F., Adams C.W., Breece T.N., Presta L.G., de Vos A.M., and Sidhu S.S. (2002). Comprehensive functional maps of the antigen-binding site of an anti-ErbB2 antibody obtained with shotgun scanning mutagenesis. J. Mol. Biol. *320*, 415–428
20. Einhauer A., and Jungbauer A. (2001). The FLAG peptide, a versatile fusion tag for the purification of recombinant proteins. J. Biochem. Biophys. Methods *49*, 455–465
21. Saviranta P., Haavisto T., Rappu P., Karp M., and Lovgren T. (1998). In vitro enzymatic biotinylation of recombinant fab fragments through a peptide acceptor tail. Bioconjug. Chem. *9*, 725–735
22. Sibler A.P., Kempf E., Glacet A., Orfanoudakis G., Bourel D., and Weiss E. (1999). In vivo biotinylated recombinant antibodies: high efficiency of labelling and application to the cloning of active anti-human IgG1 Fab fragments. J. Immunol. Methods *224*, 129–140
23. Howarth M., and Ting A.Y. (2008). Imaging proteins in live mammalian cells with biotin ligase and monovalent streptavidin. Nat. Protoc. *3*, 534–545

24. Hirsch J.D., Eslamizar L., Filanoski B.J., Malekzadeh N., Haugland R.P., Beechem J.M., and Haugland R.P. (2002). Easily reversible desthiobiotin binding to streptavidin, avidin, and other biotin-binding proteins: uses for protein labeling, detection, and isolation. Anal. Biochem. *308*, 343–357
25. Chattopadhaya S., Tan L.P., and Yao S.Q. (2006). Strategies for site-specific protein biotinylation using in vitro, in vivo and cell-free systems: toward functional protein arrays. Nat. Protoc. *1*, 2386–2398
26. Starck S.R., Green H.M., Alberola-Ila J., and Roberts R.W. (2004). A general approach to detect protein expression in vivo using fluorescent puromycin conjugates. Chem Biol. *11*, 999–1008
27. Agafonov D.E., Rabe K.S., Grote M., Voertler C.S., and Sprinzl M. (2006). C-terminal modifications of a protein by UAG-encoded incorporation of puromycin during in vitro protein synthesis in the absence of release factor 1. Chembiochem. *7*, 330–336
28. Tan L.P., Lue R.Y., Chen G.Y., and Yao S.Q. (2004). Improving the intein-mediated, site-specific protein biotinylation strategies both in vitro and in vivo. Bioorg. Med. Chem. Lett. *14*, 6067–6070
29. Yin J., Lin A.J., Golan D.E., and Walsh C.T. (2006). Site-specific protein labeling by Sfp phosphopantetheinyl transferase. Nat. Protoc. *1*, 280–285
30. Los G.V., and Wood K. (2007). The HaloTag: a novel technology for cell imaging and protein analysis. Methods Mol. Biol. *356*, 195–208

Chapter 14

High-Throughput Insect Cell Protein Expression Applications

Mirjam Buchs, Ernie Kim, Yann Pouliquen, Michael Sachs, Sabine Geisse, Marion Mahnke, and Ian Hunt

Summary

The Baculovirus Expression Vector System (BEVS) is one of the most efficient systems for production of recombinant proteins and consequently its application is wide-spread in industry as well as in academia. Since the early 1970s, when the first stable insect cell lines were established and the infectivity of baculovirus in an *in vitro* culture system was demonstrated *(1, 2)*, virtually thousands of reports have been published on the successful expression of proteins using this system as well as on method improvement. However, despite its popularity the system is labor intensive and time consuming. Moreover, adaptation of the system to multi-parallel (high-throughput) expression is much more difficult to achieve than with *E. coli* due to its far more complex nature. However, recent years have seen the development of strategies that have greatly enhanced the stream-lining and speed of baculovirus protein expression for increased throughput via use of automation and miniaturization. This chapter therefore tries to collate these developments in a series of protocols (which are modifications to standard procedure plus several new approaches) that will allow the user to expedite the speed and throughput of baculovirus-mediated protein expression and facilitate true multi-parallel, high-throughput protein expression profiling in insect cells. In addition we also provide a series of optimized protocols for small and large-scale transient insect cell expression that allow for both the rapid analysis of multiple constructs and the concomitant scale-up of those selected for on-going analysis. Since this approach is independent of viral propagation, the timelines for this approach are markedly shorter and offer a significant advantage over standard baculovirus expression approach strategies in the context of HT applications.

Key words: Baculovirus; Deep-well block protein expression; Virus titration; High-throughput

14.1. Introduction

Baculovirus mediated insect cell expression has become one of the most popular vehicles for the production of large quantities of recombinant protein for structural and functional studies of therapeutically relevant bio-molecules. To date, hundreds

Sharon A. Doyle (ed.), *Methods in Molecular Biology: High Throughput Protein Expression and Purification, vol. 498*

Book doi: 10.1007/978-1-59745-196-3

of proteins from a variety of organisms have been expressed using the system, rendering it a highly valuable tool in the protein production laboratory. The adaptation of the system to multi-parallel expression is much more difficult to achieve than with *E. coli* due to its far more complex nature *(3, 4)*. However, recent years have seen the development of strategies that have greatly enhanced (at least in part) the stream-lining and speed of baculovirus mediated insect cell expression system for increased throughput via use of automation and miniaturization. We therefore describe a series of optimized protocols which have enabled us to improve the speed and throughput of baculovirus expression in the context of supporting drug discovery efforts at Novartis. It is important to stress, that whilst the protocols have made significant (positive) impact on our daily work, a truly fully automated strategy is still some way off. Indeed, many bottlenecks and problems still remain and these will be discussed in some detail also.

As mentioned above one of the drawbacks of baculovirus expression is the generation of viral particles sufficient in quantity and infectivity prior to protein expression. Additionally, high protein expression rates are only achievable if the infectious process is optimized with respect to multiplicity of infection, cell densities and duration of infection. To take advantage of insect cell lines as hosts for recombinant protein production while circumventing the tedious stages of virus generation and characterization, attempts have been made to apply transient transfection technologies to insect cells. This approach potentially offers a highly attractive vehicle for high-throughput eukaryotic expression screening. We therefore also provide a series of optimized protocols for small and large-scale transient insect cell expression that allow for both the rapid analysis of multiple constructs and the concomitant scale-up of those selected for on-going analysis. Since this approach is independent of viral propagation, the timelines for this approach are markedly shorter and offer a significant advantage over a standard baculovirus expression approach with regards to the analysis of large numbers of constructs and/or conditions.

14.2. Materials

14.2.1. General Reagents

1. Sf21, Sf9 and Hi5™ (BTI-TN-5B1-4) cell lines (Invitrogen, Carlsbad, CA, USA).
2. Ex-Cell 420 serum-free (SFM) cell culture medium (SAFC-Biosciences, Milwaukee, WI, USA). For cultivation of Sf cell lines addition of 1% (v/v) fetal calf serum (SAFC Biosciences) and 1% (v/v) L-glutamine (200 mM stock solution, Invitrogen) to the medium is recommended, as these additives improve cell growth.

3. Standard tissue culture plastic ware: 6-well plates, 35 mm Petri-dishes, tissue culture flasks (e.g. Nalge-Nunc, Rochester, NJ, USA; BD Biosciences, Franklin Lakes, NJ, USA; Greiner bio-one, Monroe, NC, USA), disposable Erlenmeyer shake flasks with vented cap (Corning, Corning, NY, USA, e.g. Cat. no. 431143 for 125 mL flasks; Cat. nos. 431147, 431255, 431253 for large-scale applications).
4. Sterile cryovials, centrifuge tubes (polystyrene or polypropylene) and 15 mL and 50 mL centrifugation tubes (e.g. BD Biosciences & Corning).
5. DMSO (Fluka), Fetal Calf Serum (FCS), Glutamine solution (200 mM, Invitrogen).
6. Cell culture incubators: Standard shaker incubators (e.g. from Infors AG; Bottmingen, CH; Kühner AG, Birsfelden, CH; New Brunswick, Edison, NJ, USA; Snijders Scientific, Tilburg, NL; GFL, Burgwedel, FRG).
7. Cell culture microscope.
8. Seed cultures: Maintained in disposable shake flasks on an orbital shaker platform at 28°C. The seed cultures are usually passaged twice a week by dilution to a density of $1–0.5 \times 10^6$ cells/mL – (*see* **Note 1**).
9. Cell density determination: manually using a hemocytometer 'Neubauer Improved' counting chamber, or automatically using Cedex™ cell counting system (Innovatis, Bielefeld, FRG) or Vi-Cell™ instrument (Beckman Coulter, Fullerton, CA). Staining solution: Trypan blue (Sigma-Aldrich, St. Louis MI, USA), 0.4 g dissolved in 100 mL of phosphate buffered saline and sterile filtered through a 0.22 µm filter to remove debris; alternatively: 0.4% ready-to-use solution (Sigma-Aldrich, Cat. no. T8154 or Bioconcept, Cat. no. 5-72F00-H).

14.2.2. High-Throughput Bacmid Propagation

1. MAX Efficiency DH10Bac chemically competent cells (Invitrogen).
2. S.O.C. medium (Invitrogen).
3. Repeater pipette with 25 and 50 mL tips.
4. Vented QTray™ with cover and 48-well divider (Genetix, New Milton, Hants. UK, Cat. no. X6029).
5. ColiRoller Plating Beads (Novagen/EMD Biosciences, Madison, WI, USA, Cat. no 71013).
6. Blue-gal (Invitrogen).
7. 1,250 µL Matrix 6 channel pipette and 1,250 µL Matrix pipette tips.
8. PerfectPrep™ BAC 96 kit (Eppendorf, Hamburg FRG).

9. 96-well E-Gel pre-cast agarose Electrophoresis system (Invitrogen).
10. 96-well V-bottom microplate (Innovative Instruments, CT, USA).
11. Optional: Freedom EVO® workstation (Tecan, Durham, NC, USA).
12. Optional: Blue hard-shell 96-well plate (Bio-Rad Hercules, CA, USA, Cat. no. HSP-3831).

14.2.3. HT-Suspension Based Insect Cell Transfection

1. Transfection reagent: Cellfectin™ (Invitrogen), Insect GeneJuice™ (Novagen/EMD Biosciences) & FuGene HD (Roche Diagnostics, IN, USA) transfection reagents all work equally well.
2. 24 deep-well round bottom blocks (Qiagen, Germantown, MD USA).
3. AirPore™ cover sheets (Qiagen).
4. Option: Eppendorf epMotion 5070 Workstation.

14.2.4. Methods of Recombinant Viral Titer Determination

1. Pluronic F68, 10% (Invitrogen).
2. Carboxynapthofluorescein diacetate (Sigma-Aldrich).
3. AlamarBlue™ (AbD Serotec, Duesseldorf, FRG).
4. Option: Agilent 2100 Bioanalyzer (Agilent Life Science Technologies, Santa Clara, CA, USA).
5. Option: CytoFluor II (Applied Biosystems, Foster City, CA, USA).

14.2.5. HT Miniaturized Deep Well Insect Cell Protein Expression

1. Incubators suitable for HT insect cell expression: Growth of insect cell cultures in deep well blocks can be achieved in either conventional shakers (Multitron II, Infor & Innova 4430, New Brunswick) or those designed specifically for that purpose following careful optimization of agitation speeds (Micro Expression Shaker, GlasCol, Terre Haute, IN & HiGro, GeneMachines, San Carlos, CA).
2. 24 deep-well, round bottom blocks (Qiagen).
3. AirPore™ cover sheets (Qiagen).
4. NP-40 (Sigma-Aldrich).
5. Complete Protease Inhibitor Cocktail Tablets (Roche Diagnostics).
6. XCell SureLock MiniCell gel electrophoresis unit (Invitrogen).
7. XCell Western Blot Module (Invitrogen).
8. NuPAGE LDS Sample Buffer (Invitrogen).
9. NuPAGE MES/MOPS SDS Running Buffer (Invitrogen).
10. NuPAGE Western Blot Transfer Buffer (Invitrogen).

11. NuPAGE Novex 4–12% Bis-Tris Precast Gels (Invitrogen).
12. Option: HT E-PAGE 96 system (Invitrogen).

14.2.6. Transient Insect Cell Expression

1. Plasmid preparation kits: Nucleospin® (Macherey & Nagel, Dueren, FRG) or Compact Prep Plasmid Midi (Qiagen) kits.
2. Expression vectors: A series of expression plasmids harboring different tags (His_6 and S-tag at N- or C-terminus and GST-fusions) are available from Novagen/EMD Biosciences. We used the expression plasmid pIEX-4 which features the S-tag and his_6-tag in tandem at the C-terminus. For alternative expression plasmids and kits *see* **Note 2**.
3. Insect cell line and culture medium: Essentially the same cell lines, culture media and conditions are applied as used for baculovirus-mediated expression approaches (*see* **Section 14.2.1**).
4. Transfection reagents: Standard transfection reagents (*see* **Section 14.2.3**) and in addition Polyethylenimine (Polysciences Inc., Warrington, PA). Preparation of a PEI stock solution at 1 mg/mL concentration (a) Dissolve the appropriate amount of PEI in distilled waster. (b) Adjust pH to 7.0 by addition of 1M HCl. (c) Filter-sterilize (0.22 μm filter), prepare aliquots and store frozen at –80°C.

14.3. Methods

The methods described below outline (1) Baculovirus mediated insect cell systems suitable for HT protein expression applications, (2) HT-bacmid propagation, (3) HT-suspension based insect cell transfection, (4) Methods of recombinant viral titer determination, (5) HT-miniaturized deep-well block insect cell expression, and (6) Transient insect cell expression.

14.3.1. Baculovirus Mediated Insect Cell Expression System for High-Throughput Applications

Several baculovirus expression systems are commercially available to drive recombinant protein expression, some of which are more amenable to high-throughput protein expression than others (**Table 14.1**). Undoubtedly, the most popular vehicle is the Bac-to-Bac™ system (Invitrogen), although it also has its problems with regards to its suitability for HT expression. HT propagation of bacmid is laborious and costly – even on a robot. Moreover, recent reports suggest that recombinant baculovirus propagated via bacmid transposition appear to be inherently unstable, with the spontaneous deletion of the transposed bacmid sequence observed with increasing passage number of the culture *(5)*. Regardless, it still remains the dominant tool in the propagation

Table 14.1
Overview on commercially available baculovirus expression systems

Baculovirus expression kits and vendors	Compatible transfer vectors	Methodology for cloning foreign gene into transfer vector	Transfer of foreign gene into Baculovirus genome	Selection/ Recombination efficiency
BacPAK™ (Clontech)	Based on homologous recombination at polyhedrin locus	Ligase dependent	Homologous recombination in insect cells	≥90%
Bac-to-Bac™ (Invitrogen)	Based on site-specific transposition	Ligase-dependent, Gateway™ adapted	Site-specific transposition in bacterial cells	Selection of recombinants by blue-white selection on agar plates
BaculoDirect™ (Invitrogen)	Based on site-specific recombination	Gateway™ adapted	Site-specific recombination in Eppendorf tube	Antibiotic selection of transfectants in insect cells
flashBac™ (OET/ NextGen Sciences)	Based on homologous recombination at polyhedrin locus	Ligase dependent	Homologous recombination in insect cells	100%
BacVector™ 1000, 2000, 3000 (EMD/ Novagen)	Based on homologous recombination at polyhedrin locus	Ligase dependent	Homologous recombination in insect cells	≥ 95%
BaculoGold™ (Cloentech)	Based on homologous recombination at polyhedrin locus	Ligase dependent	Homologous recombination in insect cells	≥ 95%
DiamondBac™ (Sigma-Aldrich)	Based on homologous recombination at polyhedrin locus	Ligase dependent	Homologous recombination in insect cells	≥ 95%

of recombinant baculovirus for the majority of scientists and as such many of the developments to expedite increased throughput have been focused on the Bac-to-Bac™ system.

Other approaches that may be more amenable to HT baculovirus expression include BaculoDirect™ (Invitrogen) and FlashBAC™

(Oxford Expression Technologies, Oxford, UK) systems. Their use is still not widespread, although FlashBAC™ in particular (which is also marketed by Novagen/EMD Biosciences as BacMagic™) has gained wider acceptance in recent years.

14.3.2. High-Throughput Bacmid Propagation

The purification of recombinant bacmid is specific to the Bac-to-Bac™ driven baculovirus protein expression system and is traditionally achieved via alkaline lysis (*see* Bac-to-Bac Baculovirus Expression System Instruction Manual; http://invitrogen.com). Other systems such as FlashBAC™ have no such step and are therefore at an immediate advantage for high-throughput baculovirus applications.

The standard bacmid purification protocol utilizes a variety of centrifugation and liquid handling steps and whilst suitable for 1–6 constructs, becomes very labor intensive and time consuming when operating at numbers in excess of 24 constructs. Whilst some commercial kits can provide more convenient approaches for the propagation of a few constructs (S.N.A.P. Midi Prep Kit, Invitrogen), they are not particularly suitable for preparation of large numbers of bacmid DNA. Here we describe a protocol using a PerfectPrep™ BAC 96 kit (Eppendorf/Brinkmann) adapted for use on a Tecan robot that is suitable for HT bacmid propagation.

14.3.2.1. High-Throughput Transposition

1. Prepare 320 mL of molten LB agar. When agar has cooled down to approximately 55°C, add 50 μg/mL kanamycin, 7 μg/mL gentamicin, 10 μg/mL tetracycline, 100 μg/mL Blue-gal and 40 μg/mL IPTG.
2. Pour the molten LB agar into a QTray™ (with 48 well divider removed), then carefully place the divider into the QTray™.
3. Let the LB-agar plate solidify at room temperature and dry the plate in bio-safety cabinet for 2 h uncovered. The plate may be stored overnight in the dark at room temperature.
4. Label 24 × 1. mL sterile microfuge tubes and place on ice.
5. Thaw 1.2 mL of DH10Bac competent cells (Invitrogen) and aliquot 50 μL volumes into each pre-chilled microfuge tube.
6. To each add 0.5 μL of purified recombinant pFastBac DNA, gently mix and incubate the cells on ice for 30 min.
7. Heat-shock the cells for 45 s at 42°C, then place the tubes on ice and chill for 2 min.
8. To each add 450 μL of room-temperature S.O.C medium and incubate at 37°C with constant shaking at 225 rpm for 4 h.
9. Aliquot 0.5 mL volumes of S.O.C medium into each of the wells of a 24 deep-well block. To each add 3.8 μL of the transformed culture.

10. Remove 5 µL from each well and dilute 100–fold into fresh S.O.C. medium in a new block.
11. Transfer 20 µL of this dilution into individual wells of the QTray™ LB agar plate (described above). After each transfer, tilt the plate in all directions to cover the entire surface in the well. This can also be achieved by using sterile ColiRoller Plating Beads (Novagen/EMD Biosciences).
12. Allow the inoculation(s) to completely dry, before incubating the plate for 48 h at 37°C.

14.3.2.2. Recombinant Bacmid Purification

1. Prepare 200 mL of 2YT medium (containing 50 µg/mL kanamycin, 7 µg/mL gentamycin and 10 µg/mL tetracycline) and dispense 1.7 mL into each well of 4 × 24 deep-well blocks.
2. Select four large white colonies from each transformation and inoculate four 24-well blocks containing 1.7 mL 2YT medium (*see* **Note 3**).
3. Seal the blocks with AirPore™ tape and incubate overnight at 37°C with constant shaking – we use a Shel Lab Shaking Incubator (Thomson Scientific, San Diego, CA), set at 400 rpm.
4. Fill 4 × 24-well blocks with fresh 1.3 mL 2YT medium (containing appropriate antibiotics) and using a 6-channel pipette, inoculate each with 200 µL from the overnight culture.
5. Grow the cultures for 4 h at 37°C with constant shaking.
6. Prepare *diluted trapping buffer* and *diluted wash buffer* according to the PerfectPrep™ BAC 96 manual (*see* http://www.eppendorfna.com).
7. Filter sterilize *elution buffer* from the PerfectPrep™ kit.
8. Centrifuge the cultures in the 24-well blocks at 2,000 × *g* for 10 min.
9. Pour off supernatants and blot the block on absorbent material to remove residual culture medium.
10. Purify bacmids using the protocols provided in the kit. This can be done manually with a vacuum manifold or adapted to a robot (e.g. Tecan Freedon EVO® workstation).
11. Analyze the recombinant bacmid DNA to verify successful transposition of your gene of interest by PCR. Briefly, set-up 25 µL reaction mixes on a 96-well plate using M13 Forward (-40) and M13 Reverse Primers (available from Invitrogen). Verify the correct molecular weight by analysis on DNA agarose gel electrophoresis; we use an Invitrogen E-Gel 96 gel (*see* **Note 4**).

This protocol allows for the rapid purification of high-quality recombinant bacmids suitable for transfection. In comparison to other approaches it is faster, easier to use and provides sufficient quantity of high quality DNA to allow the user to perform multiple transfection experiments. DNA yields appear to be fairly consistent between plates, each with a success rate of 90% based on PCR verification. Unlike some BAC kits, the PerfectPrep™ kit does not require alcohol precipitation steps, allowing a 96 plate to be processed in less than 1 h. Since both vacuum filtration and centrifugation protocols are available, the kit can be used with standard equipment found in most labs. Indeed, we have successfully integrated the PerfectPrep™ BAC 96 kit onto our existing plasmid purification robot (Tecan Freedom EVO® workstation), simplifying the procedure further.

14.3.3. HT-Suspension Based Insect Cell Transfection

This protocol is modified from a protocol first published by McCall et al *(6)* and has been adapted to an Eppendorf epMOTION 5070 Workstation. The protocol routinely generates viral stocks with titers of at least one order of magnitude higher than conventional technologies.

1. Preparation of DNA transfection reagent: Using a multichannel pipette, dispense 150 µL per well of serum-free medium (SFM) into a sterile flat-bottomed 96-well plate and add 8 µL of transfection reagent. Then, dispense 2 µg recombinant bacmid DNA per well, followed by a gentle mixing of the well content.
2. Incubate for 30 min at room temperature for efficient complex formation. It is important not to use serum-containing medium, as serum may inhibit complex formation. Plates are covered with lids to prevent evaporation and contamination.
3. Separate mid-logarithmic insect cells and culture medium containing 1% FCS by centrifugation at 1,000 × *g* for 5 min and gently resuspend cells in fresh serum free medium to a density of 2.9×10^6 cells/mL.
4. Dispense 0.85 mL mid-logarithmic insect cells per well into a fresh sterile 24 deep-well plate using a multi-channel pipette (*see* **Note 5**).
5. Transfer the transfection mix into each well of the 24 deep-well plate.
6. Seal plate with AirPore™ cover sheet (*see* **Note 6**).
7. Incubate the 24 deep-well plate for 5 h at 28°C on an orbital shaker at 250 rpm.
8. Carefully remove the AirPore™ tape sheet and discard it. Add 4 mL of fresh culture medium containing 10% FCS (v/v) to each well resulting in 5.0 mL cell suspension at a density of 0.5×10^6 cells/mL.
9. Seal plate with new AirPore™ cover sheet.

10. Shake the 24 deep-well plate for 7 days at 28°C on an orbital shaker at 250 rpm.
11. After 7 days, measure a sample from each well using the Vi-Cell™/Cedex™ cell counter, monitoring cell density, cell viability and cell diameter to check for signs of infection.
12. Separate cells and culture supernatant by centrifugation at 3,890 × *g* for 10 min (4,000 rpm using a JS-5.3 rotor and a Avanti G-20XPi centrifuge, Beckman Coulter Inc., Fullerton, CA, USA) and store virus-containing supernatant at 4°C ready for amplification and protein expression analysis.

In comparison to standard adherent culture techniques, the transfection of insect cells in suspension generates viral stocks with titers of at least one order of magnitude higher than conventional technologies. This negates the time consuming amplification of the viral stocks, whilst generating sufficient virus to enable both initial expression tests to be performed in 24-well plates, plus amplification of the selected constructs from the same master stock. Moreover, by combining this powerful transfection protocol with automation (Eppendorf epMOTION 5070 Workstation), the number of constructs processed simultaneously can be greatly enhanced for a single operator, thereby further expediting high throughput construct screening and protein expression analysis (*see* **Note 7**).

14.3.4. Methods of Recombinant Viral Titer Determination

Traditionally as part of the initial optimization of baculovirus infection and concomitant protein expression, the titer of the recombinant viral stock is determined, thereby allowing the calculation of MOI and ensuring cross-referencing and reproducibility in subsequent experiments. The titer of a recombinant baculovirus stock can be determined by plaque assay *(7)*, end point dilution *(8)* and immunoassay *(9, 10)* – *see* **Note 8** for further information.

All of these approaches have their own respective merits, however they all share an overriding problem in that they take considerable periods of time to complete and/or can be difficult to interpret and give rise to high variability between users. Therefore, despite the popularity of the baculovirus expression system in protein production laboratories worldwide, there is no fast, reliable and inexpensive method of virus titer determination. Here we describe a series of protocols currently used in our groups to facilitate rapid titer determination.

14.3.4.1. High Throughput Baculovirus Titration Assay Using GFP Co-Expression

This automated method for the determination of baculovirus titer uses GFP-linked co-expression plasmids similar to those published *(11, 12)* and the Agilent 2100 Bioanalyzer (Agilent Life Science Technologies) to generate fast, highly reproducible viral titer estimates.

1. Plasmid generation: GFP co-expression plasmids amenable to Bac-to-Bac™ mediated baculovirus expression are not commercially available and thus must be generated in-house. This is easily achieved with the following protocol:
   (a) An *Xho*I – *Kpn*I flanking GFP fragment can be generated by standard PCR methodology and sub-cloned into the p10 multiple cloning site (MCS) of pFastBac Dual vector (Invitrogen) to create a GFP co-expression plasmid compatible with the Bac-to-Bac™ Expression System – the remaining MCS under the control of PolH promoter is then retained for use in expressing the gene of interest.
   (b) Transform DH10Bac (Invitrogen) competent cells with GFP pFastBac Dual plasmid and isolate recombinant bacmid following the manufacturer's instructions (Invitrogen). Once verified by PCR transfect the bacmid into Sf21 cells to generate a high titer recombinant viral stock.
2. Infect a 25–50 mL insect cell culture with 1 mL of your GFP expressing recombinant baculovirus (in addition to your gene of interest).
3. Following incubation for 48 h at 27°C, count cells, harvest and resuspend at 1×10^6 cells/mL in HEPES buffered saline supplemented with 0.05% Pluronic acid (Invitrogen).
4. Stain for 15 min with 0.5 µM of the live cell dye Carboxynaphthofluorescein Diacetate (CBNF), centrifuge (500 × *g*, 5 min) and resuspend in cell buffer (2 × 106 cells/mL).
5. Load 10 µL onto an Agilent cell fluorescence LabChip™ and analyze using the 2100 Bioanalyzer following the manufacturer's instructions.
6. In order to determine the percentage of GFP expressing cells, live cells in the CBNF-positive population are simply cross-gated onto the GFP histogram. In the example given in **Fig. 14.1**, 13.3% of live CBNF-positive cells can be cross-gated with GFP fluorescence. Knowing the number of CBNF-positive cells and the number of GFP expressing cells allows the calculation of the viral titer using the equation described by Berns and Giraud *(13)*. The calculated value determined from data shown in **Fig. 14.1** is 1.26×10^6 IU/mL and compares favorably with the viral titer calculated (for the same viral stock) using the traditional plaque assay (4.1×10^6 pfu/mL) or the BacPAK™ immunoassay (6.4×10^6 IU/mL), respectively.

$$\text{Viral Titer (IU / mL)} : \frac{\#\text{GFP Gated Cells} \times \text{Dilution Factor} \times \#\text{CNBF-Positive cells}}{\text{Volume of Virus Added (mL)}} \quad (14.1)$$

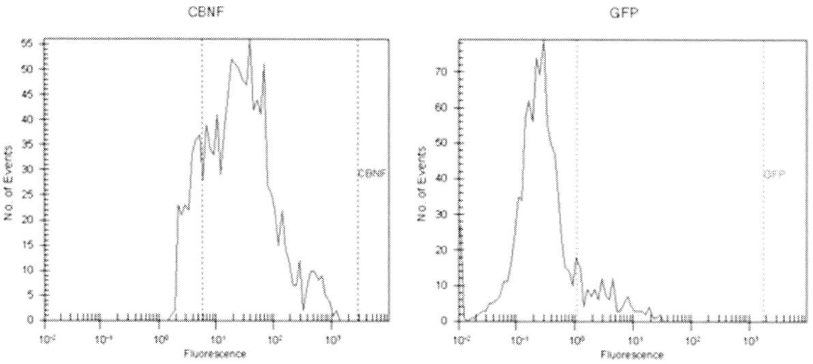

Marker	#Events	% Gated
Total	1189	-
Gated by CBNF	983	-
Gated by GFP	131	13.3

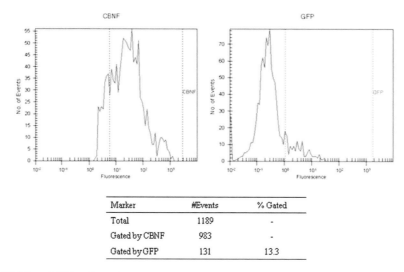

Fig. 14.1. GFP infected Sf21 cells analyzed by the 2100 Bioanalyzer. Cell culture volumes of 50 mL (0.5×10^6 cells/mL) were infected with 1 mL of GFP virus and incubated for 48 h, at 90 rpm, 27°C. Following incubation cells were counted, harvested and resuspended at 1×10^6 cells/mL in HBSS with 0.05% (v/v) pluronic acid (Molecular Probes). Cells were then stained for 15 min at room temperature with 0.5 μM live cell dye carboxy-naphthofluorescin diacetate [CBNF] (Molecular Probes), pelleted by centrifugation (500 × *g*, 5 min), and resuspended in cell buffer (2×10^6 cells/mL). 10 μL of viral sample was then loaded onto a cell fluorescence LabChip™ and analyzed on the Agilent 2100 Bioanalyzer.

This method offers several advantages over alternative approaches for viral titer calculation. Firstly, since data collection is automated by the 2100 Bioanalyzer, it does not rely upon the operator to differentiate between infected and non-infected cells. Automation of data collection therefore removes user-to-user variability and thereby one of the largest sources of error associated with other viral titer determination methods. Secondly, the method is very quick and simple to perform. Forty-eight hours post-infection, insect cells infected with GFP-containing virus can be harvested and within 90 mins a value for viral titer determined. This compares favorably with the more time-consuming and labor intensive plaque assay and immunoassay which both require extensive washing and fixing of cells. Whilst at first glance this strategy requires the use of an expensive machine (Agilent 2100 Bioanalyzer), the protocol could also be used in concert with other flow cytometry systems (FACS or MoFlo), thus making this approach widely applicable to many laboratories. For further details *see (14)*.

14.3.4.2. High Throughput Baculovirus Titration Assay Using Alamarblue™

An alternate method of viral titer determination method uses AlamarBlue™ to monitor early insect cell growth arrest induced by a 24 h baculovirus infection cycle *(15)*. Briefly, AlamarBlue™ when added to the cells is reduced by the mitochondrial enzyme activity of viable insect cells, causing a change in emitted color and shift in fluorescence. Therefore, viral infectivity can easily be monitored by a change in color (as the viral infection proceeds). The approach does not require the construction of GFP co-expression plasmids or purchase of costly equipment. Automation of this assay has been shown to greatly improve both intra-assay and inter-assay standard deviation (compared to manual pipetting). Therefore the protocol described below utilizes an Eppendorf epMOTION 5070 Workstation (*see* **Fig. 14.2**).

1. Dispensing viral solutions and media control: Manually dispense 200 µL of an undiluted viral solution (working virus stock) and medium control to the wells in column 1 as illustrated in **Fig. 14.2a**. Transfer the plates to the workstation. Fill 20 mL of fresh medium into a sterile "tubs" container and dispense 0.1 mL medium into the wells in column 2–11.
2. Viral dilutions: Perform automated serial dilutions by transferring 0.1 mL of the virus solution from one column to the next one, starting in column 1 and finishing in column 12. Program the "mixing after dispensing" module set at "Speed 4" to four (0.1 mL) cycles and fix the height of aspiration/dispensing to 1 mm above the well bottom. Exchange tips after each dilution step to ensure a reliable dilution process.

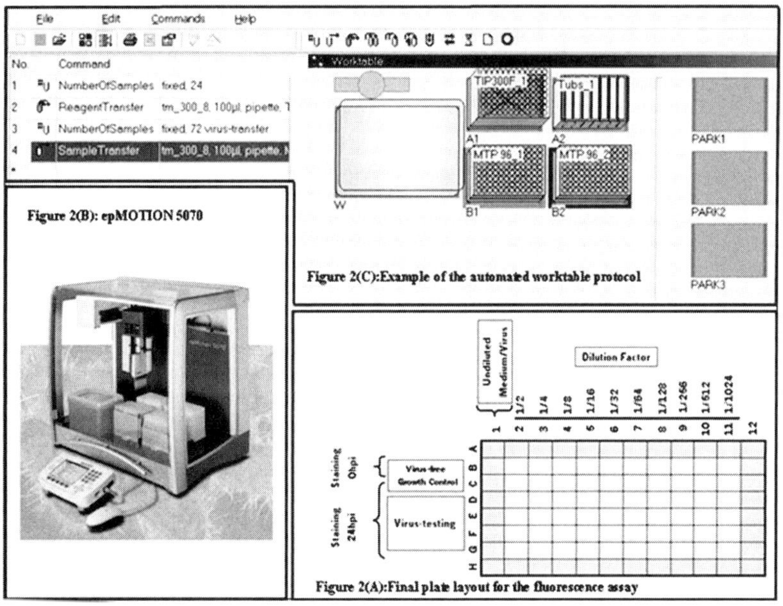

Fig. 14.2. epMOTION 5070 and automated worktable protocol. The worktable is composed of two 96-well micro-titer plates (MTP 96_1 and 2), tubs holder for medium and cell suspension (Tubs_1), filter tip box for single and multi-channel pipetting (TIP300F_1), and a waste box (W). The final set up for the 96-well micro-titer plates is shown in **a.**

3. Prepare 50 mL of Sf21 cell suspension at exactly 3×10^4 cells/mL for dispensing into two 96 well plates (MTP_96_1 & MTP_96_2 respectively) (*see* **Note 5 and 9**).
4. Cell dispensing: For each 96-well plate, transfer 20 mL cell suspension in a sterile "tubs" and start immediately the automated addition of 0.1 mL cell suspension per well. Set the "mixing before aspirate" modules to three cycles of 0.3 mL with speed 4. Set the "mixing after dispensing" modules to four cycles of 0.1 mL with speed 3, and fix the height of aspiration/dispensing to 1 mm above the well bottom. Exchange tips before new aspiration of the cell suspension.
5. AlamarBlue™ dispensing: Replace the "tubs" holder with the "tube" holder. Prepare the required volume of AlamarBlue™ solution (Serotec) in a 1.5 mL Eppendorf tube. Perform the automated addition of 20 µL dye per well as follows: set the dispensing modus to multi-dispense "from top". Exchange tips before new aspiration of the AlamarBlue™ solution. Incubate the plate at 28°C for 5 h and then read fluorescence to obtain the t_0 values. Re-incubate the plate at 28°C for a further 19 h to complete the 24h infection cycle. Determine the t_{24} values by adding 10% AlamarBlue™ to the remainder of the wells and read its fluorescence 5 h later. Fluorescence measurements are performed using a fluorescence multiwell plate reader (CytoFluor II from Applied Biosystems). The excitation wavelength is set to 530 nm and the emission wavelength to 590 nm.
6. Determination of viral titer:
   (a) Calculate t_{24}–t_0 values for each dilution and for the virus-free medium control. This virus-free medium control value represents 100% growth performance, i.e. undisturbed cell growth. Next, determine the percentage of Growth Inhibition (GI) for each virus dilution using the following equation:

$$GI(\%) = \frac{t_{24}virus - t_0control}{t_{24}control - t_0control} * 100 \qquad (14.2)$$

   (b) To determine the virus dilution corresponding to 50% growth inhibition (TCLD50), analyze data with the Origin 7 software (OriginLab Corporation, MA) using the following equation of a sigmoid curve;

$$y = A_1 + \frac{A_2 - A_1}{1 + e^{-(D-D_0)p}} \qquad (14.3)$$

where y is the growth inhibition, A_1 is the minimum growth inhibition (undisturbed cell growth), A_2 is the maximum growth inhibition (100% infected cells), D is the dilution factor, D_0 is the dilution at which the growth inhibition was 50% ($1/TCLD_{50}$) and p is a slope factor.

(c) Convert the obtained D_0 value to $TCLD_{50}/mL$ by applying the following equation:

$$TCLD_{50}\ /\ mL = \frac{1}{(D_0 V)} \tag{14.4}$$

where V is the volume of virus solution added per well.

We find the protocol to be a simple and extremely reliable high-throughput (semi-automated) method of determining the titer of a recombinant baculovirus stock. The data collection from the fluorescent plate reader is fast, convenient and is not subject to user specific interpretation as is the case for viral plaque or foci counting (*see* **Fig. 14.3**). Hence, within 3 h, it is possible to set up four 96-well plates for the assay on the Eppendorf epMOTION 5070, allowing the titration of eight virus stocks in

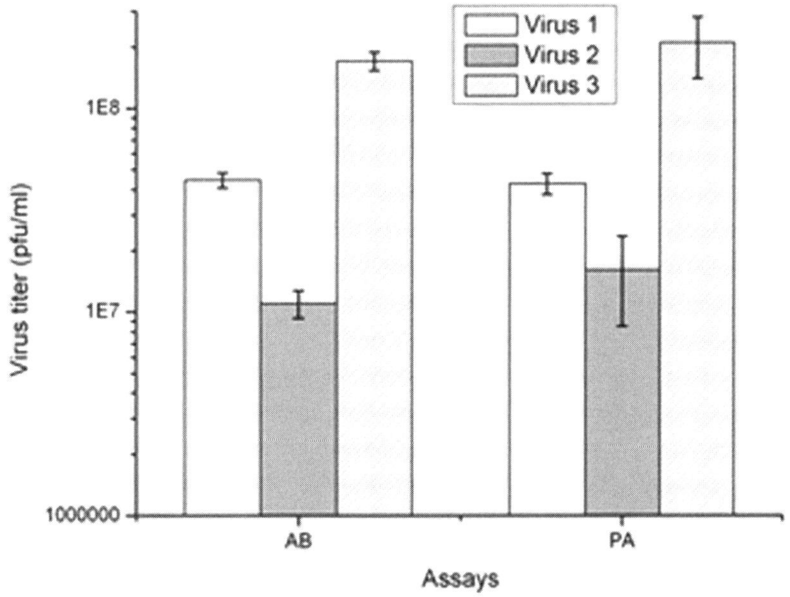

Fig. 14.3. Titration of three different virus stocks with AlamarBlue fluorescent assay (AB) in comparison to plaque assay (PA). Virus titrations were carried out on three different recombinant working virusstocks generated in Sf21 cells. Titers were calculated from $TCLD_{50}$ and a standard curve and compared to titers obtained by classical plaque assay. The graph shows the comparison of mean values and standard deviations of data obtained from three independent AlamarBlue and Plaque Assays.

parallel. Moreover, with an overall assay time of 30 h the AlamarBlue fluorescent titration assay is one of the fastest titration assays available, increasing its attractiveness for rapid high throughput applications. For more information *see (16)*.

14.3.4.3. Use of Cedex Cell Counter to Determine Cell Diameter and Viral Infectivity

This method developed by scientists at Roche *(17)* is an appealing method to titer viruses using the insect cell diameter information obtained from a CEDEX™ cell counter. Since there is a quantitative relationship between the viral concentration and the increase in average cell diameter of an infected culture this information can be used to determine a value for virus titer. It is quick, simple and appears to generate very accurate and reproducible results. By using serial dilutions of a viral stock to infect a series of cell cultures, a dose response plot of viral concentration vs. average cell diameter can be generated 24 h post infection. The titer of an unknown virus can be determined by the magnitude of the shift between the dose response curve of an unknown virus and that of a reference standard. While the titers obtained by this 'cell swell' assay do not correspond precisely with the titers from a plaque assay, the titers are reproducible with a relatively low standard deviation of approximately 10%. Determining the value of the curve shift requires a sigmoidal curve fitting statistical analysis. There is no interpretation of data by the user for this method, thus eliminating the user-to-user variability found in other titer methods. There is also little handling by the user making this a very low time investment, with results within 24 h. Unfortunately, this method is limited by the number of samples that can be processed at once, thus making it useful for low to mid-throughput applications. Lastly, a well kept tissue culture facility is required to maintain cells with the necessary uniform cell morphology to ensure accurate titers over a long period of time.

14.3.5. HT Miniaturized Deep-Well Block Insect Cell Protein Expression

Once a high-titer baculovirus stock has been propagated, a myriad of optimization experiments are typically required before large-scale expression and purification can commence. These studies are far more labor intensive and time consuming than those employed in *E. coli* multi-parallel protein expression, since for each recombinant baculovirus, optimal growth and infection conditions require careful characterization. These experiments include varying the time of infection (48–96 h) and the volume of virus added to the viable cell number (expressed as multiplicity of infection). Typically therefore, for any given protein, several different constructs and tags, each requiring individual recombinant viral propagation, may need optimization. To address these issues, many laboratories have developed strategies that make use of 24 deep-well blocks to conduct initial optimization screening in a similar fashion to those described for *E. coli (18, 19, 6)*. The use of deep-well blocks to rapidly scout a multitude of conditions also results in a reduction in lead-time before commencement of large-scale bioreactor experiments.

14.3.5.1. Insect Cell Growth Optimization in Deep-Well Blocks

When embarking on deep-well block studies it is advisable to first optimize insect cell growth for the incubator you have chosen to perform your studies in, since evaporation, optimal volumes and agitation rates will be different for individual incubators. Here we describe the use of a Micro Expression Shaker (GlasCol Terre Haute, IN, USA) routine growth of insect cells in deep-well blocks.

1. Individual wells of a 24 deep-well block (Qiagen) are seeded with 2–7 mL of a 0.5–1.0 × 10^6 cell/mL insect cell master stock. For each planned time point, triplicate wells are set-up for accuracy.
2. Seal blocks with an adhesive AirPore™ sheet.
3. Incubate at 27°C with constant shaking (300 rpm) in a Micro Expression Shaker. Remove 1 mL aliquots at 24, 48 and 72 h and analyze for cell density, viability and cell diameter using a Cedex™ cell counter.

14.3.5.2. Protein Expression Studies Using Deep-Well Blocks

1. Seed 24 deep-well blocks with the appropriate volume (2–7 mL; as determined above) of 1 × 10^6 cells/mL of insect cells.
2. Seal with an AirPore™ sheet and incubate overnight at 27°C with constant shaking.
3. Check that cell viability and growth have been maintained, and then infect cells with 0.05–1 mL volumes of recombinant baculovirus generated as described above (**Section 14.2.1**).
4. Seal with an AirPore™ sheet and incubate at 27°C with constant shaking for a further 24–48 h, checking for cell viability on a regular basis.
5. Harvest cells by centrifugation (13,200 × *g*, Eppendorf 5415R bench top centrifuge), then resuspend in 1 mL lysis buffer [50 mM Tris pH8, 10% (v/v) glycerol, 10 mM DTT, 0.4M NaCl, 0.2% (v/v) NP40 supplemented with a Complete Protease Inhibitor™ (Roche Applied Sciences) tablet per 50 mL] and incubate on ice for 30 min (*see* **Note 10**).
6. Pellet the insoluble fraction and cellular debris by centrifugation for 30 min at 20,000 × *g* and 4°C.
7. The supernatant containing the soluble fraction is then finally prepared by the addition of LDS loading buffer (Invitrogen) supplemented with 50 mM DTT and analyzed for expression levels of recombinant protein by SDS-PAGE (NuPAGE; Invitrogen) and Western blot analysis.

Figure 14.4 shows the growth profiles of 1–10mL suspension cultures maintained in 24 deep-well blocks over a period of 24–72 h when grown in an Infors Multitron II incubator. At small (1 mL) volumes, despite the presence of the AirPore™ tape sheet and a humidified incubator, the losses in volume from

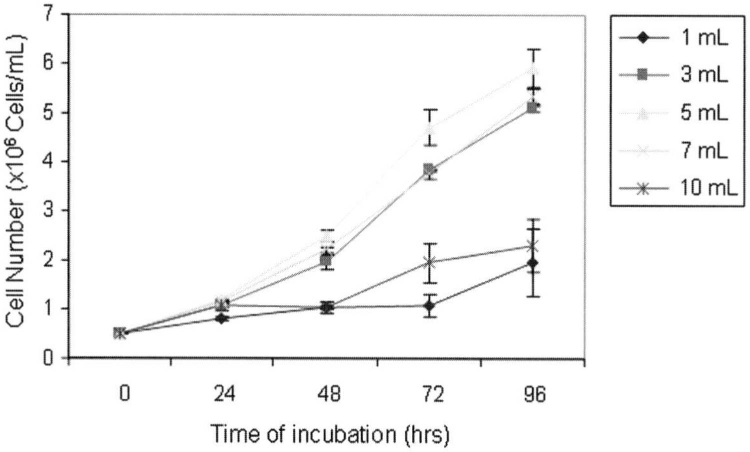

Fig. 14.4. Growth kinetics of *Sf*21 cells grown in deep-well blocks. *Sf*21 cells were seeded in a 24-deep well block at an initial density of 0.5×10^6 cells/mL in culture volumes ranging from 1 to 10 mL. The culture volumes were set up in triplicate to allow an independent growth measurement to be taken at 24 h intervals over a three day period. Cells were counted using a Neubauer haemocytometer. All data points are means ± S.E. of triplicate measurements from a single well.

evaporation were sufficiently large to have a profound effect on the growth of the insect cells. In addition, at the largest volume analyzed (10 mL), cells appear to settle to the bottom of the wells, reflecting poor mixing. Taken together, these data suggest that the optimal volume for continuous and logarithmic growth is between 3 and 7 mL when shaken at 250 rpm. Cells seeded at higher densities ($1–2 \times 10^6$/mL) appear to work equally well using similar conditions (data not shown). These data appears to be consistently reproducible in a variety of incubators and suggests that media volumes of 3–7 mL are optimal in most cases. Moreover, the growth kinetics of insect cells grown in deep-well blocks agree well with those maintained under standard cell culture conditions (*see* **Note 11**).

The major advantage of our approach over traditional methods lies in the scale on which the experiments are performed. By performing the optimization in 2–7 mL culture volumes in standard 24 deep-well plates, we not only reduce the footprint of the equipment required to perform multiple experiments in parallel, but we also reduce the cost and labor associated with expression optimization, not to mention the amount of baculovirus required for assay formats. Taken together, these results clearly show the advantages of using a 24 deep-well block format to rapidly characterize different expression conditions (e.g. scouting of different fusion partners (GST, MBP, His_6) and/or constructs and cell lines).

14.3.6. Transient Insect Cell Protein Expression

Insect cells play an important role as hosts for the generation of recombinant proteins and it is fair to say that the Baculovirus/insect cell system is one of the most popular expression systems nowadays. However, despite the many improvements made in recent years (some of which are described above), this system inherently calls for a double-track strategy: the successful generation of sufficient recombinant virus and the optimization of conditions necessary to express the protein of choice to high levels. For recombinant protein production using mammalian cells as hosts transient transfection technologies are gaining increasing importance. Very efficient transfection technologies and reagents give rise to high abundance of plasmid DNA introduced transiently into the nucleus of cells, which results in a burst of recombinant protein produced within short time frames. To take advantage of insect cell lines as hosts for recombinant protein production while circumventing the tedious process steps of virus generation and characterization, attempts have been made to use stably transfected insect cells. Apart from the well-known Drosophila Schneider cell lines *(20, 21)*, *Spodoptera frugiperda* and *Bombyx mori* descendants (Sf9, Sf21, Bm) have been used to generate stable recombinant cell lines following transfection and selection with appropriate antibiotics *(22, 23, 24)*. Yet, the overall production rates obtained were not convincing enough to give rise to widespread application of this technology. A very recently published review summarizes excellently different approaches using stable transformants for production *(25)*.

Alternatively, a transient transfection process is now frequently applied to mammalian expression trials in HEK293 or CHO cell systems *(26, 27, 28)* and could be envisaged using insect cells. Not much information is available in the public domain on such expression trials, even though this possibility is supported by the commercial availability of suitable expression plasmids and kits *(26, 29, 30)*; see also links to websites/manuals in reference section). When working with commercial transfection reagents based on e.g. lipofection, the expression trials are usually restricted to small scale levels due to price constraints. An attractive alternative is polyethylenimine (PEI) as carrier of foreign genetic material. Since its first description in 1995 by Boussif et al. *(31)* PEI has been described as a successful and cost-effective transfection reagent in in-vitro cell systems as well as in gene therapy trials in a multitude of publications *(26, 27, 28, 32)*. The application of PEI to insect cell transfection has, however, not been demonstrated.

We describe in the following an optimized protocol for transient insect cell expression on small and large scale using lipofection and PEI for gene transfer. High transfection efficiencies and yields allow the rapid analysis of expression in case of multiple construct screening and the option to scale this approach to production/purification levels (*see* **Note 12**).

14.3.6.1. Plasmid Preparation

The recombinant plasmids used for transfection are propagated in the *E.coli* strain DH5α and purified using conventional plasmid preparation kits (e.g. Nucleospin® (Macherey & Nagel) or Compact Prep Plasmid Midi (Qiagen) kits.

14.3.6.2 Cell Cultivation

1. Seed cultures are maintained in disposable shake flasks on an orbital shaker platform at 94 rpm; however, shaking speed may vary for individual cell lines (*see* **Note 13**.). The shaker is installed in a CO_2 free environment at 28°C. The seed cultures are usually passaged twice a week by dilution to a density of 3×10^5 cells/mL.
2. Importantly, the cells need to be in the exponential growth phase prior to transfection, i.e. at densities between 1.5 and 3.5×10^6 cells/mL. It is advisable to assess the growth behavior of the insect cells by establishing individual growth curves for each cell line before starting with transfections. In our hands, Sf cells usually exhibit a population doubling time of 18–24 h and a continuous viability >95%.
3. To ensure that cells are in the exponential growth phase at transfection, prepare a shake flask containing 1.5×10^6 cells/mL the day before transfection. Total number of cells and culture volume of this pre-culture are dependent on the number of transfections planned.

14.3.6.3. Transient Insect Cell Transfection

1. Transient transfections can be done using either suspension cultures grown in shake flasks or adherent cultures grown in e.g. 6-well tissue culture plates.
2. Performing transient transfections in suspension culture facilitates the sampling at different time points post transfection (usually after 48 and 72 h to monitor cell density, cell viability and protein expression) vs. adherent cultures, especially if the protein is expressed in the cytosol of cells.
3. If 6-well plates are chosen as culture vessels for transfection, the seed culture may also be kept in suspension and distribution into the wells should be done at an early time-point to allow the cells to settle at the bottom of the well prior to transfection. Alternatively, cells can be taken from seed cultures growing adherently on tissue culture flasks; these are detached by vigorous tapping of the flask (do not use trypsin for detachment).
4. The following protocol for suspension cultures is derived from Novagen/EMD Biosciences InsectDirect™ expression protocol (http://www.merckbiosciences.co.uk/product/22001) which was modified and further optimized as described.

14.3.6.4. Small-Scale Transfection Protocol in Shake Flasks

In this protocol only PEI-mediated transfection is outlined in detail, but the same protocol can be applied to other commercially available transfection reagents. Detailed quantities of reagents, cells and plasmid DNA are listed in **Table 14.2.**

The quantities in this table refer to a culture volume of 15 mL with a final cell density of 1.5×10^6 cells/mL.

1. Preparation of DNA: PEI complexes: Add 500 µL of serum-free medium to a sterile 15 mL polystyrene tube and add 15 µg of plasmid DNA. Similarly, mix 500 µL serum-free medium with 30 µg of the PEI solution in a second tube (DNA: PEI ratio = 1:2 (µg: µg), *see* also **Note 14**). Gently mix the contents of both tubes by shaking and leave at room temperature for 15 min (*see* **Note 15**). For multiple transfections, a 48-well polystyrene plate can be used instead of the 15 mL tubes for preparation of the mixes.
2. Add the PEI solution to the tube with the DNA solution, gently mix the tube and incubate for another 15 min at room temperature to allow for complex formation. It is important not to use serum-containing medium for these two steps, as serum may inhibit complex formation.
3. During incubation of the transfection mix the cell culture is prepared: 7 mL of culture at a density of 3.2×10^6 cells/mL are added to a 125 mL disposable shake flask.
4. Gently re-mix the transfection mix before adding it to the cells.
5. Incubate the shake flask for 30 min at 28°C on an orbital shaker at 94 rpm.

Table 14.2
Quantities of reagents and medium needed for small scale transfection in suspension culture

Cell culture	Plasmid DNA diluted in SFM medium	Transfection Reagent diluted in SFM medium	Medium needed to adjust volume (ExCell 420 SFM)
Sf21 at 3.2×10^6 cells/mL (7 mL)	15 µg in 500 µL	Insect GeneJuice™ 30 µL in 500 µL	7 mL
Sf21 at 3.2×10^6 cells/mL (7 mL)	15 µg in 500 µL	Cellfectin™ 45 µL in 500 µL	7 mL
Sf21 at 3.2×10^6 cells/mL (7 mL)	15 µg in 500 µL	PEI (1 µg/µL) 30 µL in 500 µL	7 mL

6. Add 7 mL of fresh culture medium to the shake flask, resulting in a 15 mL culture at a density of 1.5×10^6 cells/mL. Continue incubation under the above-mentioned conditions.
7. After 48 and/or 72 h take samples for protein analyses.
8. The protocol for transfections on enlarged scale implies the same process steps as outlined for small scale suspension transfections. For details on volumes etc. *see* **Table 14.3.** If the total volume of the transfection mix exceeds 50 mL, two 50 mL sterile centrifuge tubes (e.g. Corning, Cat. no. 430304) are used and the concentrations of DNA and PEI are divided by two. Again, final cell densities of 1.5×10^6 cells/mL after transfection are obtained.

14.3.6.5. Transfection of Adherently Growing Cell Cultures in 6-Well Plates

1. Seed 2mL of cell culture at a density of 7.5×10^5 cells/mL per well and incubate plate at 28°C for a minimum of 30 min, but not longer than 4 h (after 4 h part of the cells proliferate already and the resulting cell density will be too high for transfection).
2. Preparation of DNA-PEI complexes: Add 100 µL of serum-free medium to a sterile 15 mL polystyrene tube and add 3 µg of plasmid DNA. Similarly, mix 100 µL serum-free medium with 6 µg of the PEI solution in a second tube (DNA: PEI ratio = 1:2 (µg: µg), *see* also **Note 14**). Gently mix the contents of both tubes by shaking and leave at room temperature for 15 min (*see* **Note 15**).
3. Add the PEI solution into the tube with the DNA solution, gently mix the tube and incubate for another 15 min at room temperature to allow for complex formation. Do not use serum-containing medium in these two steps, as serum may inhibit complex formation.

Table 14.3
Transfections on large scale in suspension culture

Cell density	Plasmid DNA diluted in SFM medium	PEI diluted in SFM medium	Medium needed to adjust volume	Culture vessel
140 mL Sf21 at 3.2 $\times 10^6$ cells/mL	0.3 mg add 10 mL	0.6 mg add 10 mL	140 mL (Final volume:300 mL)	1 L shake flask Cat. no.431147
275 mL Sf21 at 3.2 $\times 10^6$ cells/mL	0.6 mg add 25 mL	1.2 mg add 25 mL	275 mL (Final volume:600 mL)	2 L shake flask Cat. no.431255
550 mL Sf21 at 3.2 $\times 10^6$ cells/mL	1.2 mg add 50 mL	2.4 mg add 50 mL	550 mL (Final volume:1,200 mL)	3 L shake flask Cat. no.431252

4. After the incubation, add 800 μL of medium to the transfection mix and gently re-mix the tube by hand.
5. Remove the medium covering the cells and add carefully drop wise the transfection mix to the cells. There is no washing step necessary to remove residual medium with FCS, as only during complex formation the absence of FCS is important.
6. Incubate the plate for 4 h in the incubator at 28°C.
7. Some transfection reagents, including PEI, may inhibit cell growth slightly and exhibit toxicity when applied in high concentrations or over a prolonged incubation time. To prevent cells from damage the medium can be exchanged with 2 mL of fresh medium after the 4 h incubation period, or alternatively the transfection mix can be diluted by addition of 2 mL of medium prior to addition to the cells (total volume is then 3 mL).

14.3.6.6. Expression Analysis

1. In most instances recombinant protein production is facilitated by expression of a tagged protein, in particular in eukaryotic cells. Thus, analyses of expression (and subsequently protein purification) are highly tag-dependent with respect to methods applied. If a secretory protein is to be expressed, detection and determination of expression levels are mostly done using affinity HPLC against e.g. an Fc-fusion protein, or in case of His_6-tagged proteins by capture on Ni-NTA-coated magnetic beads or columns, followed by LC-MS analyses.
2. In case of proteins expressed intracellularly or on the cell surface, a reliable (yet time-consuming) analysis is the traditional Western Blot. Despite being semi-quantitative with respect to yields, it not only allows monitoring of the integrity of the recombinant protein; additionally, the solubility of a protein expressed in the cytosol of the cells can be addressed, if the lysate is fractionated into the soluble and insoluble portion (*see* **Note 16, 17** for a protocol).

Table 14.4 summarizes results obtained during our optimization experiments carried out with different transfection reagents. It shows the effect of the reagents on cell growth and toxicity, as well as transfection efficiencies as judged by visualization of expression of the Green Fluorescent Protein (GFP). All transfections were carried out in Sf21 cells. Read-outs were done 72 h post-transfection by visual inspection (for adherent cultures in 6-well-plates) and for suspension cultures by FACS analyses and cell count determination using the CEDEX™ cell counting instrument.

In brief, our experiments revealed that the transfection efficiency (as judged by GFP fluorescence) can be increased by using higher concentrations of the transfection reagents; unfortunately,

Table 14.4
Comparative analyses of different transfection reagents

Reagent	Cellfectin™	Insect GeneJuice™	PEI
Cell growth	Significant inhibition	Inhibition only at high concentrations	Inhibition only at high concentrations
GFP expression (fluorescence)	50–70% of cells (weaker intensity)	70–90% of cells (variable intensities)	>90% of cells (medium to high intensities)
Cost of reagent (per 1 L culture)	576 US$	248 US $	0.052 US $

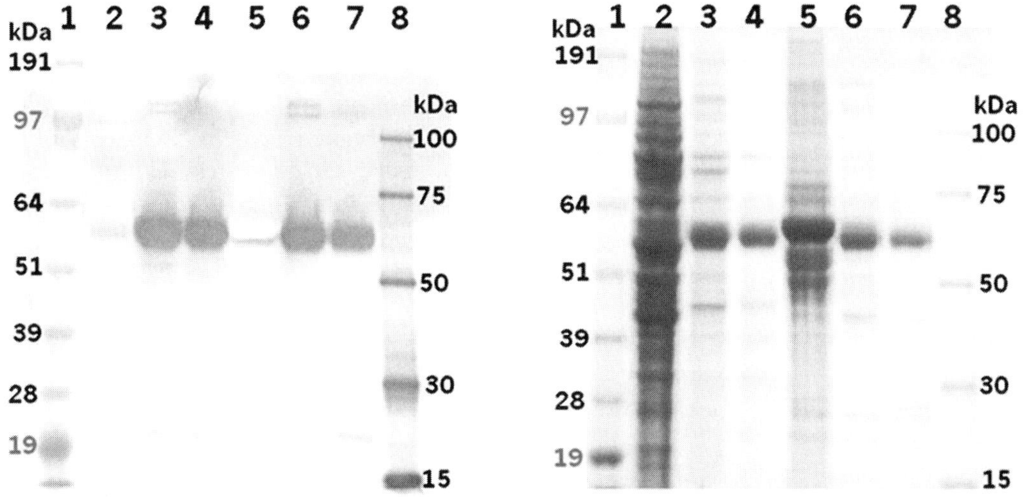

Fig. 14.5. Comparative expression of a cytokine in Hi5™ cells infected with recombinant Baculovirus vs. Sf9 cells transfected with recombinant plasmid DNA and PEI as carrier. Transfected cells (lanes 2–4): 1.5 L culture harvested 72 h days post transfection. Baculovirus-infected cells (lanes 5–7): 2 L culture harvested 72 h post infection. Both preparations were subjected to protein purification using Ni-NTA and SPX chromatography. Lanes 2 and 5 show Ni-NTA eluates, lanes 3, 4, 6 and 7 fractions collected after size-exclusion chromatography. *Right panel* shows analysis by SDS-PAGE, *left panel* analysis of the same fractions by anti-his Western Blot. Each lane contains approximately 1.5 μg of total protein. Yields: Baculovirus-derived protein: 4.5 mg/L; protein derived from transfection: 7.3 mg/L culture.

high concentrations of reagent also lead to retardation of cell growth and toxicity.

Another example (**Fig. 14.5**) shows expression of a cytokine by transient means using PEI as transfection reagent on the multi-liter scale in shake flask and subsequent purification steps. For comparison, the same gene was expressed using the conventional Baculovirus-mediated expression approach. The results in terms of purified protein match very well and show that transient insect cell expression is also a viable approach for the generation of multi-milligrams of protein.

14.4 Notes

1. For a more complete review of insect cell culture techniques, please refer to reference *(33)*.
2. The expression vector pIB/V5-His6 is part of the InsectSelect™ expression kit available from Invitrogen. The Canadian company CytoStore (www.cytostore.com) offers the Triple Express™ Insect Expression system (Cat. no. ASA-200) including the expression plasmid pIE/153A (V4). The vectors differ with respect to the insect promoters/enhancers used. In pIB/V5-His6 (InsectSelect™) transcription is based on the constitutive promoter OpIE2, derived from the Baculovirus *Orgyia pseudotsugata* multicapsid nuclear polyhedrosis virus *(34)*. The Triple Express™ expression plasmid features the actinA3 promoter in conjunction with the *Bombyx mori*-derived ie-1 transcription factor and hr3 homologous repeat region *(35, 36)*. The pIEx vectors (InsectDirect™ system) employ transcriptional elements derived from *AcNPV*, namely the ie-1 promoter and hr5 element *(2)*. In our initial studies we compared the Invitrogen expression plasmid pXinsect-DEST39 (featuring the actin promoter and ie-1/hr3 enhancer) to the EMD Biosciences/Novagen plasmid IEx and found no striking superiority; thus, we conducted all further experiments using the InsectDirect™ plasmid IEx-4. The pXinsect-DEST39 plasmid mentioned above is meanwhile no longer marketed by Invitrogen.
3. In the protocol described above, white (recombinant) colonies from transposition reactions were picked manually; this can be adapted to a colony picker to facilitate the process (e.g. Genetix).
4. Although it is possible to confirm recombinant bacmids by identification of high molecular weight species by agarose gel electrophoresis, this is not at all reliable since it is difficult to correctly identify DNA of this size. For more precise identification and quantification of bacmid DNA it is recommendable to use quantitative PCR methods.
5. Insect cells need to be in the exponential growth phase prior to transfection, i.e. at densities between $1.5–3.5 \times 10^6$ cells/mL. A population doubling time of 18–24 h and a continuous viability >95% are prerequisites for successful transfection and virus titration.
6. AirPore™ tape sheets can be used only once and have to be discarded after removal in order to avoid cross-contaminations between wells.
7. Whilst the ability to perform HT transfections utilizing 24-well blocks is technically feasible and has been demonstrated

(6) the subsequent propagation of large volumes of viral stocks through the P2 (and if needed, P3) stages is not technically feasible in a HT format. This represents one of the most formidable challenges in the implementation of HT processes in the baculovirus expression system.

8. The most accurate and most established viral titer determination method is the plaque assay. However, the disadvantage of this method is that it is difficult to perform and that it requires a long processing time (approximately 1 week). Furthermore, visualizing plaques can be difficult and requires an expert, well-trained eye. A number of vectors carrying an infection marker have therefore been developed to overcome problems associated with the identification of recombinant plaques and infection. Typically these incorporate the β-galactosidase gene and require the addition of substances such as X-gal *(37, 38)*, although green fluorescent protein (GFP) has also been used with some success in this and other laboratories *(11, 12, 14)*. Immunoassay procedures are marketed by Clonetech s and EMD Biosciences/Novagen under the trade names BacPAK™ and FastPlax™.
9. Various cell lines were tested during assay set-up and optimal results were obtained with the Sf21 cell line described. It is therefore advisable to use this cell line or to compare the available in-house cell lines before starting larger experiments. It was observed that cell passage number can influence the quality of the assay; therefore using a culture beyond 30 passages should be avoided.
10. A variety of approaches for cell lysis and protein expression analysis exist, many of which are easily amenable to automation. Of particular note are those approaches that obviate the need for centrifugation by virtue of in-well lysis (e.g. FastBreak Cell Lysis Reagent (Promega) and Insect Pop Culture (Novagen/EMD Biosciences).
11. Similar growth kinetic characterization studies to those described above were also performed for other commonly used insect cell lines such as Sf9 and Hi5™ cells with broadly similar effects. Whilst insect cells can be cultivated successfully in 24-well format in volumes of 2–7 mL, adaptation to a 48- and 96-well format, however, is somewhat more difficult to achieve, and has yet to be reliably demonstrated.
12. The described protocols are based on the Sf21 cell line and ExCell 420 culture medium, but other commercially available *Spodoptera frugiperda* or *Trichoplusia ni* (trade-name High-Five™) derived cell lines can equally be used. Similarly, the ExCell 420 culture medium can be replaced by other serum-free cultivation media such as SF900 II (Invitrogen)

or ExCell 405 (SAFC Biosciences) or Express Five (Invitrogen) for High Five™ cells.

13. In suspension culture most insect cell lines are agitated at a speed of 90–100 rpm on an orbital shaker device. High Five™ cells require a speed of 120–125 rpm to avoid the formation of cell aggregates (clumping). Some insect cell culture protocols recommend addition of heparin (tissue culture grade) to avoid cell aggregation; however, apart from being expensive the addition of heparin to the culture may negatively impact on protein production. The maximal culture volume in a shake flask should not exceed 30% of the total shake flask volume to ensure sufficient oxygen exchange.
14. The optimal DNA: PEI ratio can vary in dependence on the cell line. For example, the Sf21 cell line sold by Clonetech shows in our hands better transfection efficiencies with a DNA: PEI ratio of 1:3 (μg: μg). It is advisable to perform some experiments to determine the optimal DNA: PEI ratio using the cell line(s) available in the lab.
15. Frequently the PEI mixture is prepared in 150 mM NaCl solution; however, no adverse effect has been found when dissolving the PEI stock solution in serum-free culture medium *(39)*. A second important aspect relates to the order of addition: the PEI solution must be added to the DNA solution. This has been shown to impact transfection rates by a factor of 10 when the solution was added drop wise *(31)*.
16. If the solubility of an intracellularly expressed recombinant protein is to be addressed, the cell pellet obtained after the first centrifugation step is resuspended in lysis buffer (50 mM Tris-HCl, pH 8.0, 100 μg/mL PMSF, 10 mM β-Mercaptoethanol, dissolved in 0.9% NaCl solution; this corresponds to a 2× concentrated buffer). Choose a volume that corresponds to 10–20% of the culture volume, e.g. if 1 mL of culture was centrifuged, add 100–200 μL of lysis buffer to the pellet. Carefully resuspend the cell pellet by vortexing, followed by 15 min incubation on ice and subsequent centrifugation at 10,000 × *g* for 3 min. The supernatant now contains the soluble fraction of the protein expressed.
17. The remaining pellet harbors the insoluble fraction of the protein. Upon removal of the supernatant this pellet is dissolved in again 100–200 μL of the following buffer: 20 mM HEPES, pH 7.4, 100 mM NaCl, 1 mM β-Mercaptoethanol, 0.1% SDS and Complete™ protease inhibitor cocktail (Roche Diagnostics). The mixture is sonicated twice for 15 s (amplitude 41% = 20 μ) prior to further analysis by SDS-PAGE and Western Blot.

Acknowledgements

The authors wish to thank Brendan Kerins and Jean-Marc Schlaeppi for technical support and James Groarke for useful discussions and comments during the preparation of this manuscript. This manuscript is dedicated to the memory of Robert Cheung

References

1. Vaughn, J. L., Goodwin, R. H., Tompkins, G. J., and McCawley, P. (1977) Establishment of 2 Cell Lines from Insect *Spodoptera-Frugiperda* (*Lepidoptera-Noctuidae*). *In Vitro-Journal of the Tissue Culture Association*. **13**, 213–217
2. Smith, G. E., Fraser, M. J., and Summers, M. D. (1983) Molecular engineering of the Autographa-Californica Nuclear Polyhedrosis-virus genome – deletion mutations within the polyhedrin gene. *Journal of Virology*. **46**, 584–593
3. Chambers, S. P. (2002) High-throughput protein expression for the post-genomic era. *Drug Discov Today*. 7, 759–765
4. Hunt, I. (2005) From gene to protein: a review of new and enabling technologies for multi-parallel protein expression. *Protein Expression and Purification*. **40**, 1–22
5. Pijlman, G. P., van Schijndel, J. E., and Vlak, J. M. (2003) Spontaneous excision of BAC vector sequences from bacmid-derived baculovirus expression vectors upon passage in insect cells. *Journal of General Virology*. **84**, 2669–2678
6. McCall, E. J., Danielsson, A., Hardern, I. M., Dartsch, C., Hicks, R., Wahlberg, J. M., and Abbott, W. M. (2005) Improvements to the throughput of recombinant protein expression in the baculovirus/insect cell system. *Protein Expression and Purification*. **42**, 29–36
7. Hink, W. F. and Vail, P. V. (2004) A plaque assay for titration of alfalfa looper nuclear polyderosis virus in cabbage looper TN-368 cell line. *Journal of Invertebrate Pathology*. **22**, 268–174
8. O'Reilly, D. R., Miller, L. K., and Luckow, V. A. (2004) *Baculovirus expression vectors: a laboratory maunal*. WH Freeman, NY
9. Kwon, M. S., Dojima, T., Toriyama, M., and Park, E. Y. (2002) Development of an antibody-based assay for determination of baculovirus titers in 10 hours. *Biotechnology Progress*. **18**, 647–651
10. Volkman, L. E. and Goldsmith, P. A. (1981) Baculovirus bioassay not dependent upon polyhedra production. *Journal of General Virology*. **56**, 203–206
11. Cha, H. J., Gotoh, T., and Bentley, W. E. (1997) Simplification of titer determination for recombinant baculovirus by green fluorescent protein marker. *Biotechniques*. **23**, 782–786
12. Eriksson, S., Raivio, E., Kukkonen, J. P., Eriksson, K., and Lindqvist, C. (1996) Green fluorescent protein as a tool for screening recombinant baculoviruses. *Journal of Virological Methods*. **59**, 127–133
13. Berns, K. I. and Giraud, C. (1996) Biology of adeno-associated virus. *Current Topics in Microbiology and Immunology*. **218**, 1–23
14. Malde, V. and Hunt, I. (2004) Calculation of baculovirus titer using a microfluidic-based Bioanalyzer. *Biotechniques*. **36**, 942–946
15. O'Brien, J., Wilson, I., Orton, T., and Pognan, F. (2000) Investigation of the Alamar Blue (resazurin) fluorescent dye for the assessment of mammalian cell cytotoxicity. *European Journal of Biochemistry*. **267**, 5421–5426
16. Pouliquen, Y., Kolbinger, F., Geisse, S., and Mahnke, M. (2006) Automated baculovirus titration assay based on viable cell growth monitoring using a colorimetric indicator. *Biotechniques*. **40**, 282–286
17. Chao-Min Lu. (2006) Simple, Overnight Titration of Baculovirus and Scale-up Production of High Titer Virus Stock. *Baculovirus Technology, Cambridge Healthtech Institute Meeting*, Boston, September 25–26th 2006
18. Bahia, D., Cheung, R., Buchs, M., Geisse, S., and Hunt, I. (2005) Optimisation of insect cell growth in deep-well blocks:development of a high-throughput insect cell expression system. *Protein Expression & Purification*. **39**, 61–70
19. Chambers, S. P., Austen, D. A., Fulghum, J. R., and Kim, W. M. (2004) High-through put screening for soluble recombinant expressed

kinases in Escherichia coli and insect cells. *Protein Expression Purification.* **36**, 40–47
20. Kim, Y. K., Shin, H. S., Tomiya, N., Lee, Y. C., Betenbaugh, M. J., and Cha, H. J. (2005) Production and N-glycan analysis of secreted human erythropoietin glycoprotein in stably transfected Drosophila S2 cells. *Biotechnology and Bioengineering.* **92**, 452–461
21. Shin, H. S. and Cha, H. J. (2002) Facile and statistical optimization of transfection conditions for secretion of foreign proteins from insect Drosophila S2 cells using green fluorescent protein reporter. *Biotechnology Progress.* **18**, 1187–1194
22. Keith, M. B., Farrell, P. J., Iatrou, K., and Behie, L. A. (1999) Screening of transformed insect cell lines for recombinant protein production. *Biotechnology Progress.* **15**, 1046–1052
23. McCarroll, L. and King, L. A. (1997) Stable insect cell cultures for recombinant protein production. *Current Opinion in Biotechnology.* **8**, 590–594
24. Pfeifer, T. A. (1998) Expression of heterologous proteins in stable insect cell culture. *Current Opinion in Biotechnology.* **9**, 518–521
25. Douris, V., Swevers, L., Labropoulou, V., Andronopoulou, E., Georgoussi, Z., and Iatrou, K. (2006) Stably transformed insect cell lines: Tools for expression of secreted and membrane-anchored proteins and high-throughput screening platforms for drug and insecticide discovery. *Insect Viruses: Biotechnological Applications.* **68**, 113–156
26. Baldi, L., Muller, N., Picasso, S., Jacquet, R., Girard, P., Thanh, H. P., Derow, E., and Wurm, F. M. (2005) Transient gene expression in suspension HEK-293 cells: Application to large-scale protein production. *Biotechnology Progress.* **21**, 148–153
27. Derouazi, M., Martinet, D., Schmutz, N. B., Flaction, R., Wicht, M., Bertschinger, M., Hacker, D. L., Beckmann, J. S., and Wurm, F. M. (2006) Genetic characterization of CHO production host DG44 and derivative recombinant cell lines. *Biochemical and Biophysical Research Communications.* **340**, 1069–1077
28. Geisse, S. and Henke, M. (2005) Large-scale transient transfection of mammalian cells: A newly emerging attractive option for recombinant protein production. *Journal of Structural and Functional Genomics.* **6**, 165–170
29. Farrell, P. and Iatrou, K. (2004) Transfected insect cells in suspension culture rapidly yield moderate quantities of recombinant proteins in protein-free culture medium. *Protein Expression & Purification.* **36**, 177–185
30. Loomis, K. H., Yaeger, K. W., Batenjany, M. M., Mehler, M. M., Grabski, A. C., Wong, S. C., and Novy, R. E. (2005) InsectDirect System: rapid, high-level protein expression and purification from insect cells. *Journal of Structural and Functional Genomics.* **6**, 189–194
31. Boussif, O., Lezoualch, F., Zanta, M. A., Mergny, M. D., Scherman, D., Demeneix, B., and Behr, J. P. (1995) A versatile vector for gene and oligonucleotide transfer into cells in culture and in-vivo – polyethylenimine. *Proceedings of the National Academy of Sciences of the United States of America.* **92**, 7297–7301
32. Park, T. G., Jeong, J. H., and Kim, S. W. (2006) Current status of polymeric gene delivery systems. *Advanced Drug Delivery Reviews.* **58**, 467–486
33. Geisse, S. (2007) Insect Cell Cultivation and Generation of Recombinant Baculovirus Particles for Recombinant Protein Production. *Metohds in Biotechnology: Animal Cell Biotechnology.* **24**, 489–507.
34. Pfeifer, T. A., Hegedus, D. D., Grigliatti, T. A., and Theilmann, D. A. (1997) Baculovirus immediate-early promoter-mediated expression of the Zeocin resistance gene for use as a dominant selectable marker in dipteran and lepidopteran insect cell lines. *Gene.* **188**, 183–190
35. Lu, M., Farrell, P. J., Johnson, R., and Iatrou, K. (1997) A baculovirus (Bombyx mori nuclear polyhedrosis virus) repeat element functions as a powerful constitutive enhancer in transfected insect cells. *Journal of Biological Chemistry.* **272**, 30724–30728
36. Chen, Y., Yao, B., Zhu, Z., Yi, Y., Lin, X., Zhang, Z., and Shen, G. (2004) A constitutive super-enhancer: homologous region 3 of Bombyx mori nucleopolyhedrovirus. *Biochemical and Biophysical Research Communication.* **318**, 1039–1044
37. Sussman, D. J. (1995) 24-hour assay for estimating the titer of beta-galactosidase-expressing baculovirus. *Biotechniques.* **18**, 50–51
38. Yahata, T., Andriole, S., Isselbacher, K. J., and Shioda, T. (2000) Estimation of baculovirus titer by beta-galactosidase activity assay of virus preparations. *Biotechniques.* **29**, 214–215
39. Kichler, A., Zauner, W., Ogris, M., and Wagner, E. (1998) Influence of the DNA complexation medium on the transfection Vefficiency of lipospermine/DNA particles. *Gene Therapy.* **5**, 855–860

Chapter 15

High-Throughput Protein Expression Using Cell-Free System

Kalavathy Sitaraman and Deb K. Chatterjee

Summary

One of the main challenges in this post genomic era is the development and implementation of efficient methods of protein synthesis. A clear understanding of the role of genes in an organism is to comprehend the biological functions of all of its proteins. Acquiring this knowledge will depend in part on the success of rapid synthesis and purification of proteins. The future of structural genomics and functional proteomics depends on the availability of abundantly expressing, soluble proteins in a high-throughput manner. Conventional cell based methods of protein expression is rather laborious, time consuming and the ways to fail are numerous including solubility, toxicity to the host and instability (e.g. proteolysis). Cell-free or *in vitro* protein synthesis, on the other hand allows the expression and analysis of protein synthesis, may solve many of these problems. It is a simple open system which lends itself for manipulations and modifications to influence protein folding, disulfide bond formation, incorporation of unnatural amino acids, protein stability (by incorporating protease inhibitors in the system) and even the expression of toxic proteins. Cell-free synthesis can also be used as a reliable screening methodology for subsequent protein expression *in vivo*. Furthermore, this technology is readily amenable to automation. Here, we present a protocol for expressing recombinant proteins with high yield in a standard 96-well plate format using *E. coli* cell-free extract in a batch mode.

Key words: Cell-free; Protein expression; High-throughput; Recombinant proteins

Sharon A. Doyle (ed.), *Methods in Molecular Biology: High Throughput Protein Expression and Purification, vol. 498*

Book doi: 10.1007/978-1-59745-196-3

15.1. Introduction

Cell-free transcription/translation system has been traditionally used for analytical purpose to characterize gene products. It makes use of highly active cellular machinery in *E. coli* S30 extract to direct the protein synthesis in the presence of a secondary energy source *(1)*. The efficiency of protein synthesis is governed by the availability of the constant supply of energy. Recent advances made in cell-free protein synthesis have widened its usefulness to various applications in molecular biology and biotechnology. Previously, the technology suffered wide spread acceptance because of low productivity and cost consideration. Fortunately, various improvements have been made in the last few years in producing highly active cellular extract and identifying more stable energy sources, making the system highly competitive with cell-based protein production.

Some of the recent improvements in *E. coli* cell-free protein expression include preparation of the cell lysate in a more concentrated form, removal of endogenous RNAs and amino acids during the processing of the extract, stabilization of energy sources, amino acids and enzymes *(2–5)* that generate the ATP necessary to sustain prolonged protein synthesis.

E. coli cell-free system is designed for coupled transcription and translation of open reading frames (ORF) cloned downstream from the T7 RNA polymerase promoter. Proteins can be synthesized by using either plasmid DNA or linear DNA template obtained by polymerase chain reaction *(6)* or by other methods containing the gene of interest. The procedures for making the key components of the reaction, S30 extract and reaction buffer are described in detail below. The synthesis directed by T7 RNA polymerase *(7)* is performed at 30°C in a thermomixer for 3 h. The synthesized protein is analyzed by SDS–polyacrylamide gel electrophoresis or by functional activity.

15.2. Materials

15.2.1. Handling the Components

The cell free transcription/translation procedure is quite sensitive and cannot tolerate the presence of any RNases. Gloves should be worn at all times while making up the solutions and setting up the reactions. Disposable sterile plastic wares or autoclaved glass wares should be used throughout. All chemicals are purchased as analytical quality RNase free grade. All solutions are prepared using MilliQ water. The freshly made solutions are then either autoclaved or filter sterilized.

15.2.2. S30 Extract

1. *Cell culture medium:* Autoclave buffered rich growth medium containing 16g of tryptone (Difco), 10 g of yeast extract (Difco), 5 g of NaCl (Sigma), 5.68 g of Na_2HPO_4 (Sigma), 2.64 g of NaH_2PO_4 per liter. Prepare 50% glucose solution (w/v), filter sterilized. Add 10 ml of 50% solution per liter of autoclaved medium.
2. *S30 Buffer:* Prepare stock solutions of 2.2 M Tris–acetate, pH 8.2, 1.5 M magnesium acetate and 3.0 M potassium acetate using MilliQ water. Store at RT or in a cold room. Prepare 1.0 M Dithiothreitol (DTT), store at –20°C. Add to a final concentration of 1 mM to the buffer immediately before use.
3. *Pre-incubation mixture:* Prepare the following stock solutions:
   (a) ATP (100 mM), store at –80°C in small aliquots.
   (b) Phosphoenolpyruvate (1.5 M), trisodium salt, adjusted pH to 7.2 and store at –80°C in small aliquots.
   (c) Pyruvate Kinase, 1,200 U/ml, store at 4°C.
   (d) Fifty millimolar of all 20 amino acids mix thoroughly as some of the amino acids are not completely soluble, store at –80°C in aliquots.
4. *Dialysis:* You will need dialysis tubing, SnakeSkin Pleated, 10,000 MWCO (Pierce) and S30 buffer stored at 4°C with 1 mM DTT added just prior to use.

15.2.3. 2.5× Cell-Free Synthesis Buffer

1. Potassium glutamate (3.0 M), filter sterilize, store at RT.
2. Ammonium acetate (7.5 M), filter sterilize, store at RT.
3. Magnesium acetate (3.0 M), filter sterilize, store at RT.
4. HEPES (1.0 M), pH 7.5, adjust pH to 7.5 with 10 N KOH, store as above.
5. NTPS (200 mM), store at –20°C in small aliquots.
6. Folinic acid (20 mg/ml), store aliquots at –20°C.
7. *E. coli* tRNA (60 mg/ml), storage at –20°C in aliquots.
8. Amino acids (50 mM), in aliquots at –80°C.
9. Spermidine (1.5 M), –20°C storage.
10. cAMP (100 mM), sodium salt, adjust the pH to 7.5 with 0.1 N KOH and store at –20°C.
11. Betaine (5.0 M), stored at RT.
12. Trehalose (50%), filter, sterilize and store at RT.

13. D-(−)-3-phosphoglyceric acid (1.5 M), trisodium salt, store at −80°C.
14. Co-enzyme A (50 mM), store at −80°C.

15.2.4. Template Purification

15.2.4.1. Plasmid Template

1. Gateway Cloning Vector kit (Invitrogen).
2. PureLink HQ Mini Plasmid Purification Kit (Invitrogen).

15.2.4.2. Linear Template

1. Platinum High fidelity Supermix (Invitrogen) or Phusion Master Mix HF (New England Biolabs).
2. Custom made gene specific primers (Operon).
3. QiaQuick PCR purification kit (Qiagen).
4. TE buffer: 10 mM Tris–HCl, pH 8.0, 0.1 mM EDTA.

15.2.5. High Throughput Cell-Free Reaction

1. S30 extract, ribosome.
2. 2.5× cell-free buffer.
3. T7 RNA polymerase (Invitrogen or any other commercial sources).
4. Protease inhibitor cocktail (Roche – Complete Mini, EDTA-free). Prepare by dissolving one tablet in 500 µl of MilliQ water. Prepare fresh each time.
5. MilliQ water.
6. Thermomixer R (Eppendorf).
7. 96-Well 0.8 ml storage conical well bottom plate (ABgene).
8. AirPore Tape Sheets (Qiagen).
9. Silverseal Greiner adhesive plate sealer (PGC Scientific).
10. Pre-cast SDS–PAGE Tris–Glycine gels 4–20% (Invitrogen).
11. BenchMark Protein Ladder (Invitrogen).
12. Laemmli Sample Buffer (Biorad).
13. β-Mercaptoethanol.
14. SimplyBlue SafeStain (Invitrogen).

15.3. Methods

15.3.1. Preparation of S30 Extract

Crude S30 extracts capable of supporting transcription and translation was first demonstrated by Zubay and coworkers in 1973 *(1)*. To date, several changes have been made to increase the activity and productivity of S30 extract. Any *E. coli* strain with rapid growth characteristics with less RNase activity could be suitable for a S30 extract preparation. We routinely use A19 strain for standard cell-free synthesis. Cells grown at 37°C to mid-log phase are lysed and centrifuged at 30,000 × *g* to prepare the extract. The following procedure describes in detail the steps involved in making the S30 extract. One can also further fortify the final S30 extract with crude ribosomes. As ribosomes where the protein synthesis occurs, presence of active and substantial amount of ribosomes in the S30 extract is extremely important.

1. Grow (strain A19 of *E. coli* K12) in phosphate rich buffer in baffled flasks at 37°C in a shaker incubator. Use a fresh culture to inoculate the medium. For large scale, fermentation growth is recommended.
2. Harvest the cells in mid-log phase (*see* **Note 1**). The cell growth should be stopped before it reaches the stationary phase. An OD_{595} of 3.0 is recommended. Centrifuge the cells at 4,000 × *g* for 15 min at 4°C. If it is not possible to use the cells immediately, freeze the cell pellet at –80°C and use them as soon as possible (*see* **Note 2**).
3. Prepare S30 buffer: It can be made in a carboy and stored at 4°C for months. Buffer is made from filter sterilized stock buffers to a final concentration of 10 mM Tris–acetate, pH 8.2, 14.0 mM magnesium acetate and 60 mM potassium acetate. Remember to add 1 μl of 1.0 M DTT per milliliter of buffer before use.
4. Thaw cells at 4°C. Thoroughly resuspend the cell pellet in S30 buffer at 1 ml buffer/g cells. The cell suspension is passed through a 60 ml syringe with an 18 G needle to make sure that there are no clumps.
5. The cells can be broken in many different ways. The equipment Emulsiflex C5 (Avestin, Canada) is quite useful for this application (*see* **Note 3**). Before use, the pressure chamber is washed extensively with MilliQ water and then with a liter of S30 buffer.
6. Transfer the cell suspension to the pre-chilled pressure chamber. Pass cell suspension and monitor the pressure gauge. Press the cells at 15,000–20,000 psi. Collect the lysate in a beaker placed on ice. Generally one pass is enough to completely lyse the cells. Alternately, sonication of the cells is quite effective. Resuspended cells are sonicated in Branson Digital sonifier at an amplitude of 65% with a large tip for 30 s. Repeat two more times. It is very useful to allow the sample to cool between cycles. Check under the microscope for the efficiency of cracking, if needed.
7. Centrifuge the cracked cell lysate at 30,000 × *g*, using Beckman SS34 rotor (or a suitable rotor) for 30 min at 4°C to remove the cell debris and chromosomal DNA.
8. In the meantime prepare a pre-incubation mix at the following concentration (**Table 15.1**). Approximately 5 ml of

Table 15.1
Pre-incubation mixture

Reagents	Volume added (ml)/10 ml	Final concentration
2.2 M Tris–acetate, pH 8.2	2.000	440 mM
3.0 M magnesium acetate	0.046	13.8 mM
100 mM ATP	1.974	19.74 mM
1.5 M PEP	0.840	126 mM
1.0 M DTT	0.066	6.6 mM
50 mM all 20 amino acids	0.012	60 μM each
Pyruvate kinase (2,500 U/ml)	0.084	10 U/ml
Water (to 10 ml)	4.978	

pre-incubation mixture is needed for 25 ml of supernatant. Store all components on ice as some of them are labile (*see* **Note 4**).

9. After the centrifugation, carefully remove the tubes so as not to disturb the layers. Three different layers are visible; the bottom cell debris, middle white inter-phase and clear supernatant at the top. Suction off the supernatant delicately. Care should be taken not to withdraw any of the middle inter-phase or the bottom layer (*see* **Note 5**).
10. Transfer the supernatant to a sterile, disposable flask. Note the volume of the supernatant. Add 5 ml of pre-incubation mix per 25 ml of supernatant. Mix well with gentle swirling.
11. Incubate at 37°C for 80 min with gentle shaking (*see* **Note 6**).
12. Once the incubation is complete, transfer the lysate to dialysis tubing (*see* **Note** 7). Dialyze 4 times with 50 volumes of S30 buffer with 1 mM DTT at 4°C for 45 min each.
13. Transfer the dialyzed extract into centrifuge tubes and centrifuge at 4,000 × *g*, for 10 min at 4°C.
14. Collect the supernatant and discard the pellet. The extract prepared by this method contains about 30–35 mg protein per milliliter.

15. Aliquot the extract into sterile cryo-vials (Nunc) and freeze over powdered dry ice. The extract is viable for several years without any loss of activity when properly stored at –80°C (*see* **Note 8**).

15.3.2. Ribosome Preparation

1. The S30 extract is centrifuged at 150,000 × *g* for 4h in a Beckman 50 Ti rotor *(8)*.
2. The ribosome pellet is resuspended in S30 buffer with 1 mM DTT to a final concentration of 30 mg/ml.
3. The suspension is stored at –80°C in small aliquots (*see* **Note 9**).

15.3.3. S30 Extract + Ribosome Mixture

1. Thaw aliquots of S30 extract and ribosomes by quickly warming them in the palm of the hand and place them on ice.
2. Mix S30 extract and ribosome at a ratio of 4:1. For example, mix 800 μl of extract and 200 μl of ribosome to a total of 1 ml at the time of setting up the reaction.

15.3.4. Preparation of 2.5× Cell-Free Synthesis Buffer

We generally prepare the final cell-free synthesis buffer at a concentration of 2.5×, so as to provide enough room for the addition of template, chaperones and other additives, if needed. This 2.5× buffer is made form 10× buffer A and 5× buffer B as described below. The buffers A, B and final 2.5× cell-free synthesis buffer can be made in large quantities and stored in aliquots at –80°C for long term storage. Working aliquots of 2.5× buffer can be stored at –20°C. Prepare the 2.5× buffer on ice using stock solutions. Thoroughly vortex each and every component before adding as some of the components, like some amino acids are not completely soluble. Again, mix the completed reaction buffer frequently while dividing into aliquots for storage.

15.3.4.1. 10× Buffer A

10× buffer is made of potassium glutamate, ammonium acetate and magnesium acetate (**Table 15.2**). Since there are no labile elements in this buffer it can be conveniently stored at 4°C.

15.3.4.2. 5× Buffer B

5× buffer contains rest of the components needed for the protein synthesis. They include NTPs, amino acids and other ingredients. Individual components are prepared as stock solutions as described in **Section 15.2**. The buffer is prepared on ice and the components are added in the order given in **Table 15.3**.

15.3.4.3. 2.5× Cell-Free Synthesis Buffer

10× buffer A and 5× buffer B are combined in such a way to make the final 2.5× cell-free synthesis buffer in the presence of 3-phosphoglycerate *(9)* as the secondary energy source (**Table 15.4**). We prefer to use 3-phosphoglycerate as the energy source because in our hands, it has been the most efficient with respect to the

Table 15.2
10× buffer A

Reagents	Volume added (ml)/10 ml	Final concentration
3.0 M potassium glutamate	7.65	2.3 M
7.5 M ammonium acetate	1.05	800 mM
3.0 M magnesium acetate	0.40	120 mM
Water to volume 10 ml	0.90	

Table 15.3
5× buffer B

Reagents	Volume added (ml)/10 ml	Final concentration
1 M HEPES (pH 7.5)	2.860	286 mM
2 M DTT	0.045	8.8 mM
200 mM ATP	0.501	10 mM
200 mM GTP	0.501	10 mM
200 mM CTP	0.216	4.3 mM
200 mM UTP	0.216	4.3 mM
20 mg/ml folinic acid	0.087	170 μg/ml
60 mg/ml *E. coli* tRNA	0.143	0.85 mg/ml
50 mM cysteine, arginine and tryptophan[a]	1.000	5 mM
1 M spermidine	0.076	7.5 mM
50 mM amino acids	1.250	6.25 mM
100 mM cAMP	0.332	3.32 mM
5 M betaine	2.500	1.25 M
Water to volume 10 ml	0.223	

[a]Cysteine, arginine and tryptophan are added slightly more than other amino acids as they are partially soluble in water

Table 15.4
2.5× cell-free-synthesis buffer

Reagents	Volume added (ml)/10 ml	Final concentration
10× buffer A	2.500	2.5×
5× buffer B	5.000	2.5×
1.5 M 3-phosphoglyceric acid	0.700	105 mM
50 mM coenzyme A	0.313	1.56 mM
50% trehalose[a]	1.000	5%
Water to 10 ml	0.487	

[a]Conventionally PEG (polyethyleneglycol) is used in the reaction. But the presence of PEG in the reaction cause gel artifacts and protein insolubility. As a result the bands get compressed at the lower portion of the gel. Use of trehalose alleviates this problem (*see* **Note 10**)

Table 15.5
Cell-free translation/transcription reaction cocktail

Components	1 reaction for 50 μl[a]	100 reactions for 50 μl each[b]
S30 extract mix	20.0 μl	2.00 ml
2.5× cell-free synthesis buffer	20.0 μl	2.00 ml
Protease inhibitor	1.0 μl	100 μl
T7 RNA polymerase	1.0 μl	100 μl

[a]Make a total volume of 50 μl with DNA solution with up to 8 μl

[b]The volume can be adjusted for more than 100 or less than 100 reactions using the ratios for one 50-μl reaction. Also, reaction volumes can be increased or decreased according to the needs

amount of protein yield *(9, 10)* as well as it is quite inexpensive than the conventional energy sources. Other energy sources such as phosphoenolpyruvate, acetyl phosphate or creatine phosphate can also be used at same concentration.

15.3.5. DNA Templates

Any of the following DNA templates may be used in *E. coli* cell-free expression system. a. Supercoiled plasmid DNA, b. Linear

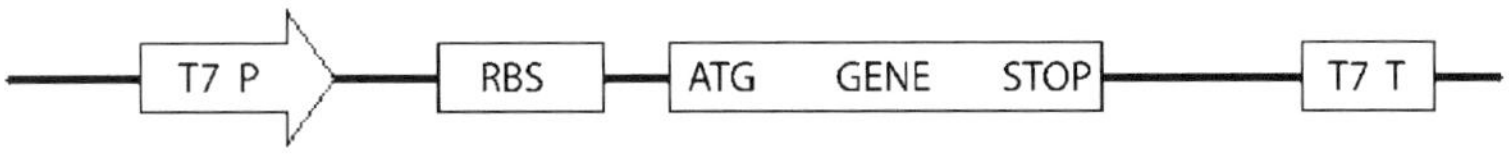

Fig. 15.1. Optimal configuration of DNA template (plasmid or PCR product) used for cell-free protein synthesis (*see* **Note 11** for detailed explanation).

DNA, c. PCR product. All these templates must contain the T7 promoter, a prokaryotic Shine-Dalgarno sequence (ribosome binding site) upstream of gene of interest and an initiation codon for efficient translation/transcription. It should also include a stop codon (**Fig. 15.1**). A transcription terminator sequence for T7 transcription termination may be used for message stability (*see* **Note 11**) and thus, efficient protein synthesis.

1. Generate PCR products using Platinum Infidelity PCR Supermix (Invitrogen) or any other suitable enzyme following manufacturer's instruction.
2. Purify the PCR product with PCR purification kit (Qiagen) or any other commercially available kit for optimal results (*see* **Note 12**).
3. To make supercoiled template, clone the gene of interest by Gateway cloning technology *(11)* (Invitrogen) or by any other methodology (*see* **Note 13**).
4. After generating the clone, isolate high quality plasmid DNA using PureLink HQ Plasmid purification kit (Invitrogen) following supplier's protocol. The concentration of the template should be kept around 500 ng to 1.0 μg/μl (*see* **Note 14**).

15.3.6. High Throughput Cell-Free Reaction

E. coli coupled cell-free transcription/translation system in a 96-well format lends itself extremely useful for high throughput applications of large scale screening of *in vitro* expression of target proteins *(12–16)*. Several parameters can be checked simultaneously for optimal protein expression. The effect of detergents, chaperones, lipids and other additives can be examined thoroughly in a single experiment.

15.3.6.1. Preparation of 96-Well Plate (0.8 ml) (see Note 15)

This plate can be used for setting up reaction volume of 10–400 μl. Accordingly the reagents can be adjusted proportionately to the desired volume. Similarly the number of reactions to be performed can also be altered. The following procedure is suitable for setting up the whole plate with a reaction volume of 50 μl/well (**Table 15.5**).

1. Calculate the amount of S30 extract mix and the amount of 2.5× buffer needed.

Number of reaction = 1 negative control + 1 positive control + X test reactions + 4 extra. More than one positive and one negative control can also be done.

Amount of extract and buffer needed = Total number of reaction × 20 μl

E.g., For 96-well plate = 96 + 4 = 100 × 20 μl = 2.0 ml (*see* **Note 16** for negative and positive control).

2. Quickly thaw S30 extract mix, invert the tubes a few times to mix thoroughly the contents and place it on ice. Do not vortex.
3. Thaw 2.5× buffer, some of the components might have precipitated. Vortex well. Remember to vortex again before adding to rest of the components.
4. Prepare cell-free protein synthesis cocktail of S30 extract and ribosome mix, 2.5× buffer, protease inhibitor, RNase out and T7 RNA polymerase as described in **Table 15.5**:
5. Dispense 42 μl of cell-free synthesis cocktail mix to each well of the 96-well microtiter plates. At this stage, the plates can be used immediately or stored for later use.
6. The plates are sealed with a single Silverseal Greiner adhesive plate sealer.
7. They can be stored at −80°C until use for several months (*see* **Note 17**).

15.3.6.2. Setting up the Reaction

1. Remove the plate from −80°C and thaw on ice for 10 min.
2. Carefully remove the adhesive seal.
3. Prepare 1.0 μg of template in nuclease free water to a total volume of 8.0 μl (*see* **Note 18**).
4. Add each template to designated wells to make up the volume to 50 μl (*see* **Note 19**).
5. Seal the plate with AirPore sealing tape.
6. Incubate the reaction at 30°C in a thermomixer fitted with 96-well plate holder for 3.0 h. with shaking at ~500 g (*see* **Note 20**).
7. When the reactions are done the expressed proteins can be analyzed immediately or stored at −20°C until ready to run the gel.

15.3.6.3. Analysis of Expression

The expressed recombinant protein samples are run on SDS–PAGE gels along with protein molecular weight standard (*see* **Fig. 15.2**). This allows the easy visualization of the expressed protein and its approximate molecular weight. The BenchMark protein ladder consists of 15 distinct protein bands in the range of 10–220 kDa.

The expressed protein as such is the total protein. The soluble proteins are obtained by centrifuging the total protein in the

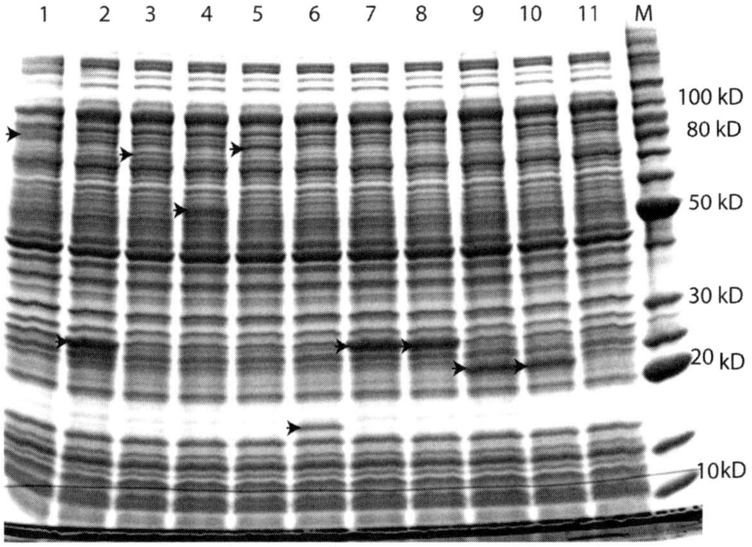

Fig. 15.2. Analysis of proteins expressed in cell-free synthesis system. A typical polyacrylamide gel electrophoresis of some of the proteins expressed in a high-throughput format in 96-well 0.8 ml microtiter plate is shown here. The reaction was performed in 50 μl volume in a thermomixer with 1,200 rpm. 1.0 μl out of 50 μl total protein synthesized was run on 4–20% Tris–glycine gel. (Lane 1) HBV-RT, (Lane 2) K-Ras, (Lane 3) Yop-D, (Lane 4) Wip1, (Lane 5) Tol, (Lane 6) SR-5, (Lanes 7 and 8) GFP, (Lanes 9 and 10) CAT, (Lane 11) No template negative control. Lanes 1–7 and 9 are synthesized using plasmid DNA as template. In lanes 8 and 10 PCR product was used as template. GFP serves as positive control.

microtiter plate at 2,200 × *g* for 15 min at 4°C. Carefully remove the supernatant, taking care not to disturb the pellet at the bottom.

1. Prepare 2× sample loading buffer by mixing 900 μl of Laemmli Sample Buffer and 100 μl of β-mercaptoethanol. Mix 500 μl of this 2× sample buffer with 400 μl of water. This is the final loading buffer to be used.
2. Add 2 μl of each total protein expressed and if desired 2 μl of soluble expressed protein sample to 18 μl of sample loading buffer.
3. Heat the samples at 70°C for 10 min and briefly centrifuge.
4. Load 10 μl of the denatured sample on SDS–PAGE gel and electrophorese at 130 V until the dye front reaches the bottom.
5. Visualize the protein by staining using SimplyBlue SafeStain following manufacturer's instructions.

15.4. Notes

1. Different cells grow at different rates. It is best to use *E. coli* with high growth rate. We generally harvest about 150–175 g of A19 cell paste from 15 L of fermentation growth in buffer rich media. The cell paste is stored in small aliquots of 25 g in sterile containers.

2. It is recommended to prepare the extract as soon as the cells are harvested to retain high specific activity. But we have used cells stored at −80°C up to couple of weeks without compromising the quality of the extract. No difference in protein synthesis was observed between the extract made from fresh and stored cell pellets.

3. The Emulsiflex-C5 temperature controlled homogenizer has an air/gas driven, high-pressure pump developed and manufactured by Avestin in Canada. It is possible to process as small as 7 ml cell suspension in this homogenizer. A high pressure pump pushes product through an adjustable homogenizing valve, homogenize, filter and extrude the lysed cells. Homogenization pressure up to 30,000 psi/207 MPa can be achieved.

4. Phosphoenolpyruvate and pyruvate kinase are used as phosphate regeneration system within the extract. Making them fresh is recommended for each extract preparation.

5. The bottom pellet is compact; but the middle whitish interphase is not. It is easy to siphon off the contaminating layer along with the supernatant. It is better to remove just two thirds of the supernatant on the top.

6. Do not shake vigorously. Care should be taken not to introduce any air bubbles into the lysate.

7. SnakeSkin dialysis tubings are suitable for large scale sample dialysis. It is open and pleated made of regenerated cellulose. It is mess free and ready to use. The sample is directly added to the tubing.

8. For throughput applications aliquots of 400 and 800 μl are convenient volumes. It enables the addition of 100 and 200 μl of ribosomes directly into these vials while setting up the reaction to make S30 extract mixture (*see* **Note 15**).

9. Ribosome is stored in aliquots of 100 or 200 μl. Avoid frequent freeze thaws as they degrade rapidly and loss activity.

10. The presence of trehalose in cell-free synthesis reaction seems to increase the solubility of synthesized protein.

11. Several commercially available T7 based expression vectors contain a T7 promoter, ribosome binding site (RBS), and

T7 terminator with the suitable spacing and sequence configuration for optimal protein expression in *E. coli* cell-free system. If you are constructing your own vector attention should be paid to the following aspects: The gene of interest should be placed down stream of T7 promoter and RBS and should contain an ATG initiation codon. For linear or PCR products, upstream of T7 promoter should contain a minimum of 4–10 nucleotides for effective promoter binding. This sequence need not be specific and it is required for linear PCR products. The sequence following the T7 promoter should contain a minimum of 15–20 nucleotides which form a potential stem and loop structure as described by Studier et al. *(7)*. Another non-specific sequence of 9–11 nucleotides between the RBS and the ATG initiation codon should be included to ensure the optimal translation efficiency. T7 terminator should be located 4–100 nucleotides down stream of gene of interest for efficient transcription termination and stability of messenger RNA.

12. PCR products can be directly used in the cell-free reaction. However, purified PCR products are more productive. Once purified using QiaQuick PCR purification kit the DNA can be resuspended in 20 or 30 μl of TE to a final concentration of ~250 ng/μl.
13. Gateway cloning technology *(11)* is a robust method of generating a variety of protein expression constructs in a multiple host system. This recombinational cloning is quite rapid and highly suitable for high throughput applications of generating large numbers of parallel clones.
14. Make sure the purified DNA is free of ethanol and excess salt. Carefully wash the ethanol precipitated DNA with 70% ethanol to remove carried over salts. The DNA should also be devoid of any RNase contamination.
15. Several microtiter plates suitable for high throughput applications are commercially available from different vendors. We examined quite a few of them and found this 0.8 ml deep well, conical bottom, nuclease free, individually wrapped microtiter plate from ABgene more suitable for reaction volumes of 25–400 μl. There is no compromise in quality of the products made between different volumes. If the volume has to be scaled up further, up to 800 μl, it can be easily done in 1.2 ml deep well storage plates. These plates can be conveniently agitated at 400 g without any spill over in a thermomixer fitted with microplate holder.
16. Inclusion of negative and positive controls is highly recommended. Negative control is the reaction without any template. This serves as the background over which the

expressed proteins can be easily identified on the gel. GFP construct is one of the examples of positive control. GFP expresses very well with a product yield of almost a mg/ml under optimal conditions and the protein can be optically visualized. This expression also gauges the condition of other proteins being synthesized as the rest of the reaction components are same as in test reactions (*see* **Fig. 15.2**).

17. The plates prepared with master cocktail mix containing extract, buffer and enzyme if stored properly at -80°C are quite stable up to 6 months. Frequent freeze thawing should be avoided. If the whole plate is not going to be used in one experiment prepare them as needed. The protein synthesized with plates stored at –80°C is as good as freshly set up reactions.
18. It is better to keep the template DNA concentration at 500 ng/µl, if possible. It enables the addition of other additives like chaperones, detergents etc., if necessary. Even though we use 1 µg of DNA/50 µl reaction it can be reduced to 0.5 µg. It is true for reactions with PCR templates as well.
19. It is preferable to make an excel spreadsheet of 96 wells with different parameters to be used for easy addition of templates and other components.
20. The microtiter plate should be vigorously shaken at 400–500 g during incubation at 30°C. We don't recommend using a water bath. The protein yield is considerably less without shaking.

Acknowledgements

This project has been funded in whole or in part with federal funds from the National Cancer Institute, National Institutes of Health, under contract N01-CO-12400. The content of this publication does not necessarily reflect the views or policies of the Department of Health and Human Services, nor does mention of trade names, commercial products, or organizations imply endorsement by the U.S. Government.

References

1. Zubay G (1973) *In vitro* synthesis of protein in microbial systems. Annu Rev Genet 7, 267–287.
2. Kim D-M and Swartz JR (1999) Prolonging cell-free protein synthesis with a novel ATP regeneration system. Biotechnol Bioeng **66**, 180–188.
3. Kim D-M and Swartz JR (2000) Prolonging cell-free protein synthesis by selective reagent additions. Biotechnol Prog **16**, 385–390.

4. Calhoun KA and Swartz JR (2005) An economical method for cell-free protein synthesis using glucose and nucleotide mono phosphates. Biotechnol Prog **21**, 1146–1153.
5. Ryabova LA, Vinokurov LM, and Shekhovtsova EA et al. (1995) Acetyl phosphate as an energy source for bacterial cell-free translation systems. Anal Biochem **226**, 184–186.
6. Lesley SA, Brow MA, and Burgess RR (1991) Use of in vitro protein synthesis from polymerase chain reaction generated templates to study interaction of *E. coli* transcription factors with core RNA polymerase and for epitope mapping of monoclonal antibodies. J Biol Chem **266**, 2632–2638.
7. Studier FW, Rosenberg AH, Dunn JJ, and Dubendorff JW (1990) Use of T7 RNA polymerase to direct expression of cloned genes. Methods Enzymol **185**, 60–89.
8. Kudlicki W, Kramer G, and Hardesty B (1992) High efficiency cell-free synthesis of proteins: refinement of the coupled transcription/translation system. Anal Biochem **206**, 389–393.
9. Sitaraman K, Esposito D, Klarmann G et al. (2004) A novel cell-free protein synthesis system. J Biotechnol **110**, 257–263.
10. Kuem J-W, Kim T-W, Park C-G et al. (2006) Oxalate enhances protein synthesis in cell-free synthesis system utilizing 3-phosphoglycerate as energy source. J Biosci Bioeng **101**(2), 162–165.
11. Hartley JL, Temple GF, and Brasch MA (2000) DNA cloning using in vitro site-specific recombination. Genome Res **10**, 1788–1795.
12. Knaust RKC, Nordlund P (2001) Screening for soluble expression of recombinant proteins in a 96-well format. Anal Biochem **297**, 79–85.
13. Shih Y-P, Kung W-M, Chen J-C et al. (2002) High throughput screening of soluble recombinant proteins. Protein Sci **11**, 1714–1719.
14. Yokoyama S. (2003) Protein expression system for structural genomics and proteomics. Curr Opin Chem Biol **7**, 39–43.
15. Busso D, Kim R, Kim S-H (2003) Expression of soluble recombinant proteins in a cell-free system using a 96-well format. J Biochem Biophys Methods **55**, 233–240.
16. Betton J-M (2004) High throughput cloning and expression strategies for protein production. Biochimie **86**, 601–605.

Chapter 16

The Production of Glycoproteins by Transient Expression in Mammalian Cells

Joanne E. Nettleship, Nahid Rahman-Huq, and Raymond J. Owens

Summary

In this chapter, protocols for the growth and transfection of Human Embryonic Kidney (HEK) 293T cells for small scale expression screening and large scale protein production are described. Transient expression in mammalian cells offers a method of rapidly producing glycoproteins with a relatively high throughput. HEK 293T cells, in particular, can be transfected with high efficiency (> 50% cell expression) and are amenable to culture at multi-litre scale. Growing cells in micro-plate format allows screening of large numbers of vectors in parallel to prioritise those amenable to scale-up and purification for subsequent structural or functional studies. The glycoform of the expressed protein can be modified by treating cell cultures with kifunensine which inhibits glycan processing during protein synthesis. This results in the production of a chemically homogeneous glycoprotein with short mannose-rich sugar chains attached to the protein backbone. If required, these can be readily removed by endoglycosidase treatment.

Key words: Glycoprotein; Mammalian; HEK 293T; Transient, Protein production; Protein purification; Glycan

16.1. Introduction

Many cell surface and secreted mammalian and eukaryotic viral proteins require glycosylation for proper folding *(1, 2)*. These glycosylation requirements have necessitated the development of expression technology using eukaryotic cells for the production of glycoproteins. The eukaryotic cell lines available include yeast, baculovirus infected insect cells and mammalian cells transfected with plasmid vectors. Currently, *Saccharomyces cerevisiae* and

Sharon A. Doyle (ed.), *Methods in Molecular Biology: High Throughput Protein Expression and Purification, vol. 498*

Book doi: 10.1007/978-1-59745-196-3

Pichia pastoris are the two most common yeasts used for protein expression. The advantages of yeast are their ease of culture, rapid growth and low production costs. Although proteins from higher eukaryotes are glycosylated in yeast cells with mannose-rich glycan structures, the protein production machinery differs from those of higher eukaryotes which can result in mis-folding of the protein-of-interest *(3)*. The construction of recombinant baculoviruses and infection of insect cells is a well established system for the expression of mammalian glycoproteins due to the processing capabilities of the host cells *(4)*. Glycosylation of the protein produced occurs with the dominant glycan forms being $GlcNAc_2Man_3$ and $GlcNAc[Fuc]GlcNAcMan_3$ (*see* **Fig. 16.1**). These sugar chains differ from the larger and more complex glycans which are added to proteins produced in mammalian cells

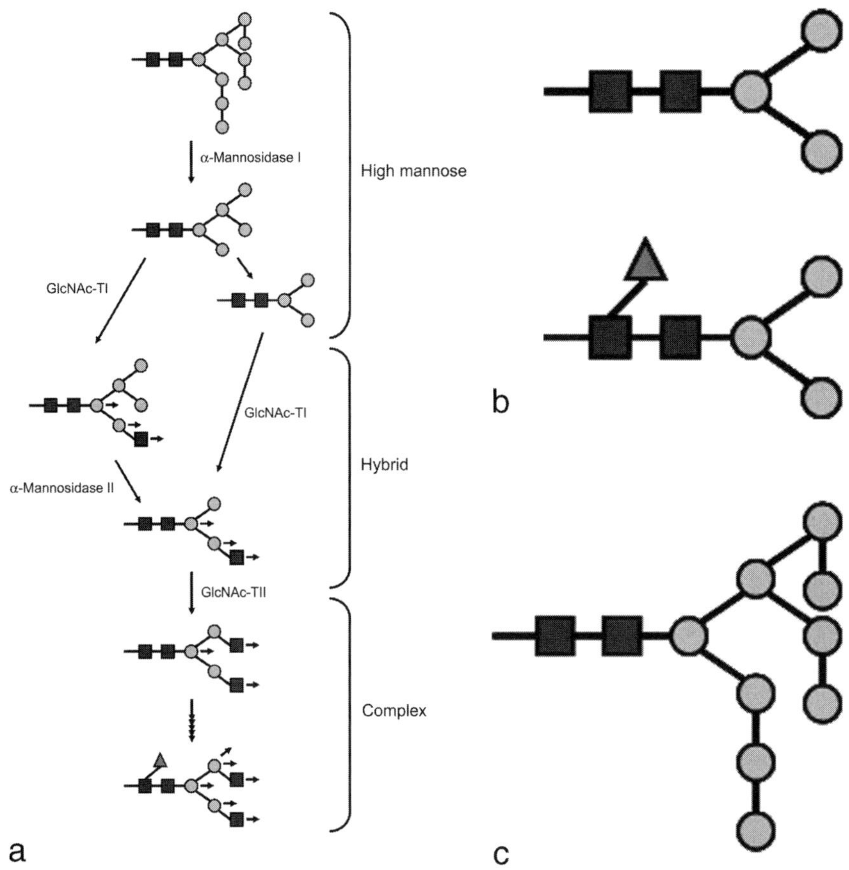

Fig. 16.1. Diagrams of the glycoforms present on proteins produced by eukaryotic cell lines. Part of the glycan formation pathway which takes place in mammalian cells showing the three subsets of N-glycans: high-mannose, hybrid and complex (**a**). The *arrows* indicate the location of branch formation in N-glycan diversification. Endoglycosidase sensitive glycoforms obtained on products produced in baculovirus infected insect cells (**b**) and in HEK 293T cells treated with kifunensine (**c**). *N*-acetylglucosamine (GlcNAc) is represented by a *black square*, mannose (Man) by a *light grey circle* and fucose (Fuc) by a *dark grey triangle.*

(*see* **Fig. 16.1**) and can readily be removed, if required, by treating the glycoproteins with a combination of Endoglycosidase (Endo) H and Endo D *(5, 6)*. However, major disadvantages of the baculovirus system are the time needed to obtain a good titre of virus for infection of the cells which is a particular problem for rapid screening of many constructs; and the relatively high cost of insect cell culture. The third option is mammalian cells which may be used for either stable expression, in which the introduced expression plasmid is stably integrated into the host genome, or transient expression involving the episomal replication of the transfected gene. If a protein-of-interest is to be expressed many times over for a number of studies or is to be used as a drug target, the use of stable cell lines is appropriate although time-consuming to set up *(7)*. However, in a high-throughput laboratory where rapid screening of many constructs is required and repeated expression experiments are not needed, the transient system is less time consuming and more convenient *(8, 9)*. In contrast to yeast and insect cells, mammalian glycoproteins expressed in mammalian cells are authentically glycosylated *(4)* resulting in a recombinant product most similar to that formed *in vivo* (*see* **Fig. 16.1**). However, if the protein is to be used for crystallisation, such complex glycoforms are heterogeneous and can form mobile regions within the protein molecule, both of which can prevent crystallisation. These problems can be overcome by blocking the glycosylation pathway within the cells with the α-mannosidase I inhibitor, kifunensine (*see* **Fig. 16.1**) *(10)*. This results in homogeneous glycans of the form $GlcNAc_2Man_9$ which can aid crystallisation of the product (*see* **Fig. 16.1**). In addition these sugars can be removed using Endo H to leave one GlcNAc moiety; thus removing mobile regions within the protein structure. Here we present detailed protocols for the transient production of glycoproteins in human embryonic kidney (HEK) 293T cells. The methods include the maintenance of HEK 293T cells (**Section 16.3.1**); small scale expression screening of constructs (**Section 16.3.2**); and large scale production and purification of glycoproteins (**Sections 16.3.3** and **16.3.4** respectively).

16.2. Materials

16.2.1. Cell Maintenance

1. HEK 293T cells (ATCC no. CRL-1573 – LGC Promochem, Teddington, UK)
2. Dulbecco's Modified Eagles Medium (DMEM) supplemented with L-Glutamine (1:100) and non-essential amino acids (1:100) plus 10% Foetal Calf Serum (FCS) (Invitrogen, Paisley, UK)

3. Trypsin-Ethylene dinitrilotetra-acetic acid (Trypsin-EDTA)
4. Phosphate buffered saline (PBS): 0.01 M phosphate buffer, 0.0027 M potassium chloride, 0.137 M sodium chloride, pH 7.4
5. Cryomedia (filter sterilised): 10% dimethyl sulphoxide (DMSO), 80% FCS, 10% Dubecco's Modified Eagle Medium (DMEM)
6. Expanded surface roller bottles (Greiner Bio-One, Stonehouse, UK)

16.2.2. Small and Large Scale Cell Transfection

1. Plasmid DNA with an A_{260}/A_{280} ratio of greater than 1.8 (*see* **Note 1**)
2. GeneJuice™ (Novagen, Nottingham, UK) or similar transfection reagent
3. Polyethylenimine (PEI) ("25 kDa branched PEI" – Sigma-Aldrich, Gillingham, UK). Prepare as a stock solution of 100 mg/ml in water. Then dilute to 1 mg/ml, neutralise with HCl, filter sterilise and store frozen in aliquots
4. Kifunensine (Toronto Research Chemicals, Canada) (*see* **Note 2**)

16.2.3. SDS Polyacrylamide Gel Electrophoresis and Western Blotting

1. NPI-10: 50 mM NaH_2PO_4, 300 mM NaCl, 10 mM imidazole, pH 8.0, in water
2. DNAse I from bovine pancreas (Sigma-Aldrich, Gillingham, UK)
3. 2× SDS–PAGE sample loading buffer: 100 mM Tris–HCl pH6.7, 40 mg/ml sodium dodecyl sulphate (SDS), 2 mg/ml bromophenol blue, 20% v/v glycerol, in water
4. SDS–PAGE MES running buffer (Invitrogen, Paisley, UK)
5. SDS–PAGE gel (NuPAGE™ – Invitrogen, Paisley, UK) and Nitrocellulose or PVDF membrane (iBlot™ stack – Invitrogen, Paisley, UK)
6. Phosphate buffered saline with Tween (PBST): 0.01 M phosphate buffer, 0.0027 M potassium chloride, 0.137 M sodium chloride, 0.05% v/v Tween 20, pH 7.4
7. Primary antibody solution: 1:500 Anti-His_6 mouse monoclonal antibody, 1% milk powder, in PBST
8. Secondary antibody solution: 1:5,000 Anti Mouse-Goat IgG-Peroxidase conjugate
9. ECL-Plus™ Western blotting detection system (GE Healthcare, Little Chalfont, UK)

16.2.4. Large Scale Glycoprotein Purification

1. 5 ml HisTrap FF Crude column (GE Healthcare, Little Chalfont, UK) (*see* **Note 3**)

2. HiLoad 16/60 Superdex 200 column or HiLoad 16/60 Superdex 75 column (GE Healthcare, Little Chalfont, UK) (*see* **Note 4**)
3. ÄKTAxpress™ purification system (GE Healthcare, Little Chalfont, UK)
4. Nickel Wash Buffer (filtered and de-gassed): 50 mM Tris–HCl, 500 mM NaCl, 30 mM imidazole, pH 7.5
5. Nickel Elution Buffer (filtered and de-gassed): 50 mM Tris–HCl, 500 mM NaCl, 500 mM imidazole, pH 7.5
6. Size Exclusion Buffer (filtered and de-gassed): 20 mM Tris–HCl, 200 mM NaCl, pH 7.5
7. Simply Blue Safe Stain (Invitrogen, Paisley, UK)
8. Endoglycosidase H (Endo H) (Sigma-Aldrich, Gillingham, UK)

16.2.5. Glycoprotein Quality Assurance by Mass Spectrometry

1. Peptide-*N*-glycosidase F (PNGase F) (Sigma-Aldrich, Gillingham, UK)
2. 20 μM myoglobin solution
3. C4 PepMap 300 μ-precolumn cartridge (Dionex, UK) (*see* **Note 5**)
4. Wash solvent: 97.3% water, 2% acetonitrile, 0.5% formic acid, 0.2% trifluoroacetic acid
5. Solvent A: 95% water, 5% acetonitrile with 0.1% formic acid
6. Solvent B: 20% water, 80% acetonitrile with 0.1% formic acid
7. Pierceable Power Seals (Greiner Bio-One, Stonehouse, UK)

16.3. Methods

16.3.1. Cell Maintenance

Maintaining healthy cell cultures is essential for obtaining a good yield of glycoprotein from a transient expression experiment. Therefore the cells need to be passaged regularly, typically twice per week and fresh cells taken from cryo-storage and resuscitated approximately once every 3 months. The viability and appearance of cells need to be monitored routinely to check for contamination and to ensure the cultures are proliferating as expected (cell doubling time is approximately 24 h) (*see* **Fig. 16.2**). In addition, it is important to maintain frozen stocks by periodically freezing down batches of cells.

16.3.1.1. Cell Resuscitation

1. All cell manipulations are carried out in a Class 2 laminar flow hood.

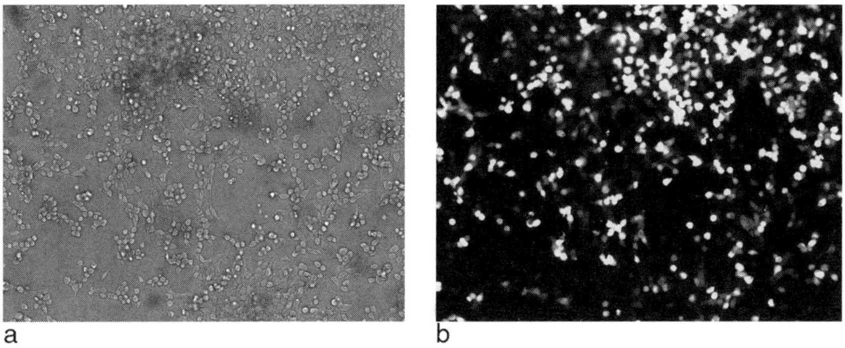

Fig. 16.2. Photographs at 10× magnification of HEK 293T cells expressing green fluorescent protein (GFP) 24 h after transfection. (**a**) is taken under white light and (**b**) under ultra-violet light. The cells can be seen to be ~60% confluent with around 55% of the cells expressing GFP (seen as *white* in (**b**)

2. Remove the cryovial containing the HEK 293T cells (usually 2×10^6 cells/ml) from the liquid nitrogen storage dewar and place in a container on ice.
3. Thaw the cells by slowly adding 0.5 ml of room temperature DMEM supplemented with 10% FCS, 1× non-essential amino acids and 1 mM glutamine. Then pipette the cells slowly up and down until they are resuspended.
4. When the cells have thawed, transfer the contents of the cryovial to a 15 ml centrifuge tube and centrifuge at 250×*g* for 10 min.
5. Discard the supernatant and re-suspend the cells slowly in 5 ml of DMEM supplemented with 10% FCS, 1× non-essential amino acids and 1 mM glutamine.
6. Transfer the cell suspension to a 175 cm^2 tissue culture flask.
7. Slowly add 20 ml of DMEM supplemented with 10% FCS, 1× non-essential amino acids and 1 mM glutamine and incubate at 37°C in a 5% CO_2/95% air environment for 24 h.
8. Remove the media from the flask and discard.
9. Add 25 ml of DMEM supplemented with 10% FCS, 1× non-essential amino acids and 1 mM glutamine and return to the incubator.

16.3.1.2. Cell Culture

1. Remove the media from a confluent 175 cm^2 tissue culture flask.
2. Add 5ml trypsin-EDTA solution and incubate at room temperature for 5 min.
3. Inactivate the trypsin-EDTA by adding 15 ml of supplemented DMEM, count and assess cell viability as follows:
   a. Resuspend the cells and take 10 μl for analysis. If the cells are fully confluent, dilute 1:10 in PBS.
   b. Add 10 μl of trypan blue, mix and place 10 μl of the resulting solution into a hemocytometer chamber.

c. Cover with a glass slide and place under a white light microscope at 10× magnification.

d. Count the number of blue cells. These are the non-viable cells as they take up the trypan blue dye.

e. Count the number of non-dyed cells which are viable. Calculate the mean number of viable and non-viable cells per ml and hence the percentage of viable cells (*see* **Note 6**).

4. Add 4 ml of the cell suspension into a new 175 cm^2 tissue culture flask and add 20 ml of supplemented DMEM. Place the flask in a static incubator at 37°C in a 5% CO_2/95% air atmosphere.

16.3.1.3. Cryo-Preservation of Cells

1. Ensure the cells are healthy with > 95% viability (**Section 16.3.1.2**).
2. Remove the media from a confluent 175 cm^2 tissue culture flask, trypsinise the cells, transfer the cell suspension to a 50 ml tube and centrifuge at 250×*g* for 5 min.
3. Remove the supernatant and discard and re-suspend the cells in cryomedia so that the final cell density is between 1.5×10^6 and 2×10^6cells/ml.
4. Pipette the cell solution into cryovials, label well with solvent resistant marker and place in a cryo-freezing container partially filled with isopropanol.
5. Store overnight in a –20°C freezer.
6. Remove the cryovials from the freezer and transfer to a liquid nitrogen store for long-term storage. For short-term storage (6 months to 1 year) the cryovials may be stored at –80°C.

16.3.2. Small Scale Expression Screening

Small scale expression screening allows for fast, effective assessment of expression level from a large number of plasmids in parallel. The level of secreted and intracellular protein expression can be assessed within a week of cloning the gene-of-interest. A useful example of parallel expression screening in a number of hosts including HEK 293T is given in the work of Berrow et al. *(11)*. A protocol for the parallelised small scale expression screening of plasmids using HEK 293T cells is given in **Section 16.3.2.1** followed by methods for sample preparation and Western blot analysis of both the secreted and intracellular products (**Sections 16.3.2.2** and **16.3.2.3** respectively). Products giving positive secreted expression can then be scaled-up for production and purification (**Sections 16.3.3** and **16.3.4**).

16.3.2.1. Transfection of Cells in 24 well Plates

1. Seed the HEK 293T cells into a 24-well plate at a density to give 75–80% confluency after 24 h (typically ~1.5×10^5 cells/ml).
2. Mix 2 µl of 1.33 mg/ml GeneJuice™ with 60 µl of DMEM supplemented with 1× non-essential amino acids and 1 mM glutamine in a V-well micro-titre plate.

3. Add ~1 µg of plasmid DNA, mix thoroughly and incubate for 30 min at room temperature.
4. Carefully aspirate the media from the cell layer in the 24-well plate and discard.
5. Make up the DNA/GeneJuice™ cocktail to 1 ml with DMEM supplemented with 2% FCS, 1× non-essential amino acids and 1 mM glutamine and add to the plated cells.
6. Incubate the cells at 37°C in a 5% CO_2/95% air environment for 96 h.

16.3.2.2. Harvest and Western Blot Analysis of Secreted Proteins

1. To analyse the expression of secreted proteins, harvest the media (containing the protein-of-interest) from the cells and centrifuge at 17,000×*g* for 10 min. Mix an equal volume of supernatant and 2× SDS–PAGE sample loading buffer. Heat the sample to 95°C for 3 min to denature the protein.
2. Run the samples on an SDS–PAGE gel and transfer to a nitrocellulose or PVDF membrane. Block the membrane for at least 30 min with 5% milk powder in PBST.
3. Remove the blocking solution and add the primary antibody solution, incubating for 1 h.
4. Wash the membrane three times in PBST for 5 min each time.
5. Remove the wash buffer and incubate the membrane for 1 h in the secondary antibody solution.
6. Wash the membrane three times in PBST for 5 min followed by two washes for 5 min in PBS (*see* **Note 7**).
7. To visualise the protein, remove excess buffer from the membrane and cover with 2 ml ECL-Plus™ reagent for 5 min.
8. Develop the Western blot by wrapping it in cling film and placing it against a sheet of Hyperfilm™ in an autoradiography cassette for 20 s and then developing the film using a Xograph imaging system (Xograph Healthcare, Tetbury, UK).

16.3.2.3. Harvest and Western Blot Analysis of Intracellular Proteins

1. To analyse the expression level of non-secreted/intracellular proteins, wash the adherent cells in 1 ml per well of PBS after removal of the media containing secreted protein (*see* **Note 8**).
2. Add 200 µl of NPI-10 supplemented with 1% v/v Tween 20 and 100 KUnits/ml DNAse I.
3. Pipette the contents of the well up and down until all the cells are detached.
4. To aid lysis of the cells, freeze the plate at –80°C and then immediately thaw it. Make sure the cells are homogeneously resuspended by pipetting up and down.
5. Take 10 µl of the resulting lysate and mix with 10 µl of 2× SDS–PAGE sample loading buffer. Heat the sample to 95°C for 3 min to denature the protein.

6. Run the samples on an SDS–PAGE gel and transfer to a nitrocellulose or PVDF membrane.
7. Follow the Western blotting protocol given in s**teps 2–8** of **Section 16.3.2.2**.

16.3.3. Large Scale Glycoprotein Production

Products identified using the small scale screen are taken forward and scaled up, usually to 1L, to produce protein for biochemical or structural studies. Using the protocols given, several products can be scaled in parallel depending on the capacity of roller bottle incubator used. As discussed previously, for structural studies it is convenient to modify the type of glycan attached to a protein. In general, we use kifunensine which results in glycans of the type $GlcNAc_2Man_9$ (*see* **Fig. 16.1**), although other drugs or modified cell lines can be used *(10)*.

16.3.3.1. Large-Scale Transient Expression

1. For a 1L culture, $4\times175\ cm^2$ flasks containing fully confluent HEK 293T cells are required.
2. Remove cells by trypsinisation and add all the cells (typically $\sim7.5\times10^5$ cells/ml) from one flask into one roller bottle with 250 ml of DMEM containing 1× non-essential amino acids and 1 mM glutamine and 2% FCS.
3. The roller bottles are then gassed with 5% CO_2/95% air for 20 s.
4. Incubate the cells for 4 days at 37°C with the bottles rolling at 30 rpm in a suitable incubator (e.g. Wheaton Science Products, NJ, USA).
5. Replace media in each roller bottle with 200 ml of DMEM supplemented with 1× non-essential amino acids and 1 mM glutamine and 2% FCS. Incubate the cells at 37°C in the roller incubator for 2 h.
6. Pipette 2 mg of plasmid DNA at 1 mg/ml (*see* **Note 9**) into 200 ml of DMEM containing 1× non-essential amino acids and 1 mM glutamine in a sterile conical flask. Mix and then add 3.5 ml of 1 mg/ml polyethyleneimine (PEI) to form the transfection cocktail.
7. Mix thoroughly and incubate at room temperature for 30 min.
8. Add 50 ml of the transfection cocktail to each of the roller bottles.
9. If control of the glycan-type is required, add 1 mg kifunensine per litre of media.
10. The bottles are then gassed with 5% CO_2/95% air for 20 s.
11. Incubate the cells for 4–5 days at 37°C with the bottles rolling at 30 rpm.
12. To harvest the secreted protein, remove the media from the cells and centrifuge at 5,000×*g* for 30 min to remove any remaining non-attached cells (*see* **Notes 10** and **11**).

13. Filter the media through a 0.2 μm bottle-top filter and store at 4°C until use.

16.3.4. Large Scale Glycoprotein Purification

One of the major challenges to purification of glycoproteins is the large volume of viscous media which needs to be processed, which can often cause difficulties with column pressure. In addition, some media components displace the His-tagged protein-of-interest during immobilised metal affinity chromatography (IMAC) of large culture volumes. Several methods have been used to overcome this problem, including dialysis or dilution of the media into a buffer such as PBS *(12)* or purification using lectin sepharose *(13)*. Dialysis removes the contaminants from the media; however it is time consuming and uses large quantities of buffer. Dilution is much more time efficient and removes the problems of column pressure and non-compatible media components, however dilution of around 3× the original volume results in a very large volume of protein-containing buffer to be purified. Lectin sepharoses are often used to purify glycoproteins and lentil lectin sepharose is of particular use as it binds mannose-rich glycans such as the ones produced on proteins expressed by HEK 293T cells when kifunensine has been used (*see* **Fig. 16.1**) *(13)*. There are two main difficulties with lentil lectin sepharose purification: firstly, the increase in pressure due to the viscosity of the DMEM media; and secondly, the relatively poor specificity of the lentil lectin sepharose for the protein-of-interest which results in loss of product.

Here we describe an automated method of sample loading onto an IMAC column and which includes brief washes of the column. The column contains an IMAC resin with highly cross-linked agarose beads that are compatible with a wide range of compounds (HisTrap FF crude column). Thus the difficulties of high pressure due to media viscosity and IMAC incompatible components in the media are addressed. In addition, as the whole protocol of IMAC followed by size exclusion chromatography is automated using the ÄKTAxpress™ purification system, the purification is labour-efficient and can be routinely run overnight.

If the product is produced using kifunensine in the media the glycans, which are $GlcNAc_2Man_9$, can be trimmed to $GlcNAc_1$ using Endo H. Short sugar chains such as these can aid crystallography as the formation of crystal contacts requires a homogeneous, rigid protein structure. It is recommended to test the quality of the purified protein before using it in other applications such as structural studies or enzymatic assays. One of the most common methods for quality assurance of protein products is mass spectrometry and a protocol is given here in **Section 16.3.4.5** *(14)*.

16.3.4.1. Programme for Glycoprotein Purification

1. The programme given in the **Appendix** is used for secreted protein purification on the ÄKTAxpress™ system. This

is programmed into the Method Editor section of the Unicorn™ software (*see* **Note 12**).

2. Briefly, 50 ml batches of media are loaded at 8 ml/min through the 5 ml HisTrap FF Crude column, each followed by 10 ml of Nickel Wash Buffer. This is automatically repeated until all the media is loaded which is detected by the integral air sensor in the system. The column is then washed with Nickel Wash Buffer before elution of the protein-of-interest in Nickel Elution Buffer. The elution peak is automatically collected in the sample loop and re-injected onto the pre-equilibrated size exclusion column. The protein is subjected to size exclusion chromatography in Size Exclusion Buffer and 2 ml fractions are collected.

16.3.4.2. Purification of the Glycoprotein

1. Using an ÄKTAxpress™ unit; Pre-equilibrate the appropriate HiLoad 16/60 Superdex column in Size Exclusion Buffer (*see* **Note 4**).
2. After equilibration of the size exclusion column, insert the inlet tubes A1 and A2 into Nickel Wash Buffer. Insert A3 into Nickel Elution Buffer and A4 into Size Exclusion Buffer.
3. Insert the outlet tube F3 into a bottle which is at least 1.5× the sample volume and place a plate into the fraction collector.
4. Screw a pre-charged 5 ml HisTrap FF Crude column into column position 1.
5. Manually wash the pumps using inlet A2.
6. Just before starting the purification programme, remove the inlet line A2 from the Nickel Wash Buffer and insert it into the bottle containing the sample (usually 1L of conditioned media with the protein-of-interest).
7. Using a Method Run in the System Control section of the Unicorn™ software, run the glycoprotein purification programme.

16.3.4.3. Analysis of Purified Product

1. Open the chromatogram in the Evaluation section of the Unicorn™ software. Analyse the A_{280} trace of the size exclusion profile, making a note of the fractions which contain protein – represented by a peak in the A_{280} trace.
2. Take 10 µl from each fraction-of-interest and mix with 10 µl of 2× SDS–PAGE sample loading buffer. Heat the sample to 95°C for 3 min to denature the protein.
3. Run the samples on an SDS–PAGE gel and when the dye reaches the bottom, wash the gel three times in water for 1 min each time at full power in a microwave. Remove the water and add 20 ml of Simply Blue Safe Stain. Microwave on full power for 1 min, and then incubate at room temperature with rocking for 10 min.

4. To de-stain the gel, remove the stain and incubate the gel in water at room temperature with rocking for 10 min or more – until a clear background is obtained.
5. Combine fractions of >95% purity by SDS–PAGE (*see* **Fig. 16.3a**).

16.3.4.4. Glycan Removal with Endoglycosidase H

1. Measure the concentration of protein in the sample using the A_{280} and the extinction coefficient. Take a 10 μl sample for SDS–PAGE analysis (*see* **step 2**) Add 1 kUnit of Endo H per mg of protein and incubate at room temperature overnight (*see* **Note 13**).

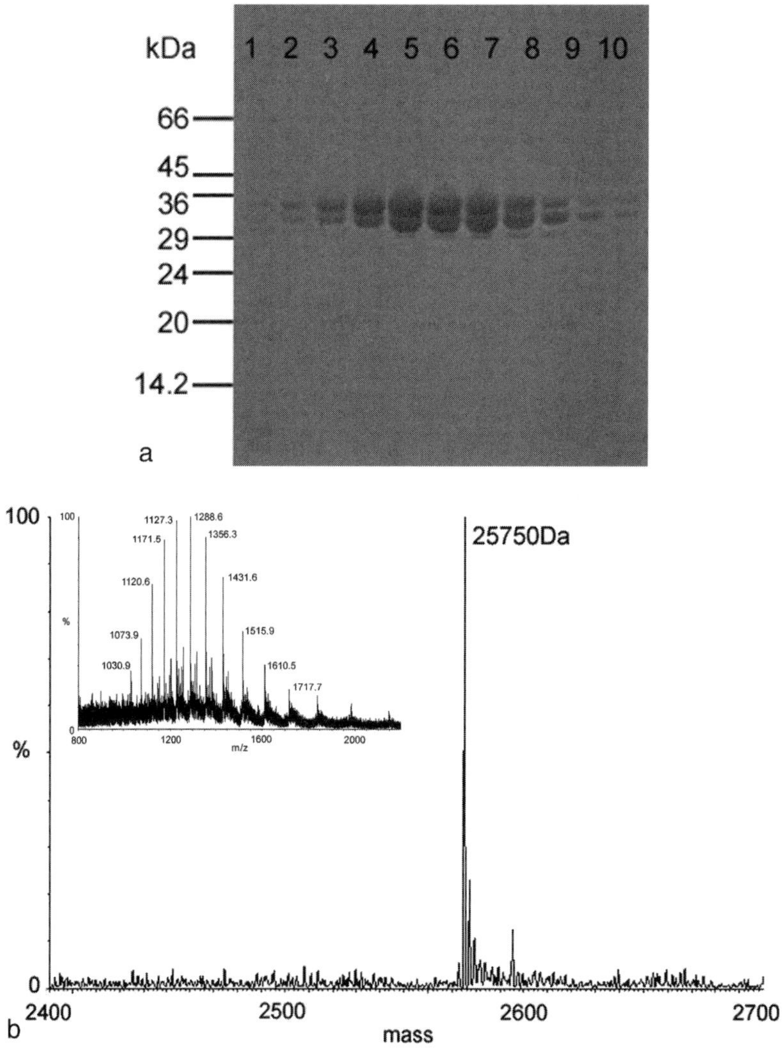

Fig. 16.3. Example of a glycoprotein purification which gave a yield of ~9 mg from 1 L of media. SDS–PAGE gel of the size exclusion fractions 1–10 of the purified glycoprotein (**a**). The two bands in (**a**) represent different occupancy of the four N-glycosylation sites. Raw data (insert) and deconvoluted mass spectra showing the quality assurance of the protein after treatment with PNGase F (expected molecular weight = 25750 Da) (**b**)

2. Take a 10 μl sample of the cleaved protein and run an SDS–PAGE gel for comparison with the un-cleaved product (*see* **step 1**) using the protocol given in **steps 2–5** of **Section 16.3.4.3**. The glycosylated band should be at a higher molecular weight and may be smeary due to the heterogeneity of the sugars; whereas the de-glycosylated band should be at a lower molecular weight and should be focussed.
3. If the reaction has not gone to completion, add more Endo H and re-incubate. In some cases sugars can be shielded from the enzyme.
4. To separate any glycosylated protein and the Endo H from the de-glycosylated protein, inject the sample onto a pre-equilibrated HiLoad 16/60 Superdex column and elute in Size Exclusion Buffer.

16.3.4.5. Glycoprotein Quality Assurance by Mass Spectrometry

1. Denature 15 μl of a 20 μM protein sample by boiling for 10 min.
2. Cool to room temperature and add 1 μl of PNGase F. Incubate the solution at 37°C for 3 h or more (*see* **Note 14**).
3. In the first well of the PCR plate, add 15 μl of the myoglobin solution which is used for the calibration of the mass spectrometer.
4. Pipette 15 μl water to the second well of the PCR plate, followed by the denatured protein sample into the third well. Add more samples into the subsequent wells as required.
5. Seal the PCR plate before automatically running the samples which can take place overnight.
6. During the automatic quality assurance of the samples, each protein sample is injected through the autosampler onto the C4 precolumn. The precolumn is then washed at 5 μl/min for 5 min with wash solvent.
7. The protein is eluted into the mass spectrometer in the reverse direction at 5 μl/min using a gradient of 5–80% solvent B over 1 min. To ensure full elution of the protein sample, the mobile phase is held at 80% Solvent B for 10 min before re-equilibration of the pre-column.
8. Analysis of samples is by MS (as oppose to MS/MS) which gives an accurate mass of the glycoprotein without its sugars and with each asparagine converted to an aspartic acid – this conversion happens during the PNGase F reaction. For analysis of the spectra, the data under the peak on the chromatogram is combined to give a raw data file showing % ions against mass/charge. To gain an average mass of the protein, the raw data is deconvoluted using the MaxEnt algorithm supplied as part of the MassLynx software (*see* **Fig. 16.3b**).

16.4. Notes

1. Good quality DNA can be obtained using a standard plasmid preparation kit such as the QIAPrep™ Spin Miniprep kit (Qiagen, Crawley, UK) or by using a parallelised protocol such as the Wizard SV96 (Promega, Southampton, UK) which can be run using a vacuum manifold.
2. Kifunensine can be added during expression of a glycoprotein in order to modify the glycan type. With this drug present, the most abundant glycan form attached to a given product is $GlcNAc_2Man_9$ which is Endo H sensitive.
3. IMAC columns from other manufacturers are suitable for this method so long as the flow properties are appropriate for viscous solutions.
4. The type of Superdex column depends on the size of the protein-of-interest: The HiLoad 16/60 Superdex 200 column separates globular proteins from 10 to 600 kDa and the HiLoad 16/60 Superdex 75 column, proteins from 3 to 70 kDa *(15)*. In practice, a protein under 25 kDa is usually purified using a Superdex 75 column, otherwise the Superdex 200 column is used.
5. C4 pre-columns are reused, however some protein samples remain stuck to the column and re-elute in all subsequent samples. In general, the precolumn is changed every 2 months.
6. Healthy cell cultures should have viability of 95% or over. If the viability is below 75% the cells should be discarded and new cells resuscitated.
7. The PBS wash after washing with PBST lowers the background signal seen on the final blot.
8. It is good practice to analyse the intracellular component of the expression screen as a problem may have occurred with secretion of the glycoprotein.
9. Good quality DNA can be obtained using a standard mega plasmid preparation kit such as the PureLink HiPure Plasmid Megaprep Kit (Invitrogen, Paisley, UK).
10. For a maximum yield, a time-expression course can be performed at small scale, where 10 µl of sample is taken each day and screened according to **Section 16.3.2.2**. The large scale culture can then be harvested after the optimal number of days.
11. Fresh media can be added to the harvested roller bottles and the cells returned to the roller bottle incubator and harvested after a further 4–5 days. Generally the yield from the second harvest is approximately 75% that of the first one.

12. The ÄKTAxpress™ programme, given in the **Appendix**, can be modified for different purification systems including any in the range of laboratory ÄKTA™ systems from GE Healthcare (Little Chalfont, UK) and purification apparatus from other manufacturers. For example, if the system has only two buffer inlets, Inlet A can be used for the Nickel Wash Buffer and Inlet B for the sample. Loading of the sample through the His-Trap FF Crude column can be performed, followed by transfer of Inlet B into Nickel Elution Buffer for the next step in the process. In this case, size exclusion chromatography is performed as a separate purification to the IMAC step.
13. Endo H has maximum efficiency at around pH 5.2, however as the majority of proteins are unstable at this pH, the reaction is carried out at pH 7.5 (Size Exclusion Buffer). Also, many proteins are unstable at 37°C for long periods of time so room temperature is preferred for the reaction. Endo H is active at room temperature and at pH 7.5; however, the reaction therefore takes longer than it would if it was run under optimal conditions.
14. It is essential that all sugars are removed by the PNGase F reaction. Therefore as the protein is already denatured, longer incubation times can be used which will ensure the reaction goes to completion. If convenient, the reaction may be left overnight.

Acknowledgements

The Oxford Protein Production Facility is supported by grants from the Medical Research Council, UK, the Biotechnology and Biological Sciences Research Council, UK and Vizier (European Commission FP6 contract: LSHG-CT-2004–511960).

References

1. Hammond, C., Braakman, I. and Helenius, A. (1994) Role of N-linked oligosaccharide recognition, glucose trimming, and calnexin in glycoprotein folding and quality control. *Proceedings of the National Academy of Sciences of the United States of America*, **91**, 913–917.
2. Mitra, N., Sinha, S., Ramya, T.N.C. and Surolia, A. (2006) N-linked oligosaccharides as outfitters for glycoprotein folding, form and function. *Trends in Biochemical Sciences*, **31**, 156–163.
3. Romanos, M.A., Scorer, C.A. and Clare, J.J. (1992) Foreign gene expression in yeast: a review. *Yeast*, **8**, 423–488.
4. Yin, J., Li, G., Ren, X. and Herrler, G. (2007) Select what you need: a comparative evaluation of the advantages and limitations of frequently used expression systems for foreign genes. *Journal of Biotechnology*, **127**, 335–347.
5. Harder, T.C. and Osterhaus, A.D.M.E. (1997) Molecular characterization and

baculovirus expression of the glycoprotein B of a seal herpesvirus (Phocid Herpesvirus-1). *Virology*, **227**, 343–352.
6. Kwong, P.D., Wyatt, R., Desjardins, E., Robinson, J., Culp, J.S., Hellmig, B.D., Sweet, R.W., Sodroski, J. and Hendrickson, W.A. (1999) Probability analysis of variational crystallization and its application to gp120, the exterior envelope glycoprotein of type 1 human immunodeficiency virus (HIV-1). *The Journal of Biological Chemistry*, **274**, 4115–4123.
7. Wurm, F.M. (2004) Production of recombinant protein therapeutics in cultivated mammalian cells. *Nature Biotechnology*, **22**, 1393–1398.
8. Durocher, Y., Perret, S. and Kamen, A. (2002) High-level and high-throughput recombinant protein production by transient transfection of suspension-growing human 293-EBNA1 cells. *Nucleic Acids Research*, **30**, E9.
9. Wurm, F. and Bernard, A. (1999) Large-scale transient expression in mammalian cells for recombinant protein production. *Current Opinion in Biotechnology*, **10**, 156–159.
10. Chang, V.T., Crispin, M., Aricescu, A.R., Harvey, D.J., Nettleship, J.E., Fennelly, J.A., Yu, C., Boles, K.S., Evans, E.J., Stuart, D.I. et al. (2007) Glycoprotein structural genomics: solving the glycosylation problem. *Structure*, **15**, 267–273.
11. Berrow, N.S., Alderton, D., Sainsbury, S., Nettleship, J., Assenberg, R., Rahman, N., Stuart, D.I. and Owens, R.J. (2007) A versatile ligation-independent cloning method suitable for high-throughput expression screening applications. *Nucleic Acids Research*, **35**, e45.
12. Aricescu, A.R., Lu, W. and Jones, E.Y. (2006) A time- and cost-efficient system for high-level protein production in mammalian cells. *Acta Crystallographica Section D*, **62**, 1243–1250.
13. McCoy, J.P. Jr. (1987) The application of lectins to the characterization and isolation of mammalian cell populations. *Cancer and Metastasis Reviews*, **6**, 595–613.
14. Nettleship, J.E., Brown, J., Groves, M.R. and Geerlof, A. (2007) In Kobe, B., Guss, M. and Huber, T. (eds.), *Structural proteomics: High-throughput methods.* Humana Press, Totowa, NJ.
15. (2002) *Gel filtration: Principles and methods.* AI ed. GE Healthcare, Uppsala, Sweden.

Appendix
ÄKTAxpress Glycoprotein Purification Programme

Main method:

(Main)

```
0.00 Base CV 5.027 {ml} HisTrap_FF_5_ml
0.00 Block Set_up_System_1
  (Set_up_System_1)
  0.00 Base SameAsMain
  0.00 ResetFracPosition
  0.00 PumpWash A1
  0.00 SystemWash 10.00 {ml}
  0.00 Alarm_Pressure Enabled 0.7 {MPa} 0.000 {MPa}
  0.00 Flow 8 {ml/min} Buffer 0.10 {ml/min} Yes
  0.00 ColumnPosition Position1
  0.00 Set_Mark "Equilibration of His Column 1"
  3.00 AutoZeroUV
  3.00 ColumnPosition Bypass
  3.00 Flow 0.00 {ml/min} Buffer 0.10 {ml/min} No
  3.00 End_Block
0.00 Watch_AirSensor Equal 1 Finish_load_1
  (Finish_load_1)
  0.00 Base Volume
  0.00 InletValve A1
```

(continued)

Appendix (continued)

```
0.00 Flow 0.00 {ml/min} Buffer 0.10 {ml/min} No
0.00 Block Air_removal_1
  (Air_removal_1)
  0.00 Base Volume
  0.00 InletValve A1
  0.00 Set_Mark "Air removal"
  0.00 InjectionValve Waste
  0.00 Flow 20 {ml/min} Buffer 0.10 {ml/min} Yes
  20.00 End_Block
0.00 Block Wash_Column_1
  (Wash_Column_1)
  0.00 Base SameAsMain
  0.00 Set_Mark "Washing column 1"
  0.00 Flow 0.00 {ml/min} Buffer 0.10 {ml/min} No
  0.00 Watch_Off AirSensor
  0.00 Alarm_AirSensor Enabled
  0.00 InjectionValve Inject
  0.00 Flow 5.00 {ml/min} Buffer 0.10 {ml/min} Yes
  10.00 AutoZeroUV
  10.00 End_Block
0.00 Block End_fractionation_load_1
  (End_fractionation_load_1)
  0.00 Base SameAsMain
  0.00 OutletValve WasteF1
  0.00 End_Block
0.00 Block Set_up_elution_1
  (Set_up_elution_1)
  0.00 Base SameAsMain
  0.00 ColumnPosition Bypass
  0.00 PumpWash A3
  0.00 OutletValve LoopFracF12
  0.00 LoopSelection Bypass
  0.00 ColumnPosition Position1
  0.00 End_Block
0.00 Block Elution_Peak
  (Elution_Peak)
  0.00 Base SameAsMain 5.027 {ml} HisTrap_HP_5_ml
  0.00 Watch_UV Greater_Than 100.000 {mAU} Peak_Collection
    (Peak_Collection)
    0.00 Base Volume
    0.00 LoopSelection LP1
    0.00 Set_Mark "Peak_Start"
    0.00 Watch_UV Less_Than 100.000 {mAU} Peak_End
      (Peak_End)
      0.00 Base Volume
      0.00 LoopSelection Bypass
      0.00 Set_Mark "Peak_End"
      0.00 End_Block
    7.50 Block Peak_End
      (Peak_End)
      0.00 Base Volume
```

(continued)

Appendix (continued)

```
      0.00 LoopSelection Bypass
      0.00 Set_Mark "Peak_End"
      0.00 End_Block
      7.50 End_Block
    5.00 Watch_Off UV
    5.00 LoopSelection Bypass
    5.00 End_Block
  0.00 Block Set_up_Gel_Filtration
    (Set_up_Gel_Filtration)
    0.00 Base SameAsMain
    0.00 Flow 0.00 {ml/min} Buffer 0.10 {ml/min} No
    0.00 ColumnPosition Bypass
    0.00 OutletValve WasteF1
    0.00 PumpWash A4
    0.00 SystemWash 10.00 {ml}
    0.00 Alarm_Pressure Enabled 0.5 {MPa} 0.000 {MPa}
    0.00 Alarm_AirSensor Enabled
    0.00 Flow 1.2 {ml/min} Buffer 0.10 {ml/min} Yes
    0.00 End_Block
  0.00 Block Equilibrate_Gel_Filtration
    (Equilibrate_Gel_Filtration)
    0.00 Base CV 120.637 {ml} HiLoad_16/60_Superdex_75_prep_grade
    0.00 ColumnPosition Position5
    0.10 AutoZeroUV
    0.10 End_Block
  0.00 Block Gel_Filtration
    (Gel_Filtration)
    0.00 Base CV 120.637 {ml} HiLoad_16/60_Superdex_75_prep_grade
    0.00 Block Inject_Peak
      (Inject_Peak)
      0.00 Base Volume
      0.00 InjectionValve Reinject
      0.00 LoopSelection LP1
      0.00 Set_Mark "Peak_injection"
      10.00 InjectionValve Inject
      10.00 End_Block
    0.30 Block Fractionation_Gel_Filtration
      (Fractionation_Gel_Filtration)
      0.00 Base CV 120.637 {ml} HiLoad_16/60_Superdex_75_prep_grade
      0.00 OutletValve FracCollF2
      0.00 Fractionation 2.000 {ml}
      0.00 End_Block
    1.10 End_Block
  0.00 End_Method
0.00 Block Fractionation_load_1
  (Fractionation_load_1)
  0.00 Base SameAsMain
  0.00 OutletValve F3
  0.00 End_Block
0.00 Block Load_Sample_1
```

(continued)

Appendix (continued)

```
  (Load_Sample_1)
  0.00 Base Volume
  0.00 Flow 8 {ml/min} Buffer 0.10 {ml/min} Yes
  0.00 ColumnPosition Position1
  0.00 Set_Mark "Loading sample 1"
  0.00 Loop (200)#No_of_loops_for_load
  0.00 InletValve A2
  (50.00)#Volume_of_part_sample InletValve A1
  60.00 Loop_End
  60.00 End_Block
0.00 Message "The programme finished but there is still sample!!" Screen "error"
0.00 End_Method
```

Chapter 17

High-Throughput Expression and Detergent Screening of Integral Membrane Proteins

Said Eshaghi

Summary

Integral membrane proteins are a major challenge within structural genomics. These proteins are not only difficult to produce in quantities sufficient for analysis by X-ray diffraction or NMR, but also require extraction from their lipid environment, which leads to a new dimension of difficulties in purification and subsequent structural analysis. To overcome these problems requires new strategies enabling screening larger number of parameters dealing with expression and purification. For this reason, we have developed high-throughput methods for screening extracting and purifying detergents as well as other purification parameters, e.g. salt and pH. The method requires standard laboratory equipments, but can also be automated.

Key words: Detergent screen; Dot blot; High-throughput; Membrane proteins; Purification

17.1. Introduction

Integral membrane proteins correspond to almost 1/3 of all proteins in a typical cell and play important roles in many biological processes. Although, over 50% of the current drug targets are integral membrane proteins, these proteins correspond to only less than 1% of all the known protein structures available in the Protein Data Bank (PDB). The main reason for this lack of structural information is the challenges associated with membrane protein production and crystallization. These challenges can be divided in three stages (1) recombinant expression of large quantities sufficient for structural analysis, (2) purification of

Sharon A. Doyle (ed.), *Methods in Molecular Biology: High Throughput Protein Expression and Purification, vol. 498*

Book doi: 10.1007/978-1-59745-196-3

high quality protein, and (3) crystallization and structural determination. The difficulties in the two latter stages are principally based on the detergent treatment of the membrane proteins. It is very important to identify the right detergent or detergent mixture in order to extract, stabilize, purify and finally crystallize the target membrane protein. It has been reported in several cases that detergent exchange or detergent mixture improves the quality of the crystals (*1–9*). There are hundreds of different detergents commercially available that make identification of the right one(s) a challenge *per se*. However, this problem can be tackled in a high-throughput manner using standard laboratory equipments, such as 96-well plates and multi-channel pipettes. Detergents, buffer composition and the purifying column material can be screened using this technique (*10,11*).

17.2. Materials

17.2.1. Protein Overexpression and Cell Lysis

1. Difco™ LB (Luri-Bertani) Broth, Miller (LB-media) (BD, Stockholm, Sweden).
2. Freshly prepared isopropyl-β-D-1-thiogalactoside (IPTG) (Anatrace, Cleveland, OH) at 100 mM dissolved in water (*see* **Note 1**) and sterile filtered.
3. Detergents (Anatrace) (Table 17.1) prepared in 10% (w/v) solutions dissolved in water and stored in 1 ml aliquots at −20°C.
4. Lysis and solubilization buffer (LS-buffer): 20 mM Tris-HCl pH 8.0 (*see* **Note 2**), 50 mM NaCl, 1 mg/ml lysozyme (Sigma-Aldrich, Stockholm, Sweden), complete protease inhibitor cocktail EDTA-free (Roche, Stockholm, Sweden), 20 μ/ml recombinant DNase I (Roche), 0.5 mM Tris[2-carboxyethyl] phosphine hydrochloride (TCEP) (Sigma-Aldrich) (*see* **Note 3**) and detergent at an extraction concentration according to Table 17.1 (*see* **Note 4**).

17.2.2. Small-Scale Protein Purification

1. 96-well filter plates 0.45 μm, 0.65 μm and 1.2 μm (Millipore, Stockholm, Sweden).
2. Ni-NTA agarose (Invitrogen, Stockholm, Sweden) in 50% slurry.
3. Preparation buffer (P-buffer): 20 mM Na-phosphate pH 7.4, 300 mM NaCl, 40 mM imidazole (*see* **Note 5**) and 0.5 mM TCEP.
4. Wash buffer (W-buffer): 20 mM Na-phosphate pH 7.4, 300 mM NaCl, 40 mM imidazole, 0.5 mM TCEP and detergent at a purification concentration according to **Table 17.1.**

Table 17.1
List of the detergents and their concentrations used in the screen

Detergent	CMC (mM)	Extraction conc. (mM)	Purification conc. (mM)
Fos-Choline-12 (FC12)	1.5	32	5
HEGA-10	7	54	21
Decyl maltoside (DM)	1.8	21	6
Dodecyl maltoside (DDM)	0.17	20	0.6
CHAPS	8	32	24
Nonyl thiomaltoside (NTM)	3.2	32	10
Cymal-5	3	30	9
LDAO	1	43	3
C8E4	8	64	24
C12E8	0.09	19	0.3
Octyl glucoside (OG)	18	68	40
Triton X-100 (TX100)	0.23	15	0.8

5. Elution buffer (E-buffer): 20 mM Na-phosphate pH 7.4, 100 mM NaCl, 400 mM imidazole, 0.5 mM TCEP and detergent at a purification concentration according to Table 17.1.

17.2.3. Dot-Blot Analysis

1. Tris buffer saline with Tween-20 (TBST): 20 mM Tris-HCl pH 7.5, 100 mM NaCl, 0.1% (w/v) Tween-20.
2. INDIA His-Probe-HRP Western blotting probe (Pierce, Nordic Biolabs, Sweden) diluted 1:10,000 in TBST (probe-solution).
3. Supersignal™ West Pico Chemiluminescence substrate (Pierce) freshly mixed.

17.3. Methods

When screening for protein expression in a number of different cells, the culturing and protein expression should be performed in 96-deep-well plates. However, to perform accurate screening of detergents and other purification parameters, it is important to start with equal amounts of cells to ensure that the variations are basically due to the differences made in the buffer composition.

Therefore, the culturing and protein expression should be done on a large batch. The batch is finally divided into smaller fractions prior to the lysis and solubilization procedures.

17.3.1. Protein Overexpression and Cell Harvest

1. *Escherichia coli* (*E. coli*) cells containing the plasmid encoding the target protein with a 6-His tag attached to either N- or C-terminus are grown in 50 ml LB-media in shake flasks. In case of protein expression screening, 1-ml cultures are grown in 96-deep-well plates. The growth is performed at 200 rpm and 37°C.
2. At OD_{600} ~ 0.8, the culture is cooled down to 20°C and thereafter the protein expression is induced with 0.1 mM IPTG at final concentration. The induction is continued overnight (15–18 h) at 200 rpm and 20°C.
3. The next day, the cultures grown in 50 ml media are aliquoted in 12 × 1-ml fractions in 96-well-deep plates. Each aliquot will later be solubilized by a certain detergent. The cells are harvested by centrifugation at 3,000 × *g*. The supernatant is removed simply by inverting the plate and discarded. To ensure efficient removal of the supernatant, place the inverted plate on a paper-tissue.
4. The harvested cells can be used directly or stored at –80°C.

17.3.2. Expression/ Detergent Screen (see Note 6)

1. 100 µl LS-buffer is added to each well containing the cell pellets (*see* **Note 7**). To resuspend, lyse and solubilize the pellets, the plate is shaken at ~ 900 rpm for 1 h. The unbroken cells and large cell debris are removed by centrifugation at 3,000×*g* for 10 min.
2. The supernatant is transferred to a 0.45 µm 96-well filter plate and filtered by centrifugation at 2,000×*g* for 10 min (*see* **Note 8**). This will ensure that smaller fragments and precipitates are also removed, yielding highly clear solution.
3. 50 µl Ni-NTA agarose slurry is added to the wells of a 1.2 µm 96-well filter plate. The resins are washed by 100 µl water. Then, 200 µl P-buffer is added to the wells to equilibrate the resins. The liquids are removed at each step by centrifugation at 100×*g* for 1 min.
4. The filtrate from **step 2** is added to the equilibrated resins and incubated for 10 min on a plate-shaker at ~450 rpm. The unbound material is removed by centrifugation at 100×*g* for 1 min.
5. The resins are washed by 3 × 200 µl W-buffer and finally eluted with 40 µl E-buffer. The eluate is collected in a 96-well plate. The washing and elution are performed by centrifugation at 100×g for 1 min.

17.3.3. Dot Blot Analysis

1. 1–2 µl of the eluate is applied to nitrocellulose membrane and allowed to dry (*see* **Notes 9 and 10**).

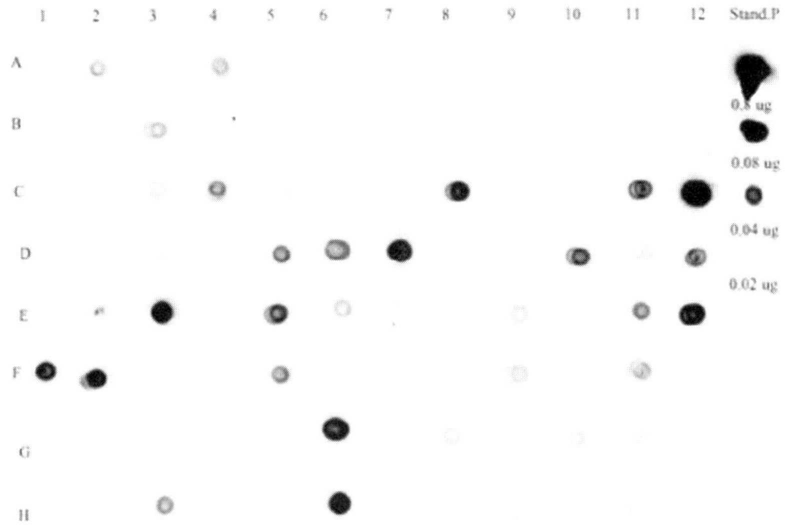

Fig. 17.1. A typical dot blot result from an expression screening of 96 different integral membrane proteins. The extraction was performed by 32 mM (1% w/v) FC12 and purification was continued with 0.6 mM (0.03% w/v) DDM in the purification buffers. Stand P. refers to the standard protein used in the experiment at various amounts.

2. The dried membrane is first wetted in TBST for a few seconds and then the INDIA His-Probe-HRP is added (*see* **Note 11**). The probe-solution should cover the membrane.
3. The membrane is incubated in the probe-solution for 1 h at room temperature while agitating.
4. After 1 h, the probe-solution is discarded and the membrane is rinsed with TBST to remove the excess probe-solution. Then, the membrane is washed 3 × 5 min with TBST.
5. TBST is removed from the membrane by holding the membrane using a pair of tweezers to let TBST drop. The final drops are removed using paper-tissue.
6. Supersignal™ West Pico Chemiluminescence substrate is then added to the membrane (*see* **Note 12**). The membrane is incubated with the substrate for 5 min. The signals are detected with a luminescence imaging system **(Fig. 17.1)**.

17.4. Notes

1. All solutions should be prepared in water that has a resistivity of 18.2 MΩ-cm and total organic content of less than five parts per billion. This standard is referred to as "water" in this text.

2. All the buffers are prepared and their pH is adjusted at room temperature. All the solutions are cooled to about 4°C before use.
3. It is recommended to use freshly prepared TCEP. However, TCEP can be stored at 4°C up to one week. Thus, a stock solution of 0.5 M can be prepared and used diluted when needed.
4. When using other detergents than mentioned in this work, it is recommended not to exceed 2% (w/v) in concentration to avoid probable damage to the column material. Also, it is recommended to use a concentration 3× CMC or higher during the course of purification.
5. Imidazole should be prepared as stock solution and pH adjusted before usage in the purification buffers to avoid pH discrepancy.
6. This method is useful for expression and detergent screening as well as screening of other parameters such as various salts and salt concentrations, imidazole concentration and different column materials, in order to obtain the optimum purification conditions.
7. When performing expression screening, it is important to use a fairly strong detergent, such as FC12, in order to prevent false negatives.
8. When the cell density is too high and the risk of clogging the filter plates is high, one can pre-filter the suspension with 1.2 μm filter plates before using 0.45 μm filters.
9. After adding the drops to the nitrocellulose membrane, the drops should be let to dry completely, preferably overnight. To reduce the time, one can blow with a normal hair drier over the membrane for about 10 min. This will increase the signal to noise ratio and speed up the whole process.
10. For a better correlation of the signals from different experiments, use a titration of a pure protein of known concentration in all the experiments.
11. There is no need of blocking the membrane with milk or BSA. The INDIA His-Probe-HRP Western blotting probe is highly specific and false positives are rare. This allows a more rapid blotting procedure. However, if an anti-histidine antibody is used instead of the INDIA His-probe, one should block the membrane with milk or BSA as a standard procedure for Western blotting.
12. To reduce the costs, the Supersignal™ West Pico Chemiluminescence substrate can be diluted with water, especially when strong signals are expected.

Acknowledgements

The author would like to thank Professor Pär Nordlund for his advice and encouragement, Dr. Marina Ignatushchenko Sabet for her useful tips and Marie Hedrén and Victoria Lieu for technical assistance. This work was supported by the European Membrane Protein Consortium (E-MeP).

References

1. Kreusch, A., Weiss, M. S., Welte, W., Weckesser, J., and Schulz, G. E. (1991) Crystals of an integral membrane protein diffracting to 1.8 Å resolution. *J. Mol. Biol.* **217,** 9–10
2. Locher, K. P., Rees, B., Koebnik, R., Mitschler, A., Moulinier, L., Rosenbusch, J. P., and Moras, D. (1998) Transmembrane signaling across the ligand-gated FhuA receptor: Crystal structures of free and ferrichrome-bound states reveal allosteric changes. *Cell* **95,** 771–778
3. Essen, L., Siegert, R., Lehmann, W. D., and Oesterhelt, D. (1998) Lipid patches in membrane protein oligomers: crystal structure of the bacteriorhodopsin-lipid complex. *Proc. Natl. Acad. Sci. USA* **95,** 11673–11678
4. Snijder, H. J., Ubarretxena-Belandia, I., Blaauw, M., Kalk, K. H., Verheij, H. M., Egmond, M. R., Dekker, N., and Dijkstra, B. W. (1999) Structural evidence for dimerization-regulated activation of an integral membrane phospholipase. *Nature* **401,** 717–721
5. Abramson, J., Larsson, G., Byrne, B., Puustinen, A., Garcia-Horsman, A., and Iwata, S. (2000) Purification, crystallization and preliminary crystallographic studies of an integral membane protein, cytochrome bo_3 ubiquinol oxidase from *Escherichia coli*. *Acta Crystallogr.* **D56,** 1076–1078
6. Koronakis, V., Sharff, A., Koronakis, E., Luisi, B., and Hughes, C. (2000) Crystal structure of the bacterial membrane protein TolC central to multidrug efflux and protein export. *Nature* **405,** 914–919
7. Jormakka, M., Törnroth, S., Abramson, J., Byrne, B., and Iwata, S. (2002) Purification and crystallization of the respiratory complex formate dehydrogenase-N from *Escherichia coli*. *Acta Crystallogr.* **D58,** 160–162
8. Lemieux, M. J., Song, J., Kim, M. J., Huang, Y., Villa, A., Auer, M., Li, X.-D., and Wang, D. -N. (2003) Three-dimensional crystallization of the Escherichia coli glycerol-3-phosphate transporter: A member of the major facilitator superfamily. *Protien Sci.* **12,** 2748–2756
9. Eshaghi, S., Niegowski, D., Kohl, A., Martinez Molina, D., Lesley, S. A., and Nordlund, P. (2006) Crystal structure of a divalent metal ion transporter CorA at 2.9 Å resolution. *Science* **313,** 354–357
10. Eshaghi, S., Hedren, M., Nasser, M. I. A., Hammarberg, T., Thornell, A., and Nordlund, P. (2005) An efficient strategy for high-throughput expression screening of recombinant integral membrane proteins. *Protein Sci.* **14,** 676–683
11. Niegowski, D., Hedren, M., Nordlund, P., and Eshaghi, S. (2006) A simple strategy towards membrane protein purification and crystallization. *Int. J. Biol. Macromol.* **39,** 83–87

Chapter 18

Cell-Free Expression for Nanolipoprotein Particles: Building a High-Throughput Membrane Protein Solubility Platform

Jenny A. Cappuccio, Angela K. Hinz, Edward A. Kuhn, Julia E. Fletcher, Erin S. Arroyo, Paul T. Henderson, Craig D. Blanchette, Vickie L. Walsworth, Michele H. Corzett, Richard J. Law, Joseph B. Pesavento, Brent W. Segelke, Todd A. Sulchek, Brett A. Chromy, Federico Katzen, Todd Peterson, Graham Bench, Wieslaw Kudlicki, Paul D. Hoeprich Jr, and Matthew A. Coleman

Summary

Membrane-associated proteins and protein complexes account for approximately a third or more of the proteins in the cell *(1, 2)*. These complexes mediate essential cellular processes; including signal transduction, transport, recognition, bioenergetics and cell–cell communication. In general, membrane proteins are challenging to study because of their insolubility and tendency to aggregate when removed from their protein lipid bilayer environment. This chapter is focused on describing a novel method for producing and solubilizing membrane proteins that can be easily adapted to high-throughput expression screening. This process is based on cell-free transcription and translation technology coupled with nanolipoprotein particles (NLPs), which are lipid bilayers confined within a ring of amphipathic protein of defined diameter. The NLPs act as a platform for inserting, solubilizing and characterizing functional membrane proteins. NLP component proteins (apolipoproteins), as well as membrane proteins can be produced by either traditional cell-based or as discussed here, cell-free expression methodologies.

Key words: Protein microarray, *In vitro* expression, Cell-free expression, Fluorescence, Protein interactions, Nanolipoprotein particles, NLP, Lipid membranes, Apolipoprotein, Nanodisc, DMPC, Bacteriorhodopsin, Membrane protein, rHDL, GPCR, Microarray

18.1. Introduction

Characterization of membrane proteins is problematic. Membrane-associated proteins and integral membrane proteins or protein complexes account for ~30% or more of the cellular proteins

Sharon A. Doyle (ed.), *Methods in Molecular Biology: High Throughput Protein Expression and Purification, vol. 498*

Book doi: 10.1007/978-1-59745-196-3

(1, 2). Membrane proteins are held within a lipid bilayer structure that consists of two opposing layers of amphipathic molecules know as polar lipids; each molecule has a hydrophilic moiety, i.e., a polar group such as, a derivatized phosphate or a saccharide group, and a hydrophobic moiety, i.e., a long hydrocarbon chain. These molecules self-assemble in a biological (largely aqueous) environment according to the thermodynamics associated with water exclusion (increasing entropy) during hydrophobic association. Membrane-bound protein complexes mediate essential cellular processes including signal transduction, transport, recognition, bioenergetics and cell–cell communication. In order to facilitate the myriad of functions, membrane proteins are arrayed in the bilayer structure. Note that some "integral" proteins span both leaflets of the bilayer and are firmly attached through hydrophobic interactions, others are anchored within the bilayer, and still others organize into loosely associated "peripheral" proteins. In general, integral membrane proteins are challenging to study because of their insolubility and propensity to aggregate when removed from their hydrophobic lipid bilayer environment. Novel approaches are needed to gain access to this large class of proteins.

Over-expression of proteins in both eukaryotic and prokaryotic cell-based systems has been quite successful for the past 20 years. However, for membrane proteins, cell-based systems have been problematic. Because membrane proteins contain large regions or structural domains that are hydrophobic (the regions that are embedded in or bound to the membrane), they can be extremely difficult to work with in aqueous systems. In fact, there are only a few general methods for rapid production of membrane proteins and membrane-associated protein complexes *(3, 4)*. Cell-free protein expression provides an alternative method that has not been fully exploited for producing recombinant proteins.

Cell-free production of proteins has become a widely accepted means to overcome bottlenecks in protein expression and purification, in support of high-throughput structural proteomics applications *(5–7)*. Cell-free protein expression is capable of overcoming the difficulties associated with obtaining recombinant proteins such as cloning, transfection, cell growth, cell lysis, and subsequent purification steps. Many proteins are inherently poorly expressed or are subject to intracellular proteolysis. Often membrane proteins may be insoluble or cytotoxic when they are over expressed. One or more of these problems can result in low or insoluble protein expression *in vivo*. Cell-free systems may be able to overcome these barriers through the use of additives since it is an "open" system. Additives including chaperonins *(8, 9)*, lipids *(10)*, redox factors *(11)*. Detergents and protease inhibitors *(3, 10, 12–14)* have also been successfully employed for aiding in the solubilization of membrane bound and membrane associated proteins. Cell-free protein synthesis has also been used to produce proteins for structural studies by NMR and X-ray crystallography since it enables the use of heavy atoms or other tags *(10, 13, 15, 16)*. In particular, these cell-free methods can be applied to integral membrane proteins in a high-throughput manner, as a variety of conditions can be rapidly tested to identify optimal expression parameters.

One particularly compelling approach for stabilization and characterization of membrane proteins has been presented by Bayburt and Sligar *(17–21)*. They described the self-assembly of a single integral membrane proteins into soluble nano-scale phospholipid bilayers, or nanolipoprotein particles (NLPs) *(18, 22, 23)*. The self-assembled discoidal particles consist of an apolipoprotein encircling a nanometer scale lipid bilayer defining a nanolipoprotein particle (NLP) (**Fig. 18.1**). Discoidal NLPs are well cited in the literature *(21, 24–27)*. The apolipoproteins used in this method for NLPs can readily be synthesized using cell-free expression to provide a scaffold for supporting membrane bound proteins. The apolipoprotein can be customized by varying the length of the amphipathic helical part of the protein which forms a belt *(28)*. Each apolipoprotein used for NLP synthesis requires requires an optimized protein to lipid ratio in order to ensure correct self-assembly of the NLPs. Using different apolipoproteins, one can customize the average diameter of the particles from 9 to 20 nm (±3%) with a height of approximately 5.5 nm, dependant on the choice of lipid chosen for the bilayer. These nano-scale bilayers offer the opportunity to understand and control the assembly of oligomeric integral membrane proteins critical to macromolecular recognition and cellular signaling *(29–32)*. Adapting NLPs to cell-free methods for obtaining membrane proteins opens up greater access to these difficult to characterize proteins, and allows isolation of pure and functional proteins.

Developing approaches to produce and characterize membrane proteins is an essential step in understanding how recep-

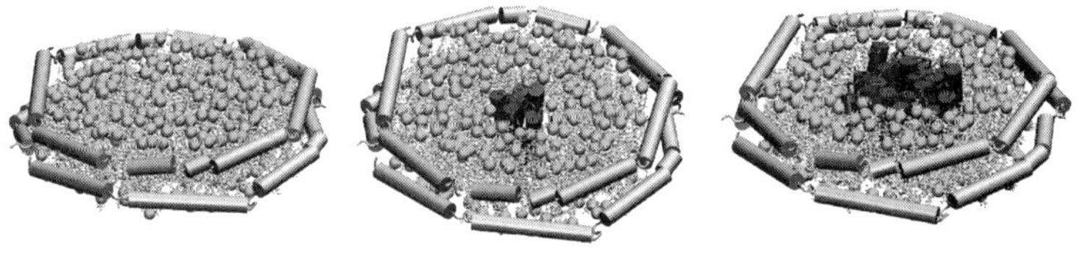

Fig. 18.1. *Models of NLPs with and without bacteriorhodopsin.* Models were constructed using molecular dynamic simulations (described elsewhere). (**A**) Model of a Nanolipoprotein particle (NLP) with a lipid bilayer in the middle and apolipoproteins encircling the hydrophobic portion of the lipids. (**B**) NLP modeled with a bacteriorhodopsin monomer inserted in the hydrophobic lipid core (**C**) NLP modeled with a bacteriorhodopsin trimer inserted in the hydrophobic lipid core (*See Color Plate 5*).

tors mediate biochemical reactions and the events associated with cellular function. Furthermore, there are no broadly applicable technologies enabling comprehensive molecular understanding of membrane-associated proteins and their respective complexes. The purpose of this method is to directly couple NLP technology to cell-free protein production for the rapid assembly and characterization of membrane proteins in discoidal NLP particles. Our method builds on available cell-free systems to enhance or bypass, many of the time-consuming steps associated with conventional expression systems that are amenable to high-throughput expression of many types of proteins.

We have used cell-free expression to screen, purify and characterize several different apolipoproteins and truncated apolipoprotein of ApoE4, ApoA1 and lipophorins (**Fig. 18.2**). Detailed description of the method in this review will focus on NLP production using a 22 kD apolipoprotein E4 fragment composed of the N-terminal domain (3 *(33, 34)* from the laboratory of Karl Weisgraber (*see* **Note 1**). Our model protein is bacteriorhodopsin (bR) from *H. salinarium*, which contains seven transmembrane (TM) spanning regions and can be produced, purified and regenerated by exogenously adding all-*trans* retinal, a required cofactor for enzyme activity *(35, 36)*. Bacteriorhodopsin is a light activated proton pump involved in energy transduction. The work presented here uses commercially available bacteriorhodopsin as a

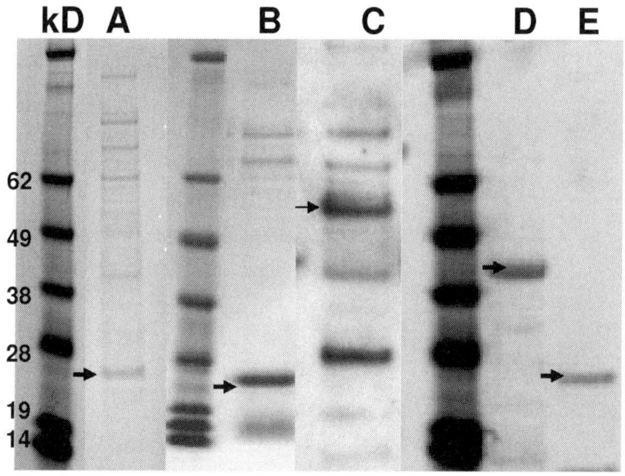

Fig. 18.2. *Lipoproteins expressed in cell-free extracts.* Lipoproteins were purified using Ni-NTA affinity chromatography and separated on a SDS–PAGE gel, stained with Coomassie Brilliant Blue (**A–B** and **D–E**) or detected by fluorescent scanning of labeled lysine residues (**C**). *Arrows* indicate apolipoprotein of interest. All proteins (A-E) are shown with SeeBlue MW marker (Invitrogen) (**A**) Full-length apolipoprotein A1 (**B**) MSP1 truncated form of ApoA1 (**C**) Full-length Apolipoprotein E4 (**D**) 22 kD truncated ApoE4-fusion protein (**E**) Thrombin cleaved truncated ApoE422k. Other Lipoproteins produced (not shown) include Apolipophorin III *B. mori,* Apolipophorin III, *M. Sexta.*

starting point for NLP formation and then takes the method one step further by combining pre-purified NLPs as an additive for the cell-free production of membrane proteins to show increased solubility of the membrane protein.

18.2. Materials

18.2.1. Cell-Free Apolipoprotein Expression and Purification

1. Apolipoprotein (ApoA1, MSP, Apo E4, lipophorinIII, or truncations Δ49ApoA1 and ApoE4 22k) clones of interest from the LLNL-IMAGE Consortium cDNA collection or as a gift from collaborating labs (*see* **Note 1**), subcloned in to an expression vector such as, pET32a thioredoxin (Novagen) *(33, 34)*, pIVEX-2.4b (Roche), or pEXP4 (Invitrogen)
2. Spectrophotometer UV–visible A_{260}/A_{280} quantification or PicoGreen dsDNA Quantification Kit (Invitrogen/Molecular Probes)
3. Cell-Free Expression System: Expressway™ Maxi Cell-Free *E. coli* Expression System (Invitrogen) or RTS 500 ProteoMaster *E. coli* HY Kit (Roche)
4. Thermomixer, Eppendorf Thermomixer R (for Roche lysates) or Incubator shaker for example New Brunswick C24 (for Invitrogen lysates)
5. Disposable fritted columns 3 mL capacity (Bio-Rad)
6. Ni-NTA Superflow resin (Qiagen)
7. Ni-NTA buffers (modified Qiagen recipes) Binding buffer: 50 mM NaH_2PO_4, 300 mM NaCl; pH 8.0. Wash Buffer: 50 mM NaH_2PO_4; 300 mM NaCl; 10 mM Imidazole; pH 8.0. Elution Buffer: 50 mM NaH_2PO_4; 300 mM NaCl; 400 mM Imidazole; pH 8.0
8. Gel electrophoresis equipment; NuPAGE 4–12% Bis–Tris SDS–PAGE gel with 1× MES-SDS running buffer (Invitrogen)
9. Protein Quantification Kit and standards, such as Bio-Rad Protein Assay (Bio-Rad)
10. Vivaspin6, ultrafiltration Devices, 10k MWCO (Sartorius Biotech) (*see* **Note 2**)
11. Centrifuge such as Eppendorf 5804R (Needs to fit 15 mL Falcon tubes)
12. Thrombin (Novagen)
13. DMPC; 1, 2-Dimyristoyl-*sn*-Glycero-3-Phosphocholine (Avanti Polar Lipids)

14. Probe or bath sonicator (*see* **Note 3**)
15. β-mercaptoethanol
16. TBS Buffer: 10 mM Tris–HCl; 0.15 M NaCl; 0.25 mM EDTA; 0.005% NaN_3 (sodium azide) adjust to pH 7.4
17. FPLC Instrument (Shimadzu SCL-10A), size exclusion column (Superdex 200 10/300 GL (GE Healthcare Life Sciences)

18.2.2. Nanolipoprotein Particle Formation and Purification

1. DMPC:1,2-Dimyristoyl-*sn*-Glycero-3-Phosphocholine (Avanti Polar Lipids)
2. Purified apolipoprotein protein or truncation (ApoE422k construct)
3. TBS Buffer: 10 mM Tris–HCl; 0.15 M NaCl; 0.25 mM EDTA; 0.005% NaN_3 (sodium azide); adjust to pH 7.4
4. 30 and 20°C water baths
5. Probe or bath sonicator (*see* **Note 3**)
6. Spin filter, 0.45 μm
7. Concentrator 50 kD MWCO, Vivaspin 2 (Sartorius, Inc.) (*see* **Note 4**)
8. FPLC Instrument (Shimadzu SCL-10A), size exclusion column (Superdex 200 10/300 GL (GE Healthcare Life Sciences)

18.2.3. Biotinylation of Membrane Protein

1. EZ-Link Sulfo-NHS-LC-Biotin (Pierce)
2. Bacteriorhodopsin (Sigma) (*see* **Note 5**)
3. Bath sonicator
4. Ultracentrifuge (Beckman-Coulter Optima TLX, TLA-120.2 fixed angle rotor)
5. 1× BupH PBS buffer (Pierce): 0.1 M NaH_2PO_4, 0.15 M NaCl; pH 7.0

18.2.4. Membrane Protein Incorporation into Nanolipoprotein Particles (MP-NLPs)

1. DMPC [1,2-Dimyristoyl-*sn*-Glycero-3-Phosphocholine] (Avanti Polar Lipids)
2. Purified apolipoprotein or truncation (ApoE4 22 kD construct)
3. TBS Buffer: 10 mM Tris–HCl; 0.15 M NaCl; 0.25 mM EDTA; 0.005% NaN_3 (sodium azide); adjust to pH 7.4
4. Sodium Cholate (Sigma) 500 mM solution in TBS
5. Biotinylated Bacteriorhodopsin (Sigma) from **Section 18.2.3**
6. 30 and 20°C and 23.8°C water baths
7. Probe Sonicator

8. Dialysis cups 10,000 MWCO (Pierce) or D-Tube Dialyzers, mini (Novagen)
9. Spin filter, 0.45 µm
10. FPLC Instrument (Shimadzu SCL-10A), size exclusion column (Superdex 200 10/300 GL (GE Healthcare Life Sciences)
11. Concentrator 50 kD MWCO, Vivaspin 2 (Sartorius, Inc.) (*see* **Note 4**)

18.2.5. Validating Protein Association by Microarray and Native Gel Electrophoresis

1. 4–20% Tris–Glycine polyacrylamide gel, 1× Tris–Glycine native running buffer, 2× Native Sample buffer, Native Mark molecular weight marker (Invitrogen)
2. Sypro Ruby Stain (Bio-Rad) light sensitive, Aqueous destain solution: 10% Methanol; 7% acetic acid
3. Fluoroimager with appropriate filter for SyproRuby stain
4. Biotinylated positive control protein such as biotinylated-bR
5. Bovine serum albumin 1 mg/ml solution
6. PBS-Tween buffer: 1.06 M KH_2PO_4; 2.97 M Na_2HPO_4; NaCl 155.1 M, 0.05% tween-20 (v/v) pH 7.4
7. 1×PBS buffer, (Gibco): 1.06 M KH_2PO_4; 2.97 M Na_2HPO_4; NaCl 155.1 M, pH 7.4
8. Cyanine-5-Strepavidin (Rockland) solution (5 µg/mL)
9. Barcoded γ-Aminopropylsilane coated glass slides (GAPS-II; Corning)
10. Robotic arrayer
11. Hybridization Chamber (Grace Bio-Labs)
12. Blocking buffer: 1 mg/mL BSA in 1× PBS, Wash Buffer: 1× PBS
13. Laser-based confocal scanner (ScanArray 5000 XL; Perkin-Elmer)
14. UV–visible plate reader (Bio-TEK Synergy HT)
15. 96 well flat bottom UV plate (Corning Costar UV Plate)

18.2.6. Membrane Protein Synthesis and Purification in a Single Step Using Cell-Free Synthesis in Conjunction with Pre-Formed Nanolipoprotein Particles

1. Membrane protein cDNA cloned into the expression plasmid pEXP4 (Invitrogen)
2. Purified pre-formed NLPs assembled with ApoE4 22k concentrated to approximately 5 mg/mL (*see above section*)
3. Cell-Free Expression System: Expressway™ Maxi Cell-Free *E. coli* Expression System (Invitrogen)
4. EasyTag(tm) L-[35S]-Methionine, 5mCi (185MBq), Stabilized Aqueous Solution, 1,175 Ci/mmol (Perkin Elmer)

5. Incubator shaker for example New Brunswick C24 (for Invitrogen lysates) or Thermomixer, Eppendorf Thermomixer R (for Roche lysates)
6. Refrigerated centrifuge (Eppendorf)
7. Borosilicate tubes, 12 × 75 mm (e.g., VWR Cat 47729–570)
8. 10% TCA (e.g., VWR Cat # JT0414–1; make a 10% (w/v) solution in water)
9. Glass fiber filters to catch precipitate (e.g., Whatman GF/C filter circle, Fisher cat# 09–874–12A)
10. Aqueous scintillation fluid (e.g., EcoLume, Fisher cat# ICN88247504)
11. Vacuum manifold (e.g., Millipore Cat #XX270550)
12. Scintillation vials and Scintillation counter

18.3. Methods

The methods described below outline (1) cell-free expression and purification of apolipoproteins, (2) NLP formation and purification, (3) biotinylation of membrane bound proteins, (4) membrane protein incorporation into NLPs (5) validating protein association by microarray, and (6) membrane protein synthesis and determination of solubility in a single step using cell-free synthesis in the presence of pre-formed NLPs.

18.3.1. Cell-Free Expression and Purification of Apolipoproteins

Cell-free protein .production is widely accepted as a method to deal with proteins that are difficult to express and purify such as membrane associated proteins *(35, 37)*. Importantly, cell-free protein expression can often bypass many of the difficulties often associated with cell-based recombinant protein expression by eliminating several steps of the process *(35)*. Cell-free systems also allow for the addition of co-factors and new strategies for protein labeling, such as fluorescent or isotope labeling. More complex mammalian proteins containing multiple post-translational modifications often require a eukaryotic environment for recombinant protein production such as a wheat germ extract or rabbit reticulocyte lysate, which are commercially available. Here we describe the cell-free production of a selected N-terminal 22 kD fragment of human apolipoprotein E4 which does not require post-translational modification (*see* **Note 1**). SDS–PAGE gels for this protein as well as other selected apolipoproteins are shown in **Fig. 18.2**.

1. Select apolipoprotein cDNAs and clone into expression vector of interest (*see* **Note 1**), such as pIVEX-2.4b (Roche Applied Science), GFP folder or pETBlue-2 (Novagen.), pET32a (thioredoxin fusion vector). Propagate plasmids by transforming into Top10 or DH5α chemically competent cells (Invitrogen) and isolate DNA using HiSpeed Plasmid Maxi or Midi Kits (Qiagen). The N-terminal truncated apolipoprotein E4–22 kD (ApoE422k) thioredoxin (trx) fusion protein construct in pET32a (ApoE422k-trx) is illustrated here.
2. Determine Midi or Maxi prepped plasmid DNA concentrations by PicoGreen dsDNA Quantification Kit (Molecular Probes) or by UV–visible spectroscopy A_{260}/A_{280}.
3. Perform cell-free protein production reactions using either the Expressway™ Maxi Cell-Free *E. coli* Expression System (Invitrogen) or RTS 500 ProteoMaster *E. coli* HY Kit (Roche) using ~15 μg of midi or maxi prepped DNA in a 1 mL reaction size.
4. Incubate reactions at 30°C shaking at 990 rpm in a thermomixer (Roche RTS ProteoMaster or Eppendorf Thermomixer R) – (Roche Lysates) or 37°C shaking at 225 rpm in a shaker incubator (New Brunswick) – (Invitrogen Lysates). All reactions are run overnight (although 4 h is sufficient). Collect a 5–10 μl sample for further analysis.
5. Purify the His-tagged apolipoprotein (ApoE422k-trx) by using Ni-NTA native affinity chromatography. Equilibrate 1 mL of the Ni-NTA slurry, equivalent to 500 μL column bed volume (Qiagen) with binding buffer and resuspend the resin to form a 50% slurry again. Add the equilibrated slurry to the cell-free post-reaction mixture and mix at 4°C for 1-2 hours.
6. Transfer the mixture to a 3-mL fritted plastic column and collected the flow through for SDS–PAGE analysis.
7. Wash the column with eight column volumes (500 μL) of native wash buffer. Fractions are collected for SDS–PAGE analysis.
8. Elute the bound apolipoprotein with six column volumes of native elution buffer.
9. All collected fractions are analyzed by denatured gel electrophoresis using a NuPAGE 4–12% Bis–Tris SDS–PAGE gel with 1× MES-SDS running buffer for 38 min at 200 V (Invitrogen). The load buffer is LDS Sample Buffer (Invitrogen). Volumes to load for SDS–PAGE gels are as follows 1 μL of total reaction and non-bound flow through, 5 μL wash fractions 1–2, 20 μL of remaining washes and all elutions. Gels are stained with Coomassie brilliant blue.

10. Elution fraction of interest determined by gel electrophoresis, are combined and concentrated and buffer exchanged into TBS using an untrafiltration device vivaspin6 (*see* **Note 2**).
11. The final protein concentration is determined by Bradford total protein concentration following the manufacturer's protocol.
12. Prepare small unilamellar vesicles of DMPC by probe sonicating 20 mg DMPC lipid into 1 mL TBS at 6 A for approximately 15 min or until optical clarity is achieved. Typically 15 min is sufficient to achieve optical clarity. An appropriate container choice is a thick walled 3 mL glass conical vial (*see* **Note 3**).
13. Transfer the sample to a 1.5 mL tube. Remove any contaminant metal from the probe by centrifugation at 13,700 RCF for 2 min in a 1.5 mL tube.
14. Remove thioredoxin fusion protein tags by incubating 2–4 mg of the produced protein with 100 μg/mL of the sonicated DMPC overnight at 24°C. Add thrombin at 1:500 w/w ratio (thrombin: apolipoprotein) and incubated at 37°C for 1 h. Halt the reaction by the addition of β-mercaptoethanol to a final concentration of 1%. Analyze 5 μg of the product by SDS–PAGE as described above and shown in **Fig. 18.2**.
15. Remove contaminant thioredoxin (trx), thrombin and β-mercaptoethanol from the apolipoprotein, ApoE422k by size exclusion chromatography using a FPLC Instrument (Akta, GE Healthcare and Life Sciences or Shimadzu SCL-10A) and size exclusion column (Superdex 200 10/300 GL) with a TBS buffer at a flow rate of 0.5 mL/min. Determine fractions of interest by gel electrophoresis combine and concentrate as above.

18.3.2 Nanolipoprotein Particle Formation and Purification

NLPs form in a self assembly process in the correct mass ratio of apolipoprotein to lipid. This ratio needs to be optimized for each different apolipoprotein. The ratio described below is for ApoE422k *(21)*. Other ratios can be found in the literature *(17, 24, 25)*:

1. Start water bath incubators. Temperatures at 30 and 20°C.
2. Probe sonicate 34 mg of DMPC into 1 mL of TBS at 6 A for approximately 15 min or until optical clarity is achieved. Centrifuge DMPC solution at 13,700 RCF for 2.5 min to remove residual metal from probe sonicator (*see* **Note 3**).
3. Transfer supernatant into new tube.
4. Combine Apo E422K with DMPC in a ratio of 1:4 by mass in TBS buffer in a 1.5 mL Eppendorf tube. Typically batches are of the 250 μL size.

5. Transition temperature procedure: Immerse tube in water bath for 10 min each 30°C (above DMPC transition temp.) followed by 20°C (below DMPC transition temp.). Repeat the procedure three times then incubate at 23.8°C overnight.
6. Filter preparation through a 0.45 µm spin filter at 13,700 RCF for 1 min.
7. Purify NLPs using size exclusion chromatography. Use a Shimadzu SCL-10A FPLC, equipped with a Superdex 200 10/300 GL column with TBS buffer, a 200 µL sample injection volume and a flow rate of 0.5 mL/min. Collect 0.5 mL fractions *see* **Fig. 18.3**.
8. Concentrate fractions using a Vivaspin 2 ultrafiltration device with a 50 k MWCO as described in **Section 18.3.1**.

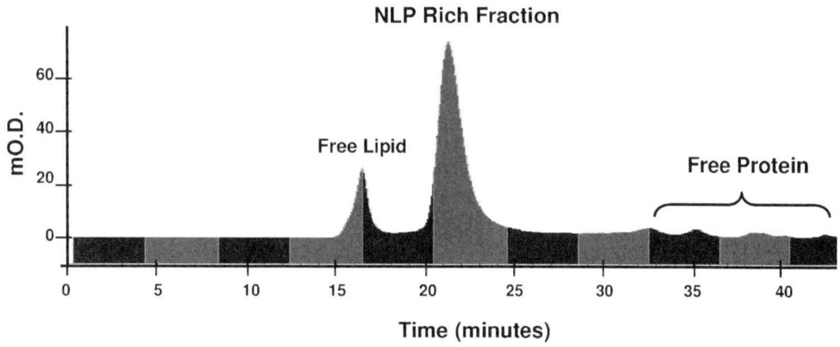

Fig. 18.3. Size exclusion chromatography separation of ApoE422k Nanolipoprotein particles (NLPs). Free lipid, and free protein denoted on graph are separated from the NLP rich fraction.

18.3.3 Biotinylation of Bacteriorhodopsin

Biotinylation of the membrane protein (MP) provides a tool for investigating the incorporation of the MP with the NLP. Biotinylation using of bacteriorhodopsin supplied in membrane sheets from Sigma selectively labels only the solvent exposed lysine residues when using EZ-Link Sulfo-NHS-LC-Biotin (Pierce) which is impermeable to membranes. Bacteriorhodopsin in membrane sheets is easily separated from the aqueous phase by centrifugation. For other membrane proteins that may be solubilized in detergent micelles removal of excess biotin solution will need to be accomplished using a desalting column or other means. Membrane proteins including bR may be expressed in a cell-free manner and biotinylated *(10, 13, 36, 38, 39)*.

1. Resuspend bacteriorhodopsin purchased from Sigma and stored as a lyophilized powder at 4°C in BupH PBS buffer in the original bottle. Avoid amine containing buffers such as TBS, due to the interaction with the biotinylation reagent. Bath sonicate the sample eight times for 1 min. each chilling the bottle on ice for 1 min. in between each burst.

2 Record UV–visible spectra to confirm the concentration of bR in solution using the molar extinction coefficient at 568 nm of 63,000 M^{-1} cm^{-1}.

3 A freshly made 10 mM solution of EZ-Link Sulfo-NHS-LC-Biotin (Pierce) is prepared according to the manufactures recommendation in ddH_2O.

4 Add the biotin solution to the bacteriorhodopsin solution in a 20-fold molar excess, and incubated on ice for 2 h.

5 Remove excess biotin by centrifugation of the solution in an ultracentrifuge at an RCF of 89,000 (although 50,000 should be sufficient) for 20 min at 4°C. Remove the supernatant and resuspend the bR pellet in TBS buffer. Repeat this process two times total (*see* **Note 6**)

6 Collect UV–visible spectra as above to calculate the concentration of the solution and the percent recovery typically around 85–90% with careful resuspension.

18.3.4. Membrane Protein Incorporation into Nanolipoprotein Particles

1. Start water bath incubators. Temperatures at 30 and 20°C.
2. Probe sonicate 34 mg of DMPC into 1 mL of TBS at 6 A for approximately 15 min or until optical clarity is achieved. Alternatively, sonicate in bath sonicator to optical clarity (*see* **Note 3**).
3. Centrifuge the solution at 13 K for 2 min to remove residual metal sloughed off from probe sonicator.
4. For a 250 μL batch in a 1.5 mL Eppendorf tube. Combine Apo E422K with DMPC in a ratio of 1:4 by mass in TBS buffer. Sodium cholate solution is then added to a final concentration of 20 mM. The biotinylated bacteriorhodopsin membrane protein is then added in a 0.67 mass ratio to the Apo E422k apolipoprotein.
5. Transition temperature procedure: Immerse tube in water bath for 10 min each 30°C (above DMPC transition temp.) followed by 20°C (below DMPC transition temp.). Repeat the procedure three times then incubate at 23.8°C overnight.
6. To remove the cholate detergent and allow for self-assembly of MP-NLPs (bR-NLPs) the sample is loaded into a pre-soaked D-Tube Dialyzers, mini (Novagen). The sample is then dialyzed against three changes each of 1 L TBS buffer over a 2–3 day period at room temperature (*see* **Note 7**).
7. Concentrate using an ultrafiltration device, Vivaspin 2 (Sartorius) MWCO 50 K to 200 μL.
8. Transfer supernatant into new tube.
9. Size exclusion chromatography is preformed using a Shimadzu SCL-10A FPLC, equipped with a Superdex 200 10/300 GL

column (GE Healthcare Life Sciences). The buffer is TBS with a 200 μL sample injection volume, a 0.5 mL/min flow rate and 0.5–1.0 mL fraction size.

10. Concentrate the fractions of interest using an ultrafiltration device, Vivaspin 2 (Sartorius) MWCO 50 K for NLP peaks.

18.3.5. Validating NLP Formation by Native Gel Electrophoresis and Confirmation of Membrane Protein Association and Functionality with NLPs by Microarray and UV–Visible Spectroscopy

Native polyacrylamide gel electrophoresis is used to validate the association of proteins of interest (apolipoprotein and/or membrane protein) with NLP fractions eluted from the size exclusion column (**Fig. 18.4**) protein identification is confirmed with mass spectrometry. We use contact microarray spotting technology to attach NLPs to an amino-silane coated glass slide in an array format for streptavidin binding studies *(37, 40, 41)*. Biotinylated bacteriorhodopsin is used to validate the incorporation of bR into NLP fractions eluted from size exclusion chromatography. Cyanine-5-Strepavidin is used for fluorescence detection

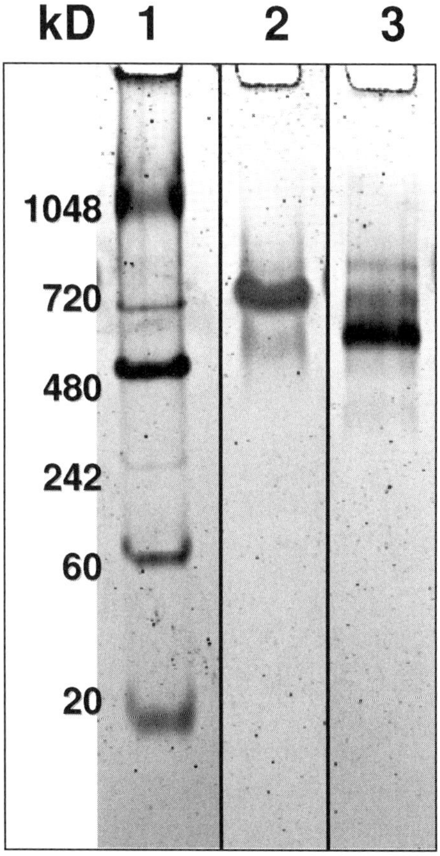

Fig. 18.4. *Native gel electrophoresis of NLPs.* (1) Native Mark molecular weight marker. (2) "Empty"-NLPs (3) Membrane protein (bacteriorhodopsin) bR-NLPs. 4–20% Tris–glycinc gel, with Tris–glycine running buffer, stained with SyproRuby Stain (BioRad) Imaged with a Typhoon scanner.

of biotinylated bacteriorhodopsin (**Fig. 18.5**). UV–visible spectroscopy of light and dark adapted bacteriorhodopsin can be used to confirm bR function and proper folding *(42)* (**Fig. 18.6**).

In-depth physical characterization of these particles is presented to demonstrate functional protein insertion. Methods such as Atomic force microscopy (AFM) and Electron microscopy (EM) address whether assembly of membrane protein-NLPs was successful. This is done by identifying insertion and localization of the membrane protein with in the NLP. AFM (**Fig. 18.7**) and Electron microscopy (**Fig. 18.8**) although not fully described here, are used to image the prepared discs and determine diameter and height measurements as well as sample heterogeneity.

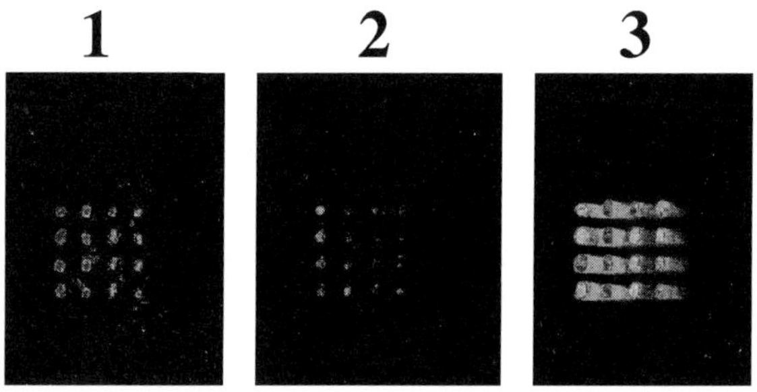

Fig. 18.5. *Protein microarray of biotinylated bR-NLPs.* (1) biotinylated-bacteriorhodopsin (2) negative control, native bacteriorhodopsin (3) biotinylated-bacteriorhodopsin associated NLPs.

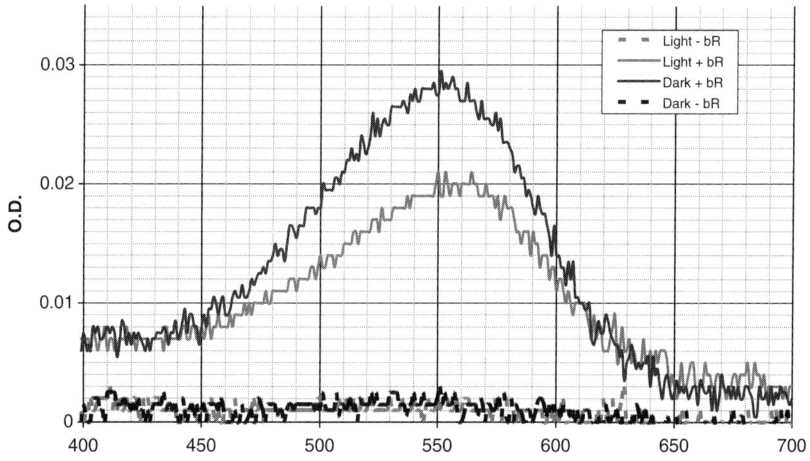

Fig. 18.6. *Light and dark adapted visible spectra of bacteriorhodopsin associated NLPs.* (*Top traces*) Light and dark adapted visible spectra of bacteriorhodopsin associated NLPs and (*bottom traces*) NLPs without membrane protein. (*Black*) Dark adapted spectra (bR λ_{max} = 550 nm). (*Grey*) Light adapted spectra (bR λ_{max} = 560 nm).

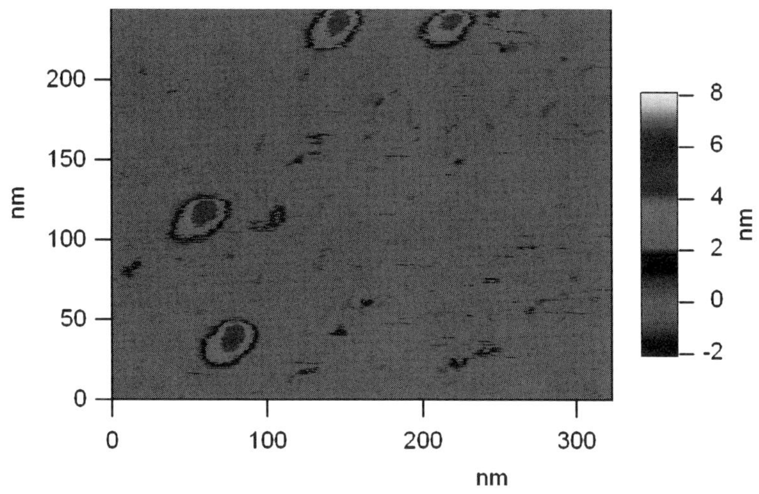

Fig. 18.7. *Atomic force microscopy of nanolipoprotein particles (NLPs).* NLPs consisting of cell-free produced apoE4 22K lipoprotein and DMPC. Particle dimensions are as follows; Height: 4.94 nm, std dev: 0.369 nm Width of top: 9.72 nm, std dev: 1.50 nm, Full width at half max: 20.4 nm std dev: 3.5 nm (*See Color Plate 6*).

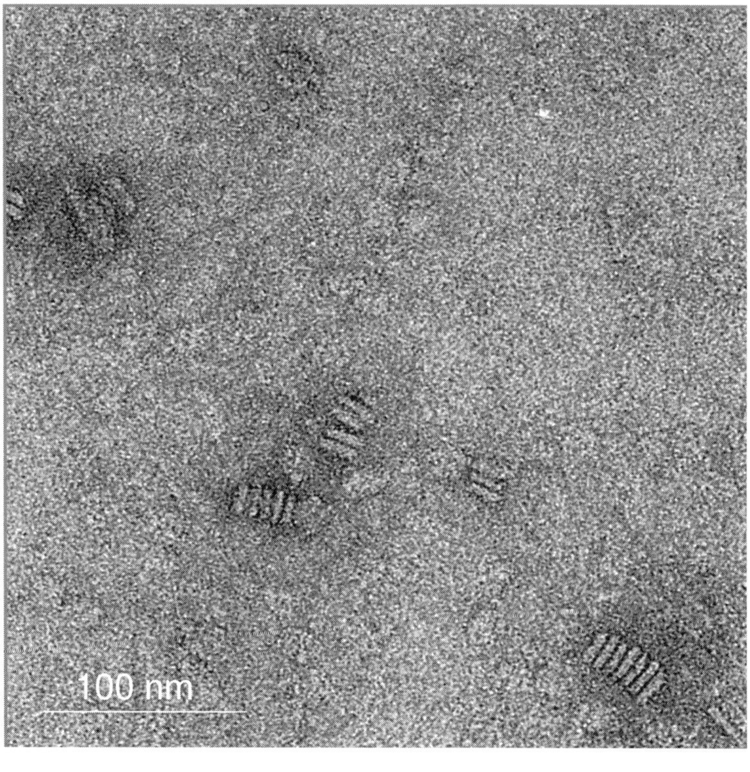

Fig. 18.8. *Electron microscopy of Nanolipoprotein particles (NLPs) showing the discoidal shape.* Magnification ×40 k.

18.3.5.1. Native Gel Electrophoresis

1. Run native-PAGE gels, 4–20% Tris–glycine with 0.75 μg total loaded protein estimated by A_{280} absorbance. Load 10 μL of molecular weight standards, Native mark (Invitrogen) diluted 20× in native sample buffer. The gel is run at 125 V for approximately 2 h.
2. Stain gels with ~150 mL of SyproRuby protein stain (Bio-Rad) following the microwave staining method: 30 s microwave, 30 s mixing on shaker table, 30 s microwave, 5 min shake, 30 s microwave, finally 23 min on shaker table at room temperature.
3. Destain gels for 1.5 h on a shaker table at room temperature.
4. Image the gel using a Typhoon Imager with appropriate filters selected for the SyproRuby fluorescence.

18.3.5.2. NLP Microarray

1. Microarray single print head is used to deposit approximately 1 nL of diluted protein solution on the slide (*see* **Note 8**). Proteins are spotted in 4×4 squares with 16 replicates of each sample, generating ~300 μm diameter spots with a spot-to-spot distance of ~350 μm.
2. Protein microarrays are spotted on GAPSII amino silane glass slides (Corning) with bacteriorhodopsin bR (non-biotinylated), biotinylated-bR, biotinylated-bR-NLPs, using a robotic arrayer. Non-biotinylated bR serves as a negative control and biotinylated-bR serves as a positive control.
3. Bacteriorhodopsin concentrations of 10 mM, as determined by UV–visible spectroscopy as described above are used for all samples.
4. Cross-link proteins to the glass slides by exposure to UV light for 5 min. Store unused slides at 4°C with out UV cross-linking.
5. Apply the hybridization chamber with a volume capacity of 950 μL to the slide carefully as to not disrupt the array. Carefully add reagents below with out injecting bubbles.
6. Block slides with BSA (1 mg/mL) for 30 min. Wash slides with 1× PBS for 15 min. Bind cyanine-5-streptavidin (5 μg/mL) for 15 min. Wash slides in 1× PBS then nanopure water each for 15 min. Dry slides by centrifugation or air dry.
7. Image protein microarrays of bR, biotinylated-bR and bR-NLPs with a laser-based confocal scanner (ScanArray 5000 XL; Perkin Elmer) using the VheNe 594 nm laser for detection of any bound Cyanine-5-streptavidin.
8. Collect images and analyze using the mean pixel intensities with Scan Array software (Perkin Elmer).

18.3.5.3. UV–Visible Spectroscopy

1. Collect UV–visible spectra in 96-well plate reader using 100 μL of sample in a UV detectable flat bottom plate. Collect dark adapted spectra after keeping the sample wrapped in foil

overnight taking care not to expose the sample prior to spectral collection. Collect light adapted spectra after exposure to a full spectrum bright lamp for 15 min.

2. A 5–10 nm visible shift between light and dark adapted spectra indicates a functional protein *(42)*.

18.3.6. Membrane Protein Synthesis and Purification in a Single Step Using Cell-Free Synthesis in Conjunction with Pre-Formed Nanolipoprotein Particles

Cell-free expression of membrane proteins has usually employed either of two possible methods; one: expression and purification in a denatured state followed by refolding in the presence of detergents and/or lipids as well as any cofactors such as all *trans*-retinal for bacteriorhodopsin or two: expression in the presence of detergents or lipids *(10, 36, 38, 39)*. Solubiliza tion of the membrane protein with detergent is generally followed by a dialysis step to return the membrane protein to a lipid bilayer vesicle. The method described here utilizes preformed NLPs as an additive to increase the membrane protein production, solubility and stabilization by incorporation into a NLP lipid bilayer (Co-translation). The procedure uses commercially available cell-free extracts with the addition of membrane protein plasmid DNA (pEXP4 expression vector (Invitrogen)) and pre-formed NLPs to synthesize folded functional membrane protein in one step. **Fig. 18.9** shows protein yield for the soluble (S) fraction based on scintillation counting of incorporated ^{35}S-Methionine in the presence and absence of added NLPs. A survey of several membrane proteins with various

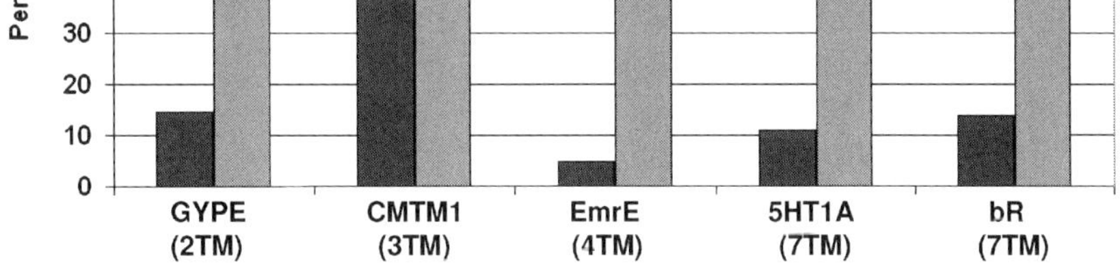

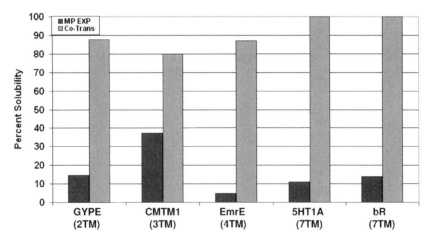

Fig. 18.9. Cell-free expression in the presence of NLPs (Co-translation) increases solubility of membrane proteins. A comparison was made between (*Black bars*) the membrane protein expressed alone or (*Grey bars*) expression of the membrane protein In the presence of pre-formed ApoA1 NLPs (Co-Translation). Membrane proteins with the number of trans membrane domain in parentheses are (GYPE) glycophorin B (MNS blood group) (2TM), (CMTM1) CKLF-like MARVEL transmembrane domain (3TM), (EmrE) *E. coli* SMR efflux transporter (4TM) (5HT1A) 5-hydroxytryptamine (serotonin) receptor (7TM), (bR) bacteriorhodopsin (7TM). In all cases the expression in the presence of NLPs increased membrane protein solubility. Solubility is determined by removing a 5 µl of the reaction supernatant after a 10 min centrifugation at 20,017 RCF and determining yield by TCA precipitation and scintillation counting as described in **Section 18.3.6**.

numbers of transmembrane (TM) segments are expressed using this method. Solubility of the membrane protein is clearly increased in the presence of pre-formed NLPs indicating association with the NLP.

1. Entry clone membrane protein cDNAs of interest are recombined by Gateway cloning into the expression plasmid pEXP4 (Invitrogen) and are propagated by transforming into TOP10 or DH5α chemically competent cells (Invitrogen). Isolate plasmid DNA using a Hipure Plasmid Maxi or MidiKits (Invitrogen).
2. Carry out cell-Free expression reactions using the Expressway™ Maxi Cell-Free *E. coli* Expression System (Invitrogen) protocols with the addition of ~15 μg of membrane protein DNA, for a 1 mL reaction, 300 μg of purified NLPs (ApoA1 assembled with DMPC *see above section*). For scintillation counting follow the manufacturer protocol for the incorporation of ^{35}S-Methionine. Reactions are scaleable to other volumes following the same ratios. Carry out control experiments with out the addition of NLPs using the same lysate batch.
3. Incubate reactions at 37°C shaking at 225 rpm in a shaker incubator (New Brunswick). Reactions continue for 1.5–2 h.
4. Retain a 5 μL aliquot of the total (T) reaction for SDS–PAGE and autoradiograms (not shown), the reaction is then centrifuged for 5 min. at 4°C and 18,000 RCF. The supernatant is collected and a 5 μL aliquot of the soluble (S) fraction placed into a 12 × 75 mm glass tube.
5. Add 100 ul of 1 N NaOH. Incubate at room temperature for 5 min.
6. Add 2 ml of cold 10% TCA (trichloroacetic acid) to the 12 × 75 mm tube. Place at 4°C for 10 min.
7. Collect the precipitate via vacuum filtration through a Whatman GF/C glass fiber filter (or equivalent). Pre-wet the filter with a small amount of 10% TCA prior to adding the sample.
8. Rinse the tube twice with 1 ml of 10% TCA and then rinse once with 3–5 ml of 95% ethanol. Pass each of the rinses through the GF/C filter.
9. Place the filter in a scintillation vial, add aqueous scintillation cocktail and count in a scintillation counter. The cpm will reflect the amount of radiolabel that was incorporated.

18.3.7. Summary

NLPs can be self-assembled into discoidal structures when apolipoproteins are combined in the proper ratio with polar lipids. Membrane protein incorporation requires the addition of a detergent and a dialysis step for self-assembly of NLPs. These membrane incorporated MP-NLPs have been used to study a variety of membrane proteins *(17, 18, 20, 22, 29, 31, 32, 43–46)* including cytochromes and receptor proteins. The soluble lipid bilayer environment provides a native-like platform for characterization and stabilization of membrane proteins. The possible utility of the NLP platform for elucidating functionality and overcoming solubility issues for multiple membrane proteins such as cell receptors, host-pathogen interactions, bioenergetics and bioremediation membrane proteins is immense. Cell-free expression of apolipoproteins will simplify the process of membrane protein incorporation into NLPs. Parallel cell-free expression of membrane bound proteins for NLP incorporation should be a future direction for expediting the characterization of membrane bound proteins.

18.4. Notes

1. Lipophorin III DNA clones (*M. sexta* and *B. mori*) were obtained from the lab of Robert Ryan at Children's Hospital Oakland Research Institute (CHORI). Truncated Apolipoprotein E4 22 kDa N-terminal domain thioredoxin fusion plasmid was obtained from Karl Weisgraber at the University of California, San Francisco. Midi or Maxi prepped plasmid DNA was prepared according to the Qiagen protocol. The 193 amino acid protein sequence of the 22 kD Apolipoprotein E4 construct is as follows, with the two initial amino acids, Gly-Ser, are left over from the thrombin cleavage site in pET32a. GSKVEQAVETEPEPELRQQTEWQSGQRW ELALGRFWDYLRWVQTLSEQVQEELLSSQVTQEL RALMDETMKELKAYKSELEEQLTPVAEETRARLSKEL QAAQARLGADMEDVRGRLVQYRGEVQAMLGQS TEELRVRLASHLRKLRKRLLRDADDLQKRLAVYQA GAREGAERGLSAIRERLGPLVEQGRVR
2. Concentration from 6 mL to 100 µL is easily achieved in ~15 min at 5,000 RCF in an Eppendorf 5804R centrifuge with a fixed angle rotor check each 3–5 min. Buffer exchange into TBS pH 7.4 requires at least three dilutions and re-concentration steps. Alternatively eluted protein can be dialyzed (spectrapor 1 MWCO 3500) against TBS buffer overnight and concentrated by immersion of the dialysis membrane in PEG 8000 (polyethylene glycol).
3. Vortex lipid solution lightly before sonicating to help get to lipid in to the buffer. Lipid should be stored at –20°C when not in use and protected from water absorption.

When sonicating lipid be sure to avoid over heating the lipids by either sonicating in a beaker of ice or cooling the sample every few minutes. The solution should be practically water clear at the end of the sonication. If the probe hits the side of the glass vessel metal will be sloughed off into the solution and the solution will become grayish. The metal can be removed by a short centrifugation at 13,700 RCF for 2 min after transferring to a 1.5 ml Eppendorf tube. Remove the supernatant and use. Any white pellet is DMPC that is not in vesicle form. Alternatively, sonicate in bath sonicator to optical clarity and skip the centrifugation step.

4. Other concentrator brands that are angled are also acceptable such as Agilent, because the NLP produced will be larger than 200 kD, a 100 kD filter may be useful.
5. Bacteriorhodopsin can also be produced in a cell-free manner and purified in the denatured state. A re-folding procedure is then employed to incorporate the retinal according to the methods of Rothschild et al. *(36, 47)*.
6. Bacteriorhodopsin in membrane sheets is extremely sticky and pellets well at the RCF listed. 85–90% recovery of bR can be achieved with careful resuspension and washing of tips and tubes. Resuspension should be in the TBS buffer used for assembly (or other buffer of interest that will be used for assembly).
7. Dialysis at 4°C could be used for unstable membrane proteins. Detergent use should be compatible with the membrane protein of interest. Adsorbent beads (Bio beads, Bio-Rad) can also be used to remove the detergent. If dialysis cups are used (Pierce) sample needs to be split into three pre-soaked dialysis cups. Care should be taken not to create bubbles or droplets on the sides of the cups.
8. We have determined that robotic spotting is best when the humidity is greater than 30%.

Acknowledgments

This work performed under the auspices of the U.S. Department of Energy by Lawrence Livermore National Laboratory under Contract DE-AC52–07NA27344, with support from the Laboratory Directed Research and Development Office (LDRD) 06-SI-003 awarded to PDH. UCRL-BOOK-235838. The authors are grateful to Drs. Karl Weisgraber and Robert Ryan for helpful discussions and providing reagents.

References

1. Drews, J. (2000) Drug discovery: a historical perspective *Science* 287, 1960–1964.
2. Wallin, E., and von Heijne, G. (1998) Genome-wide analysis of integral membrane proteins from eubacterial, archaean, and eukaryotic organisms *Protein Sci* 7, 1029–1038.
3. Klammt, C., Schwarz, D., Eifler, N., Engel, A., Piehler, J., Haase, W., Hahn, S., Dotsch, V., and Bernhard, F. (2007) Cell-free production of G protein-coupled receptors for functional and structural studies *J Struct Biol* 158, 482–493.
4. Schwarz, D., Klammt, C., Koglin, A., Lohr, F., Schneider, B., Dotsch, V., and Bernhard, F. (2007) Preparative scale cell-free expression systems: New tools for the large scale preparation of integral membrane proteins for functional and structural studies *Methods* 41, 355–369.
5. Kantardjieff, K. A., Hochtl, P., Segelke, B. W., Tao, F. M., and Rupp, B. (2002) Concanavalin A in a dimeric crystal form: revisiting structural accuracy and molecular flexibility *Acta Crystallogr D Biol Crystallogr* 58, 735–743.
6. Kigawa, T., Yabuki, T., and Yokoyama, S. (1999) [Large-scale protein preparation using the cell-free synthesis] *Tanpakushitsu Kakusan Koso* 44, 598–605.
7. Sawasaki, T., Hasegawa, Y., Tsuchimochi, M., Kamura, N., Ogasawara, T., Kuroita, T., and Endo, Y. (2002) A bilayer cell-free protein synthesis system for high-throughput screening of gene products *FEBS Lett* 514, 102–105.
8. Frydman, J., and Hartl, F. U. (1996) Principles of chaperone-assisted protein folding: differences between in vitro and in vivo mechanisms *Science* 272, 1497–1502.
9. Tsalkova, T., Zardeneta, G., Kudlicki, W., Kramer, G., Horowitz, P. M., and Hardesty, B. (1993) GroEL and GroES increase the specific enzymatic activity of newly-synthesized rhodanese if present during in vitro transcription/translation *Biochemistry* 32, 3377–3380.
10. Klammt, C., Lohr, F., Schafer, B., Haase, W., Dotsch, V., Ruterjans, H., Glaubitz, C., and Bernhard, F. (2004) High level cell-free expression and specific labeling of integral membrane proteins *Eur J Biochem* 271, 568–580.
11. Kim, D. M., and Swartz, J. R. (2004) Efficient production of a bioactive, multiple disulfide-bonded protein using modified extracts of Escherichia coli *Biotechnol Bioeng* 85, 122–129.
12. Ishihara, G., Goto, M., Saeki, M., Ito, K., Hori, T., Kigawa, T., Shirouzu, M., and Yokoyama, S. (2005) Expression of G protein coupled receptors in a cell-free translational system using detergents and thioredoxin-fusion vectors *Protein Expr Purif* 41, 27–37.
13. Klammt, C., Schwarz, D., Fendler, K., Haase, W., Dotsch, V., and Bernhard, F. (2005) Evaluation of detergents for the soluble expression of alpha-helical and beta-barrel-type integral membrane proteins by a preparative scale individual cell-free expression system *Febs J* 272, 6024–6038.
14. Martin, T. A., Harding, K., and Jiang, W. G. (2001) Matrix-bound fibroblasts regulate angiogenesis by modulation of VE-cadherin *Eur J Clin Invest* 31, 931–938.
15. Keppetipola, N., and Shuman, S. (2006) Mechanism of the phosphatase component of Clostridium thermocellum polynucleotide kinase-phosphatase *RNA* 12, 73–82.
16. Yokoyama, S. (2003) Protein expression systems for structural genomics and proteomics *Curr Opin Chem Biol* 7, 39–43.
17. Bayburt, T. H., Carlson, J. W., and Sligar, S. G. (1998) Reconstitution and imaging of a membrane protein in a nanometer-size phospholipid bilayer *J Struct Biol* 123, 37–44.
18. Bayburt, T. H., and Sligar, S. G. (2002) Single-molecule height measurements on microsomal cytochrome P450 in nanometer-scale phospholipid bilayer disks *Proc Natl Acad Sci USA* 99, 6725–6730.
19. Bayburt, T. H., and Sligar, S. G. (2003) Self-assembly of single integral membrane proteins into soluble nanoscale phospholipid bilayers *Protein Sci* 12, 2476–2481.
20. Civjan, N. R., Bayburt, T. H., Schuler, M. A., and Sligar, S. G. (2003) Direct solubilization of heterologously expressed membrane proteins by incorporation into nanoscale lipid bilayers *Biotechniques* 35, 556–560, 562–563.
21. Chromy, B. A., Arroyo, E., Blanchette, C. D., Bench, G., Benner, H., Cappuccio, J. A., Coleman, M. A., Henderson, P. T., Hinz, A. K., Kuhn, E. A., Pesavento, J. B., Segelke, B. W., Sulchek, T. A., Tarasow, T., Walsworth, V. L., and Hoeprich, P. D. (2007) Different Apolipoproteins Impact Nanolipoprotein Particle Formation *J Am Chem Soc* 129, 14348–14354.
22. Bayburt, T. H., Grinkova, Y. V., and Sligar, S. G. (2006) Assembly of single bacteriorhodopsin trimers in bilayer nanodiscs *Arch Biochem Biophys* 450, 215–222.

23. Shaw, A. W., McLean, M. A., and Sligar, S. G. (2004) Phospholipid phase transitions in homogeneous nanometer scale bilayer discs *FEBS Lett* 556, 260–264.
24. Gursky, O., Ranjana, and Gantz, D. L. (2002) Complex of human apolipoprotein C-1 with phospholipid: thermodynamic or kinetic stability? *Biochemistry* 41, 7373–7384.
25. Jayaraman, S., Gantz, D., and Gursky, O. (2005) Structural basis for thermal stability of human low-density lipoprotein *Biochemistry* 44, 3965–3971.
26. Jonas, A. (1986) Reconstitution of high-density lipoproteins *Methods Enzymol* 128, 553–582.
27. Jonas, A., Kezdy, K. E., and Wald, J. H. (1989) Defined apolipoprotein A-I conformations in reconstituted high density lipoprotein discs *J Biol Chem* 264, 4818–4824.
28. Peters-Libeu, C. A., Newhouse, Y., Hatters, D. M., and Weisgraber, K. H. (2006) Model of biologically active apolipoprotein E bound to dipalmitoylphosphatidylcholine *J Biol Chem* 281, 1073–1079.
29. Whorton, M. R., Bokoch, M. P., Rasmussen, S. G., Huang, B., Zare, R. N., Kobilka, B., and Sunahara, R. K. (2007) A monomeric G protein-coupled receptor isolated in a high-density lipoprotein particle efficiently activates its G protein. *Proc Natl Acad Sci USA* 104, 7682–7687.
30. Tufteland, M., Pesavento, J. B., Bermingham, R. L., Hoeprich, P. D., Jr., and Ryan, R. O. (2007) Peptide stabilized amphotericin B nanodisks. *Peptides* 28, 741–746.
31. Cruz, F., and Edmondson, D. E. (2007) Kinetic properties of recombinant MAO-A on incorporation into phospholipid nanodisks. *J Neural Transm* 114, 699–702.
32. Boldog, T., Grimme, S., Li, M., Sligar, S. G., and Hazelbauer, G. L. (2006) Nanodiscs separate chemoreceptor oligomeric states and reveal their signaling properties. *Proc Natl Acad Sci USA* 103, 11509–11514.
33. Forstner, M., Peters-Libeu, C., Contreras-Forrest, E., Newhouse, Y., Knapp, M., Rupp, B., and Weisgraber, K. H. (1999) Carboxyl-terminal domain of human apolipoprotein E: expression, purification, and crystallization. *Protein Expr Purif* 17, 267–272.
34. Morrow, J. A., Arnold, K. S., and Weisgraber, K. H. (1999) Functional characterization of apolipoprotein E isoforms overexpressed in Escherichia coli. *Protein Expr Purif* 16, 224–230.
35. Coleman, M. A., Lao, V. H., Segelke, B. W., and Beernink, P. T. (2004) High-throughput, fluorescence-based screening for soluble protein expression. *J Proteome Res* 3, 1024–1032.
36. Sonar, S., Patel, N., Fischer, W., and Rothschild, K. J. (1993) Cell-free synthesis, functional refolding, and spectroscopic characterization of bacteriorhodopsin, an integral membrane protein. *Biochemistry* 32, 13777–13781.
37. Segelke, B. W., Schafer, J., Coleman, M. A., Lekin, T. P., Toppani, D., Skowronek, K. J., Kantardjieff, K. A., and Rupp, B. (2004) Laboratory scale structural genomics. *J Struct Funct Genomics* 5, 147–157.
38. Klammt, C., Schwarz, D., Lohr, F., Schneider, B., Dotsch, V., and Bernhard, F. (2006) Cell-free expression as an emerging technique for the large scale production of integral membrane protein. *FEBS J* 273, 4141–4153.
39. Sonar, S., Marti, T., Rath, P., Fischer, W., Coleman, M., Nilsson, A., Khorana, H. G., and Rothschild, K. J. (1994) A redirected proton pathway in the bacteriorhodopsin mutant Tyr-57→Asp. Evidence for proton translocation without Schiff base deprotonation. *J Biol Chem* 269, 28851–28858.
40. Camarero, J. A., Kwon, Y., and Coleman, M. A. (2004) Chemoselective attachment of biologically active proteins to surfaces by expressed protein ligation and its application for "protein chip" fabrication. *J Am Chem Soc* 126, 14730–14731.
41. Rao, R. S., Visuri, S. R., McBride, M. T., Albala, J. S., Matthews, D. L., and Coleman, M. A. (2004) Comparison of multiplexed techniques for detection of bacterial and viral proteins. *J Proteome Res* 3, 736–742.
42. Wang, J., Link, S., Heyes, C. D., and El-Sayed, M. A. (2002) Comparison of the dynamics of the primary events of bacteriorhodopsin in its trimeric and monomeric states. *Biophys J* 83, 1557–1566.
43. Bayburt, T. H., Leitz, A. J., Xie, G., Oprian, D. D., and Sligar, S. G. (2007) Transducin activation by nanoscale lipid bilayers containing one and two rhodopsins. *J Biol Chem* 282, 14875–14881.
44. Baas, B. J., Denisov, I. G., and Sligar, S. G. (2004) Homotropic cooperativity of monomeric cytochrome P450 3A4 in a nanoscale native bilayer environment. *Arch Biochem Biophys* 430, 218–228.
45. Leitz, A. J., Bayburt, T. H., Barnakov, A. N., Springer, B. A., and Sligar, S. G. (2006) Functional reconstitution of Beta2-adrenergic receptors utilizing self-assembling Nanodisc technology. *Biotechniques* 40, 601–602, 04, 06, passim.

46. Nath, A., Atkins, W. M., and Sligar, S. G. (2007) Applications of phospholipid bilayer nanodiscs in the study of membranes and membrane proteins. *Biochemistry* 46, 2059–2069.

47. Coleman, M., Nilsson, A., Russell, T. S., Rath, P., Pandey, R., and Rothschild, K. J. (1995) Asp 46 can substitute Asp 96 as the Schiff base proton donor in bacteriorhodopsin. *Biochemistry* 34, 15599–15606.

Chapter 19

Expression and Purification of Soluble His_6-Tagged TEV Protease

Joseph E. Tropea, Scott Cherry, and David S. Waugh

Summary

This chapter describes a simple method for overproducing a soluble form of the tobacco etch virus (TEV) protease in *Escherichia coli* and purifying it to homogeneity so that it may be used as a reagent for removing affinity tags from recombinant proteins by site-specific endoproteolysis. The protease is initially produced as a fusion to the C-terminus of *E. coli* maltose binding protein (MBP), which causes it to accumulate in a soluble and active form rather than in inclusion bodies. The fusion protein subsequently cleaves itself *in vivo* to remove the MBP moiety, yielding a soluble TEV protease catalytic domain with an N-terminal polyhistidine tag. The His-tagged TEV protease can be purified in two steps using immobilized metal affinity chromatography (IMAC) followed by gel filtration. An S219V mutation in the protease reduces its rate of autolysis by approximately 100-fold and also gives rise to an enzyme with greater catalytic efficiency than the wild-type protease.

Key words: Maltose-binding protein; MBP; Immobilized metal affinity chromatography; IMAC; His-tag; Affinity tag; Affinity chromatography; Tobacco etch virus protease; TEV protease; Fusion protein

19.1. Introduction

The use of genetically engineered fusion tags has become a widespread practice in the production of recombinant proteins for various applications. Although originally designed to facilitate the detection and purification of proteins, subsequently it has become clear that some tags can also increase the yield of their fusion partners, protect them from intracellular proteolysis, enhance their solubility and even facilitate their folding *(1)*. However, all tags, whether large or small, have the potential to interfere with the

Sharon A. Doyle (ed.), *Methods in Molecular Biology: High Throughput Protein Expression and Purification, vol. 498*

Book doi: 10.1007/978-1-59745-196-3

biological activity of a protein and may impede its crystallization (e.g., *2–5)*. For this reason, it is generally advisable to remove the tag(s) at some stage.

Although both chemical and enzymatic reagents have been used to remove tags from recombinant proteins *(6)*, only proteases exhibit enough specificity to be generally useful for this purpose. Traditionally, the enzymes that have been used most commonly to cleave fusion proteins at designed sites are Factor Xa, enterokinase (enteropeptidase) and thrombin. However, all of these proteases have frequently been observed to cleave proteins at locations other than the designed target site *(7, 8)*.

Recently it has become clear that certain viral proteases, such as that encoded by the tobacco etch virus (TEV), have much greater stringency *(9)*, making them particularly useful for this application. TEV protease recognizes the amino acid sequence ENLYFQ/G with high efficiency and cleaves between Q and G. Hence, in contrast to Factor Xa and enterokinase, which ostensibly have no specificity for the P1′ residue in their respective recognition sites, TEV protease typically leaves a single non-native glycine residue on the N-terminus of a recombinant protein when it is used to remove an N-terminal fusion tag. It should be noted, however, that the P1′ specificity of TEV protease is relatively relaxed and many different residues can be substituted for glycine in the P1′ position of its recognition site with little or no impact on the efficiency of cleavage *(10)*, making it possible to produce many proteins with no non-native residues appended to their N-terminus.

Unfortunately, TEV protease is poorly soluble when it is overproduced in *Escherichia coli (11, 12)*. Yet, we have found that this problem can be overcome by producing the enzyme in the form of a MBP fusion protein that cleaves itself *in vivo* to generate a TEV protease catalytic domain that is highly soluble and active. Fusion to MBP somehow enables the enzyme to fold into its native conformation and avoid accumulating as insoluble aggregates *(11)*. Another problem with TEV protease is that it cleaves itself at a specific site, giving rise to a truncated enzyme with greatly diminished activity *(13)*, but this has been overcome by the introduction of a mutation (S219V) that renders the enzyme virtually impervious to autoinactivation and also increases its catalytic activity by approximately twofold *(14)*.

Here, we describe a method for the large-scale production of S219V mutant TEV protease in *E. coli* and its purification to homogeneity. The presence of a polyhistidine (His6-tag) on the N-terminus of the protease facilitates not only its purification but also its separation from the digestion products of a His-tagged fusion protein *(15)*. Although not discussed here, the polyarginine tag on the C-terminus of the catalytic domain can be used in a similar manner.

19.2. Materials

19.2.1. Overproduction of His_6-TEV(S219V)-Arg_5 Protease in E. coli

1. A glycerol stock of *E. coli* BL21(DE3) CodonPlus-RIL cells (Stratagene, La Jolla, CA, USA) containing the TEV protease expression vector pRK793 (*see* **Note 1**).
2. LB broth and LB agar plates containing 100 μg/ml ampicillin (for pRK793 selection) and 30 μg/ml chlorampenicol (for pRIL selection). For LB broth, add 10 g bacto tryptone, 5 g bacto yeast extract and 5 g NaCl to 1 l of H_2O and sterilize by autoclaving (121°C, 15 psi, 20 min, slow exhaust). Let cool to room temperature and add 1 ml of sterile 100 mg/ml ampicillin and 1 ml of 30 mg/ml chloramphenicol. Prepare a 100 mg/ml ampicillin solution by mixing 10 g of ampicillin, sodium salt, ultra with H_2O to a final volume of 100 ml. Filter sterilize and store at –20°C. Prepare a 30 mg/ml chloramphenicol solution by mixing 3 g of chloramphenicol with absolute ethanol to a final volume of 100 ml. Store at –20°C. Sterile glucose can be added to the LB-antibiotic broth to a concentration of 0.2% to increase the bacteria biomass. Prepare a stock solution of 20% (w/v) of D(+)-glucose monohydrate by mixing 100 g of D(+)-glucose monohydrate with H_2O to a final volume of 500 ml. Filter sterilize and store at room temperature. Add 10 ml 20% sterile glucose per 1 l of LB broth to achieve a 0.2% concentration. For LB agar add 12 g of Bacto agar per 1 l of LB broth before autoclaving (as above). To prepare plates, allow the 1 l of LB-agar mixture to cool until the flask or bottle can be held in hands without burning (approximately 50–60°C). Add 1 ml of 100 mg/ml sterile ampicillin stock solution (*see* above) and 1 ml of 30 mg/ml chloramphenicol stock solution (*see* above) to the LB-agar, mix by gentle swirling and pour or pipet ca. 30 ml into sterile 100 × 15 mm Petri dishes. Let plates cool to room temperature and store at 4°C.
3. Isopropyl-β-D-thiogalactopyranoside (IPTG), analytical. Prepare a fresh 200 mM stock solution by mixing 477 mg of IPTG with H_2O to a final volume of 10 ml. Five ml of 200 mM IPTG is required per 1 l of culture to achieve a final concentration of 1 mM.
4. A 500 ml and several 4 l baffle-bottom flasks. Sterilize by autoclaving (as above, fast exhaust).
5. An autoclave with fast and slow exhaust setting.
6. A temperature-controlled shaker/incubator that can accommodate 500 ml and 4 l flasks and can be set at either 30 or 37°C.
7. A high speed centrifuge (e.g., Sorvall refrigerated centrifuge).
8. A spectrophotometer and cuvette that can measure absorbance at 600 nm.

19.2.2. Purification of His_6-TEV(S219V)-Arg_5 Protease

1. Cell lysis/IMAC equilibration buffer: 50 mM sodium phosphate (pH 8.0), 200 mM NaCl, 10% glycerol, 25 mM imidazole. Prepare 2 l of buffer by mixing 14.2 g sodium phosphate dibasic, 23.38 g NaCl, 200 ml glycerol and 3.41 g imidazole with H_2O to a volume of 1980 ml. Adjust the pH to 8 using concentrated hydrochloric acid. Adjust the volume to 2 l with H_2O and check the pH. Adjust if necessary. Filter through a 0.22 µm polyethersulfone membrane (Corning Incorporated, Corning, NY, USA, or the equivalent) and store at 4°C.
2. A mechanical device to disrupt E. coli cells (e.g., a sonicator, french press, or cell homogenizer) (see Note 2)
3. A solution of 5% (w/v) polyetheleneimine, pH 8. Mix 50 ml of 50% (w/v) polyethylenimine with H_2O to a volume of 450 ml. Adjust the pH to 8 with concentrated HCl and let to cool to room temperature. Adjust the volume to 500 ml with H_2O and check the pH. Adjust if necessary. Filter through a 0.22 µm polyethersulfone filtration unit (Corning Incorporated, Corning, NY, USA, or the equivalent). The solution is stable for at least 3 years when stored at 4°C.
4. A spectrophotometer and cuvette that can measure absorbance at 280 nm.
5. ÄKTA Explorer chromatography system (Amersham Biosciences, Piscataway, NJ, USA), or the equivalent.
6. Ni-NTA Superflow resin (Qiagen Incorporated, Valencia, CA, USA).
7. Column XK 26/10 (Amersham Biosciences, Piscataway, NJ, USA).
8. IMAC equilibration and elution buffers are: (a) 50 mM sodium phosphate (pH 8.0), 200 mM NaCl, 10% glycerol, 25 mM imidazole, and (b) 50 mM sodium phosphate (pH 8.0), 200 mM NaCl, 10% glycerol, 250 mM imidazole, respectively. For preparation of IMAC (Ni-NTA Superflow) equilibration buffer (see Step 1 in Section 19.2.2). Prepare 1 l of elution buffer by mixing 7.1 g sodium phosphate dibasic, 11.69 g NaCl, 100 ml glycerol and 17.02 g imidazole, with H_2O to a volume of 950 ml. Adjust the pH to 8 using concentrated hydrochloric acid. Adjust the volume to 1 l with H_2O, let to cool to room temperature and check the pH. Adjust if necessary. Filter through a 0.22 µm polyethersulfone membrane (Corning Incorporated, Corning, NY, USA, or the equivalent) and store at 4°C.
9. A 0.5 M ethylenediaminetetraacetic acid (EDTA), pH 8 stock solution.
10. A 1 M stock solution of 1,4-dithio-DL-threitol (DTT). Prepare 10 ml by mixing 1.55 g of DTT with H_2O to a final

volume of 10 ml. Place solution on ice. Use immediately or store at −20°C.

11. Polyethersulfone filtration unit (0.22 and 0.45 μm, Corning Incorporated, Corning, NY, USA, or the equivalent).
12. An Amicon Stirred Ultrafiltration Cell concentrator and YM10 ultrafiltration membranes (Millipore Corporation, Bedford, MA, USA).
13. A HiPrep 26/60 Sephacryl S-100 HR column (Amersham Biosciences, Piscataway, NJ, USA).
14. Gel filtration buffer: 25 mM sodium phosphate (pH 7.5), 100 mM NaCl, 10% glycerol. Prepare by mixing 7.1 g sodium phosphate dibasic, 11.69 g NaCl and 200 ml glycerol with H_2O to a volume of 1980 ml. Adjust the pH to 7.5 using concentrated hydrochloric acid. Adjust the volume to 2 l with H_2O and check the pH. Adjust if necessary. Filter through a 0.22 μm polyethersulfone membrane (Corning Incorporated, Corning, NY, USA, or the equivalent) and store at 4°C.
15. 0.2 μm syringe filter (Gelman, Acrodisc Supor membrane, Pall Corporation, Ann Arbor, MI, USA).
16. A Dewar flask filled with liquid nitrogen.

19.3. Methods

19.3.1. Overproduction of Soluble His_6-TEV(S219V)-Arg5 Protease in E. coli

The induction of pRK793 with IPTG produces an MBP fusion protein (*see* **Fig. 19.1**) that self-cleaves *in vivo* to generate a soluble His_6-TEV(S219V)-Arg_5 protease. Virtually all the protease remains soluble after intracellular processing if the temperature is reduced from 37 to 30°C after the addition of IPTG.

1. Inoculate 50–150 ml of LB broth containing 100 μg/ml ampicillin and 30 μg/ml chlorampenicol in a 500 ml baffle-bottom shake flask from a glycerol stock of pRK793 transformed *E. coli* BL21(DE3) CodonPlus-RIL cells. Place in an incubator and shake overnight at 250 rpm and 37°C.
2. Add 25 ml of the saturated overnight culture to each 1 l of fresh LB broth containing 100 μg/ml ampicillin, 30 μg/ml chloramphenicol and 0.2% glucose in a 4 l baffle-bottom shake flask. To ensure that there will be an adequate yield of pure protein at the end of the process, we ordinarily grow 4–6 l of cells at a time.
3. Shake the flasks at 250 rpm and 37°C until the cells reach mid-log phase ($OD_{600\,nm}$ ~0.5); approximately 2 h.

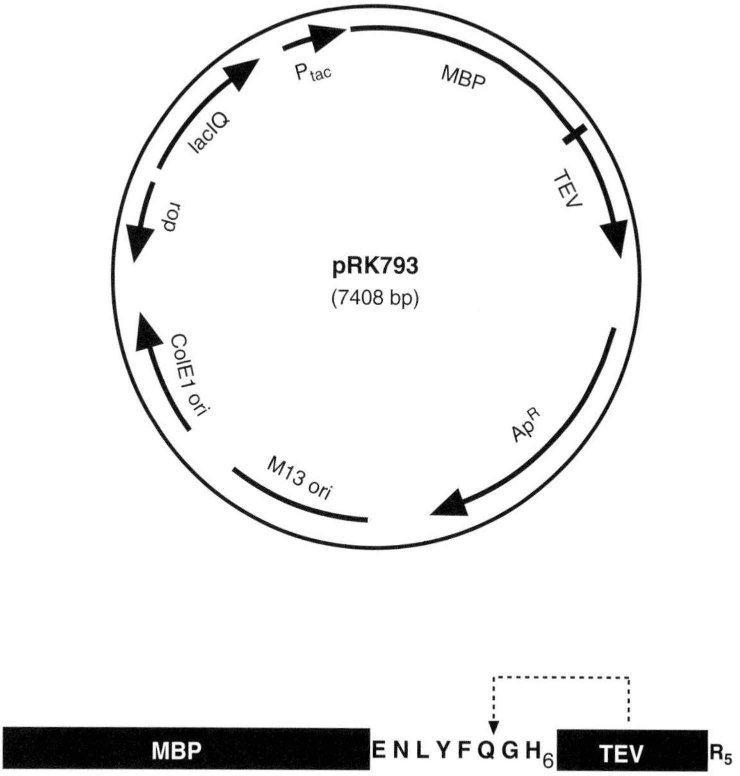

Fig. 19.1. Schematic representation (not to scale) of the TEV protease expression vector pRK793 and its fusion protein product. Further information about this plasmid can be found at http://mcl1.ncifcrf.gov/waugh_tech.html.

4. Shift the temperature to 30°C and induce the culture(s) with IPTG at a final concentration of 1 mM (5 ml of 200 mM IPTG stock solution per liter of culture). Continue shaking at 250 rpm for 4–6 h. Place cultures at 4°C.
5. Recover the cells by centrifugation at 5,000×g for 10 min at 4°C, and store at –80°C. A 6 l preparation yields 30–40 g of cell paste.

19.3.2. Protein Purification

His_6-TEV(S219V)-Arg_5 protease can be purified to homogeneity in two steps: immobilized metal affinity chromatography (IMAC) using Ni-NTA resin followed by size exclusion chromatography. An example of a purification monitored by SDS–PAGE is shown in **Fig. 19.2** (*see* **Note 3**):

1. All procedures are performed at 4–8°C. Thaw the cell paste from 6 l of culture on ice and suspend in ice-cold cell lysis/IMAC equilibration buffer (10 ml/g cell paste).
2. Lyse the cell suspension (see Note 2) and measure the volume using a graduated cylinder. Add polyethylenimine to a final

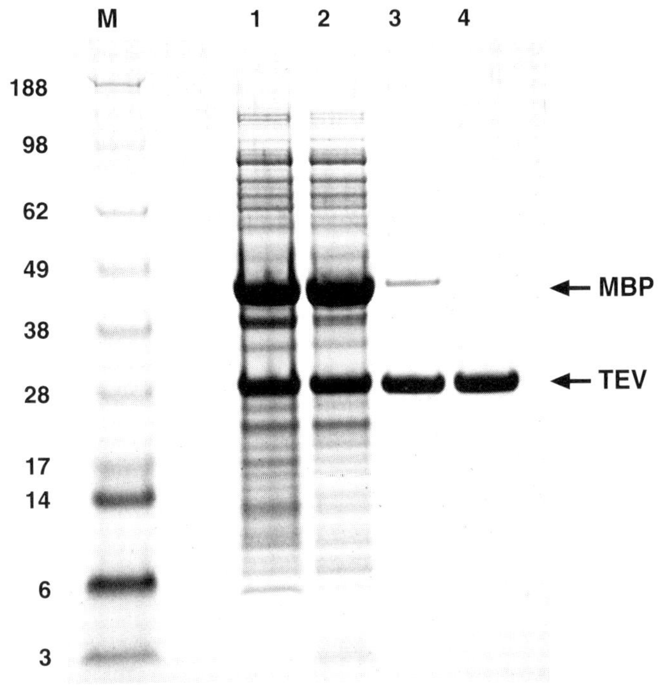

Fig. 19.2. Purification of His_6-TEV(S219V)-Arg_5 protease monitored by SDS–PAGE (NuPage 4–12% gradient MES gel). M: molecular weight standards (kDa). *Lane 1*: Total intracellular protein after induction. *Lane 2*: Soluble cell extract. *Lane 3*: Pooled peak fractions after IMAC. *Lane 4*: Pooled peak fractions after gel filtration and concentration.

concentration of 0.1% (a 1:50 dilution of the 5% stock solution at pH 8) and mix gently by inversion. Immediately centrifuge at 15,000×g for 30 min and filter (*see* **Note 2**).

3. Apply the supernatant to a 50 ml Ni-NTA superflow column equilibrated in cell lysis/IMAC equilibration buffer (*see* **Note 4**). Wash the column with equilibration buffer until a stable baseline is reached (approximately 7 column volumes) and then elute the bound His_6-TEV(S219V)-Arg_5 with a linear gradient to 100% elution buffer over ten column volumes.
4. Pool the peak fractions containing the protease and measure the volume. Add EDTA to a final concentration of 2 mM (a 1:250 dilution of the 0.5 M EDTA, pH 8 stock solution) and mix well. Add DTT to a final concentration of 5 mM (a 1:200 dilution of the 1M DTT stock solution) and mix well.
5. Concentrate the sample approximately tenfold using an Amicon stirred ultrafiltration cell fitted with a YM10 membrane. Remove the precipitation by centrifugation at 5,000×*g* for 10 min. Estimate the concentration of the partially pure protein solution spectrophotometrically at 280 nm

using a molar extinction coefficient of 32,290 M^{-1} cm^{-1}. The desired concentration is between 5 and 10 mg/ml.

6. Apply 5 ml of the concentrated sample onto a HiPrep 26/60 Sephacryl S-100 HR column equilibrated with gel filtration buffer. The volume of sample loaded should be no more than 2% of the column volume and contain no more than 50 mg of protein.
7. Pool the peak fractions from the gel filtration column(s) of pure His_6-TEV(S219V)-Arg_5 protease and concentrate to 1–5 mg/ml (*see* **Step 5** in **Section 19.3.2**). Filter through a 0.2 µm syringe filter (Gelman, Acrodisc Supor membrane, Pall Corporation, Ann Arbor, MI, USA), aliquot and flash freeze with liquid nitrogen. Store at –80°C.

19.3.3. Cleaving a Fusion Protein Substrate with TEV Protease

The standard reaction buffer for TEV protease is 50 mM Tris–HCl (pH 8.0), 0.5 mM EDTA and 1 mM DTT, but the enzyme has a relatively flat activity profile at pH values between 4 and 9 and will tolerate a range of buffers, including phosphate, MES and acetate. TEV protease activity is not adversely affected by the addition of glycerol or sorbitol up to at least 40% (w/v). The enzyme is also compatible with some detergents *(16)*. TEV protease activity is not inhibited by PMSF and AEBSF (1 mM), TLCK (1 mM), Bestatin (1 mg/ml), pepstatin A (1 mM), EDTA (1 mM), E-64 (3 mg/ml), or "complete" protease inhibitor cocktail (Roche). However, zinc will inhibit the activity of the enzyme at concentrations of 5 mM or greater and reagents that react with cysteine (e.g., iodoacetamide) are potent inhibitors of TEV protease.

The duration of the cleavage reaction is typically overnight. A good rule of thumb is to use 1 OD_{280} of TEV protease per 100 OD_{280} of fusion protein for an overnight digest. TEV protease is maximally active at 34°C *(17)*, but we recommend performing the digest at 4°C. The results of a typical TEV protease digest of a fusion protein substrate (MBP-NusG) are shown in **Fig. 19.3.**

Some fusion proteins are intrinsically poor substrates for TEV protease. This may be due to steric occlusion when the protease cleavage site is too close to ordered structure in the passenger protein, or when the fusion protein exists in the form of soluble aggregates. Sometimes this problem can be mitigated by using a large amount of TEV protease (we have occasionally used up to 1 OD_{280} of TEV protease per 5 OD_{280} of fusion protein) and/or performing the reaction at higher temperature (e.g., room temperature). Failing that, the addition of extra residues between the TEV protease cleavage site and the N-terminus of the target protein is advised. We have used polyglycine, polyhistidine and a FLAG-tag epitope with good results.

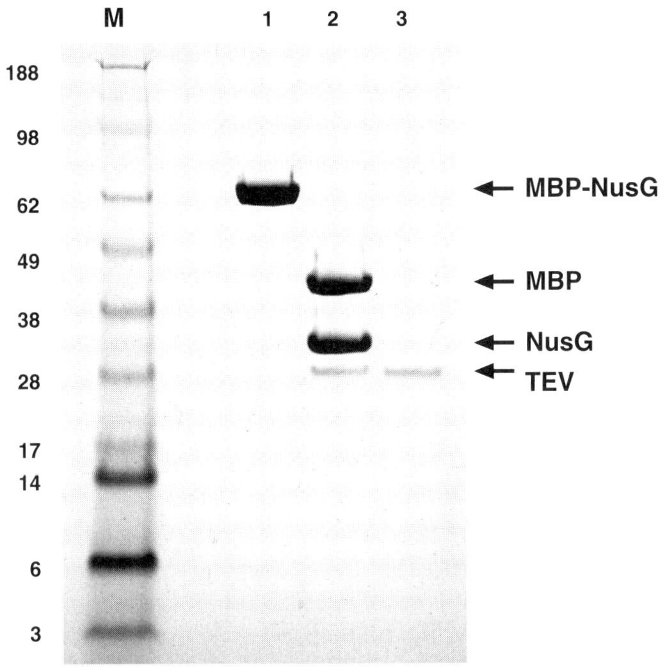

Fig. 19.3. Digestion of a fusion protein substrate by His_6-TEV(S219V)-Arg_5 protease. 22 µg of the substrate, a fusion between *E. coli* maltose-binding protein (MBP) and *Aquifex aeolicus* NusG with a canonical TEV protease recognition site (ENLYFQG) in the linker region *(10)* was incubated for 1 h at room temperature in 50 µl of standard reaction buffer (*see* **Section 19.3.3**) in the absence (*Lane 1*) or presence (*Lane 2*) of 1.3 µg His_6-TEV(S219V)-Arg_5 protease. The reaction products were separated by SDS–PAGE (NuPage 4–12% MES gradient gel) and visualized by staining with Coomassie brilliant blue. *Lane 3* contains an equivalent amount of pure His_6-TEV(S219V)-Arg_5 protease. M: molecular weight standards (kDa).

19.4. Notes

1. *E. coli* BL21(DE3) CodonPlus-RIL cells containing pRK793 can be obtained for a nominal shipping and handling fee from the non-profit distributor of biological reagents AddGene, Inc., Cambridge, MA, USA (http://www.addgene.org) or from the American Type Culture Collection (ATCC catalog number MB-145). The pRIL plasmid is a derivative of the p15A replicon that carries the *E. coli argU*, *ileY* and *leuW* genes, which encode the cognate tRNAs for AGG/AGA, AUA and CUA codons, respectively. pRIL is selected for by resistance to chloramphenicol. Due to the presence of several AGG and AGA codons in the TEV protease coding sequence, the presence of pRIL dramatically increases the yield of TEV protease.
2. We routinely break cells using an APV-1000 homogenizer (Invensys, Roholmsvej, Denmark) at 10–11,000 psi for 2–3

rounds. Other homogenization techniques such as French press, sonication, or manual shearing should yield comparable results. Centrifugation of the disrupted cell suspension for at least 30 min at 30,000×*g* is recommended. Filtration through a 0.45 μm polyethersulfone (or cellulose acetate) membrane is helpful to remove residual particulates and fines prior to chromatography.

3. We find it convenient to use precast gels for SDS–PAGE gels (e.g., 1.0 mm × 10 well, 4–12% NuPage gradient), running buffer and electrophoresis supplies from Invitrogen (Carlsbad, CA, USA).
4. We use an ÄKTA Explorer chromatography system (Amersham Biosciences, Piscataway, NJ, USA) and a Ni-NTA Superflow column (Qiagen, Valencia, CA, USA). A properly poured 50 ml Ni-NTA Superflow column (in an Amersham Biosciences XK26/20 column) can be run at 4–6 ml/min (backpressure less than 0.4 MPa) and will bind up to 400 mg of His_6-TEV(S219V)-Arg5 protease. If a chromatography system is not available, the IMAC can be performed using a peristaltic pump or manually by gravity. If the latter is used, Ni-NTA agarose should be substituted for Superflow and the elution performed with step increases of imidazole in 25 mM increments. Binding and elution profiles can be monitored spectrophotometrically at 280 nm and by SDS–PAGE. Care must be taken to properly zero the spectrophotometer because imidazole has significant absorption in the UV range.

Acknowledgment

This research was supported by the Intramural Research Program of the NIH, National Cancer Institute, Center for Cancer Research.

References

1. Waugh, D. S. (2005) Making the most of affinity tags. *Trends Biotechnol* **23**, 316–320.
2. Goel, A., Beresford, G. W., Colcher, D., Pavlinkova, G., Booth, B. J., Baranowska-Kortylewicz, J., and Batra, S. K. (2000) Relative position of the hexahistidine tag effects binding properties of a tumor-associated single-chain Fv construct. *Biochim Biophys Acta* **1523**, 13–20.
3. Bucher, M. H., Evdokimov, A. G., and Waugh, D. S. (2002) Differential effects of short affinity tags on the crystallization of *Pyrococcus furiosus* maltodextrin-binding protein. *Acta Crystallogr D Biol Crystallogr* **58**, 392–397.
4. Fonda, I., Kenig, M., Gaberc-Porekar, V., Pristovaek, P., and Menart, V. (2002) Attachment of histidine tags to recombinant tumor necrosis factor-alpha drastically changes its properties. *Sci World J* **2**, 1312–1325.
5. Chant, A., Kraemer-Pecore, C. M., Watkin, R., and Kneale, G. G. (2005) Attachment

of histidine tag to the minimal zinc finger protein of the *Aspergillus nidulans* gene regulatory proteiin AreA causes a conformational change at the DNA-binding site. *Protein Expr Purif* **39**, 152–159.

6. Arnau, J., Lauritzen, C., Petersen, G. E., and Pedersen, J. (2006) Current strategies for the use of affinity tags and removal for the purification of recombinant proteins. *Protein Expr Purif* **48**, 1–13.
7. Choi, S. I., Song, H. W., Moon, J. W., and Seong, B. L. (2001) Recombinant enterokinase light chain with affinity tag: expression from Saccharomyces cerevisiae and its utilities in fusion protein technology. *Biotechnol Bioeng* **75**, 718–724.
8. Jenny, R. J., Mann, K. G., and Lundblad, R. L. (2003) A critical review of the methods for cleavage of fusion proteins with thrombin and factor Xa. *Protein Expr Purif* **31**, 1–11.
9. Parks, T. D., Leuther, K. K., Howard, E. D., Johnston, S. A., and Dougherty, W. G. (1994) Release of proteins and peptides from fusion proteins using a recombinant plant virus protease. *Anal Biochem* **216**, 413–417.
10. Kapust, R. B., Tozser, J., Copeland, T. D., and Waugh, D. S. (2002) The P1″ specificity of tobacco etch virus protease. *Biochem Biophys Res Commun* **294**, 949–955.
11. Kapust, R. B., and Waugh, D. S. (1999) *Escherichia coli* maltose-binding protein is uncommonly effective at promoting the solubility of polypeptides to which it is fused. *Protein Sci* **8**, 1668–1674.
12. Lucast, L. J., Batey, R. T., and Doudna, J. A. (2001) Large-scale purification of a stable form of recombinant tobacco etch virus protease. *Biotechniques* **30**, 544–546.
13. Parks, T. D., Howard, E. D., Wolpert, T. J., Arp, D. J., and Dougherty, W. G. (1995) Expression and purification of a recombinant tobacco etch virus NIa proteinase: biochemical analyses of the full-length and a naturally occurring truncated proteinase form. *Virology* **210**, 194–201.
14. Kapust, R. B., Tozser, J., Fox, J. D., Anderson, D. E., Cherry, S., Copeland, T., and Waugh, D. S. (2001) Tobacco etch virus protease: mechanism of autolysis and rational design of stable mutants with wild-type catalytic efficiency. *Protein Eng* **14**, 993–1000.
15. Tropea, J. E., Cherry, S., Nallamsetty, S., Bignon, C., and Waugh, D. S. (2007) A generic method for the production of recombinant proteins in *Escherichia coli* using a dual hexahistidine-maltose-binding protein affinity tag. *Methods Mol Biol* **363**, 1–19.
16. Mohanty, A. K., Simmons, C. R., and Wiener, M. C. (2003) Inhibition of tobacco etch virus protease activity by detergents. *Protein Expr Purif* **27**, 109–114.
17. Nallamsetty, S., Kapust, R. B., Tozser, J., Cherry, S., Tropea, J. E., Copeland, T. D., and Waugh, D. S. (2004) Efficient site-specific cleavage of fusion proteins by tobacco vein mottling virus protease in vivo and in vitro. *Protein Expr Purif* **38**, 108–115.

Chapter 20

High-Throughput Protein Concentration and Buffer Exchange: Comparison of Ultrafiltration and Ammonium Sulfate Precipitation

Priscilla A. Moore and Vladimir Kery

Summary

High-throughput protein purification is a complex, multi-step process. There are several technical challenges in the course of this process that are not experienced when purifying a single protein. Among the most challenging are the high-throughput protein concentration and buffer exchange, which are not only labor-intensive but can also result in significant losses of purified proteins. We describe two methods of high-throughput protein concentration and buffer exchange: one using ammonium sulfate precipitation and one using micro-concentrating devices based on membrane ultrafiltration. We evaluated the efficiency of both methods on a set of 18 randomly selected purified proteins from *Shewanella oneidensis*. While both methods provide similar yield and efficiency, the ammonium sulfate precipitation is much less labor intensive and time consuming than the ultrafiltration.

Keywords: Ammonium sulfate precipitation; Buffer exchange; Desaltation; High-throughput; Protein concentration; Ultrafiltration

20.1. Introduction

Despite considerable progress in protein purification technologies in recent years, protein purification remains a complex, multi-step process. Significant amounts of low molecular weight materials (salts, affinity eluents) are introduced into the protein sample during this process. Removing these materials is important not only for performing the next purification step but also for storing the purified protein.

Sharon A. Doyle (ed.), *Methods in Molecular Biology: High Throughput Protein Expression and Purification, vol. 498*

Book doi: 10.1007/978-1-59745-196-3

Gel filtration, dialysis, ultrafiltration and protein precipitation are the four most-used approaches to perform desalting and buffer exchange during protein purification. Gel filtration and dialysis still dominate in a large-scale, in low-throughput protein purification, because of their well-defined outcome and scalability. However, in a high-throughput mode of protein purification, ultrafiltration and protein precipitation are more attractive. Not only they are less labor intensive and time consuming than gel filtration and dialysis, but they also can be used to efficiently concentrate proteins in addition to the buffer exchange.

Ultrafiltration is increasingly becoming the preferred method for protein concentration and buffer exchange *(1)*. Ultrafiltration membranes are cast from a variety of polymers such as polysulfone, polyethersulfone and others. Among them, the regenerated cellulose and new composite-regenerated cellulose membranes appear to have low protein adsorption and inherent chemical and thermal stability compared to other materials. Besides large-scale ultrafiltration devices with mL and L volumes, a broad selection of tube-based microdevices in a wide range of molecular weight cut-offs for volumes <1 mL is currently commercially available from multiple vendors (Millipore, Pall Corporation, Fisher Scientific). Automation-friendly 96-well ultrafiltration devices such as MultiScreen Filter Plate with UltraCell®-10 membrane (Millipore, Bedford, MA) are the most suitable for high-throughput protein concentration and buffer exchange.

Salting out of proteins using high concentrations of salts such as ammonium sulfate is one of the oldest methods of protein concentration. It can also be used for protein fractionation. However, salting out can be used as a protein concentration and buffer exchange method only if the residual ammonium sulfate will not interfere with the next step of the purification procedure or protein use. If it does, ammonium sulfate must be removed by dialysis.

In this article the application of membrane ultrafiltration and ammonium sulfate precipitation for high-throughput protein concentration and buffer exchange is described. Yield and efficiency of both methods are compared on a set of 18 randomly selected and purified recombinant proteins from *S. oneidensis* (**Table 20.1**).

20.2. Materials

20.2.1. Ultrafiltration

1. Microcon YM-10 Centrifugal Filter Devices (Amicon Bioseparations, Bedford, MA), 10,000 NMWL (*see* **Note 1**)
2. Dialysis buffer: 20 mM HEPES, 150 mM sodium chloride, pH 7.5

Table 20.1
Recovery and yield of proteins was similar for ultrafiltration (UF) and ammonium sulfate precipitation (ASP) in high-throughput protein concentration and buffer exchange using a test set of recombinant *Shewanella oneidensis* proteins

Name	Gene ID	Mol. weight (Da)	UF initial (μg)	UF final (μg)	UF recovery (%)	ASP final (μg)	ASP recovery (%)
Glutathione S-transferase-like protein	SO0012	27,859	116.66	87.4	75.0	90.1	77
Hypothetical chorismate pyruvate lyase	SO0131	24,083	216.93	241.0	111.1	130.1	60
General secretion pathway protein E	SO0167	60,379	62.23	30.0	48.3	4.8	8
Hypothetical translation initiation inhibitor	SO0337	15,561	40.17	46.3	115.2	33.3	83
Hypothetical protein	SO0338	15,127	7.06	4.1	58.0	6.9	98
Hypothetical protein	SO0342	44,379	53.39	39.4	73.8	51.0	96
Putative endori-bonuclease L-PSP	SO0358	16,429	42.56	31.7	74.5	37.9	89
Hypothetical protein	SO0620	32,259	14.51	13.2	91.0	13.9	96
Putative melanin biosyn-thesis protein TyrA	SO0740	38,479	432.33	430.4	99.6	332.3	77
Hypothetical protein	SO1068	18,250	11.62	9.6	82.8	5.2	45
Hypothetical protein	SO1287	18,646	57.51	50.8	88.3	51.5	90
Hypothetical protein	SO1306	16,006	148.19	158.2	106.8	140.6	95
Anaerobic dimethyl sulfoxide reductase, B subunit	SO1430	27,122	371.57	434.4	116.9	415.9	112
Hypothetical protein	SO1550	52,959	38.51	20.7	53.8	33.3	87
Hypothetical protein	SO1583	39,577	7.36	−0.8	−11.2	2.6	35
Hypothetical protein	SO1611	16,594	129.02	107.9	83.7	118.9	92
Putative glycine cleavage system transcriptional repressor	SO1878	23,369	723.76	633.9	87.6	664.5	92

(continued)

Table 20.1 (continued)

Hypothetical protein	SO1738	16,767	1.10	0.5	41.6	1.1	95
Average (μg)			145.49	137.6	77.6	125.5	79
Total yield (μg)			2,473.4	2,338	94.5	2,133	86
Total recovery (%)					94.5		86.2

The proteins were affinity-purified using our high-throughput procedure *(3)*. We observed no correlation between the molecular weight of proteins and their recovery using either protein concentration and buffer exchange. There is little (20%) correlation between the methods – some proteins that do not recover well using one method recovered quite well using the other method and vice versa. Therefore, we assume that the methods could be used complementary to each other. Although the total yield looks slightly better for UF (94.5%) compared to ASP (86.2%), UF is much more laborious and accident prone (especially using filtration microplates) than ASP

20.2.2. Ammonium Sulfate Precipitation

1. Ammonium sulfate
2. Saturated ammonium sulfate solution (*see* **Note 2**)
3. Dilution Buffer: 20 mM HEPES, 150 mM NaCl, pH 7.5

20.2.3. Protein Assay

1. Coomassie Plus Protein Assay Reagent Kit (Pierce, Rockford, IL)
2. 96-well standard plate (Costar EIS/RIA 96-well plates) Corning 9017 (Corning, Corning, NY)
3. Dilution buffer: 20 mM HEPES, 150 mM NaCl pH 7.5

20.3. Methods

20.3.1. Ultrafiltration

1. Transfer 200 μL of purified protein to Microcon spin column, spin at 14,000×*g* at 4°C for 30 min to reduce to ~20 μL and add 400 μL dialysis buffer. Alternatively, a 96 well ultrafiltration plate may be used (*see* **Note 1**).
2. Spin at 14,000×*g* for 30 min at 4°C to reduce the volume to 100 μL.
3. Add 300 μL of the Dialysis buffer. Spin as above to reduce to ~100 μL. This should decrease the final concentration of imidazol to <1 mM from the starting concentration of 500 mM (*see* **Note 3**).
4. Determine the final protein concentration using the Coomassie Plus Protein Assay Kit, according to the manufacturer's instructions.

20.3.2. Ammonium Sulfate Precipitation

1. Transfer 200 μL to a 1.5-mL centrifuge tube (a deep-well microplate could also be used). Add 800 μL saturated ammonium sulfate solution. Mix by inverting tube.
2. Place into a refrigerator and incubate overnight at 4°C. *Do Not Rotate*. Spin at 14,000 rpm maximum speed (35,000×*g*) for 1 h at 4°C.
3. Discard ammonium sulfate, re-suspend pellet in the Dilution buffer (*see* **Note 4**).
4. Determine final protein concentration using the Coomassie Plus Protein Assay Kit, according to the manufacturer's instructions (*see* **Notes 5 and 6**).

20.4. Notes

1. High-throughput ultrafiltration can be performed conveniently using a 96-well Filter Plate MultiScreen with Ultracell®-10 Membrane following the same procedure as above but using a floor centrifuge and swinging bucket rotors instead of the bench-top centrifuges with fixed-angle rotors.
2. If you prepare your own saturated ammonium sulfate, always set the solution pH to 7.4.
3. Some proteins form large aggregates that clog the ultrafiltration membrane. These are difficult or impossible to concentrate/buffer-exchange by ultrafiltration. The ammonium sulfate precipitation may still work fine, but dialysis is usually the best buffer exchange in this case.
4. Be careful when taking the supernatant from ammonium sulfate precipitates. Many protein precipitates are transparent, and at low quantities it is difficult to spot them in a semi-transparent polyethylene tube or a microplate, and they could mistakenly be pipetted away with the ammonium sulfate. Therefore, mark the position of the tube (microplate) in the centrifuge and take the supernatant from the opposite side of the tube (microplate well) relative to the centrifugal force.
5. Some proteins could irreversibly precipitate with ammonium sulfate. In this case, other concentration/buffer exchange methods must be used.
6. We use both protein concentration methods routinely in our process of high-throughput mapping of protein interaction networks via exogenous pull down method *(2)* for concentration and buffer exchange of affinity-purified proteins before their storage. Proteins (including the above representative set) obtained by our high-throughput affinity purification with Ni^{2+}

magnetic beads *(3)* assembled into functional protein complexes regardless of the method used for their concentration and buffer exchange. Therefore, we assume that the native protein structure and function are typically preserved after performing either of the concentration/buffer exchange methods.

Acknowledgments

This work was supported by the U.S. Department of Energy Genomics: GTL Program, Grant #KP1 102010. This manuscript has been authored by Battelle Memorial Institute, Pacific Northwest Division, under Contract No. DE-AC05–76RL0 1830 with the U.S. Department of Energy. The publisher, by accepting the article for publication, acknowledges that the United States Government retains a non-exclusive, paid-up, irrevocable, world-wide license to publish or reproduce the published form of this manuscript, or allow others to do so, for United States Government purposes.

References

1. van Reis, R., and Zydney, A. (2001) Membrane separations in biotechnology. *Curr Opin Biotechnol* **12**(2), 208–211.
2. Markillie, L. M., Lin, J. T., Adkins, J. N., Auberry, D. L., Hill, E. A., Hooker, B. S., Moore, P. A., Moore, R. J., Shi, L., Wiley, H. S., and Kery, V. (2005) Simple protein complex purification and identification method for high-throughput mapping of protein interaction networks. *J Proteome Res* **4**(2), 268–274.
3. Lin, C. T., Moore, P. A., Auberry, D. L., Landorf, E. V., Peppler, T., Victry, K. D., Collart, F. R., and Kery, V. (2006) Automated purification of recombinant proteins: combining high-throughput with high yield. *Protein Expr Purif* **47**(1), 16–24.

INDEX

A

B

C

D

E

F

G

H

I

J

K

L

M

N

O

P

T

U

V

Printed in the United States of America